STUDENT'S SOLUTIONS MANUAL

JAMES LAPP

A SURVEY OF MATHEMATICS WITH APPLICATIONS

ELEVENTH EDITION

Allen R. Angel
Monroe Community College

Christine D. Abbott
Monroe Community College

Dennis C. Runde
State College of Florida

Cover Credit: Buså Photography/Moment/Getty Images

Reproduced by Pearson from electronic files supplied by the author.

7 2022

ISBN-13: 978-0-13-574045-3
ISBN-10: 0-13-574045-2

Table of Contents

Chapter 11: Probability

Chapter 12: Statistics

Chapter 13: Graph Theory

Chapter 14: Voting and Apportionment

Chapter One: Critical Thinking Skills

Section 1.1: Inductive and Deductive Reasoning

1. Natural
3. Counterexample
5. Inductive
7. Deductive
9. $5 \times 3 = 15$
11. 1 $\underset{(1+4)}{5}$ $\underset{(4+6)}{10}$ $\underset{(4+6)}{10}$ $\underset{(1+4)}{5}$ 1
13.
15. 
17. 9, 11, 13 (Add 2 to previous number.)
19. 5, –5, 5 (Alternate 5 and –5.)
21. $\frac{1}{5}, \frac{1}{6}, \frac{1}{7}$ (Increase denominator value by 1.)
23. 21, 28, 36 $(15+6=21,\ 21+7=28,\ 28+8=36)$
25. 34, 55, 89 (Each number in the sequence is the sum of the previous two numbers.)
27. There are three letters in the pattern. $39 \times 3 = 117$, so, the 117th entry is the second R in the pattern. Therefore, the 118th entry is Y.
29. a) 36, 49, 64
    b) Square the numbers 6, 7, 8, 9 and 10.
    c) $8 \times 8 = 64$, $9 \times 9 = 81$; 72 is not a square number since it falls between the two square numbers 64 and 81.
31. Blue: 1, 5, 7, 10, 12; Purple: 2, 4, 6, 9, 11; Yellow: 3, 8
33. a) \$3700
    b) We are using observation of specific cases to make a prediction.
35. 
37. a) You should obtain the original number.
    b) You should obtain the original number.
    c) Conjecture: The result is always the original number.
    d) $n \to 3n \to 3n+6 \to \frac{3n+6}{3} = \frac{3n}{3} + \frac{6}{3} = n+2 \to n+2-2 = n$
39. a) You should obtain the number 5.
    b) You should obtain the number 5.
    c) Conjecture: The result is always the number 5.
    d) $n \to n+1 \to n+(n+1) = 2n+1 \to 2n+1+9 = 2n+10 \to \frac{2n+10}{2} = \frac{2n}{2} + \frac{10}{2} = n+5 \to n+5-n = 5$
41. $3 \times 5 = 15$ is one counterexample.
43. Two is a counting number. The sum of 2 and 3 is 5. Five divided by two is $5/2$, which is not an even number.
45. One and two are counting numbers. The difference of 1 and 2 is $1-2=-1$, which is not a counting number.
47. a) The sum of the measures of the interior angles should be $180°$.
    b) Yes, the sum of the measures of the interior angles should be $180°$.
    c) Conjecture: The sum of the measures of the interior angles of a triangle is $180°$.
49. Inductive reasoning: a general conclusion is obtained from observation of specific cases.

51. 129; The numbers in the positions of each inner 4×4 square are determined as follows.

$$\begin{matrix} a & b \\ c & a+b+c \end{matrix}$$

53. c

Section 1.2: Estimation Techniques

Answers in this section will vary depending on how you round your numbers. The answers may differ from the answers in the back of the textbook. However, your answers should be something near the answers given. All answers are approximate.)

1. Estimation

3. $26.9+67.3+219+143.3$
   $\approx 30+67+220+143=460$

5. $197,500\div 4.063\approx 200,000\div 4.000=50,000$

7. $1776\times 0.0098\approx 1800\times 0.01=18$

9. $23.97-7.05\approx 24-7=17$

11. $22\%\times 9116\approx 20\%\times 9000=0.20\times 9000$
    $=1800$

13. $\frac{\$91.35}{3}\approx\frac{\$90}{3}=\$30$

15. $12 \text{ months}\times\$47\approx 12\times\$50=\$600$

17. $\$7.99+\$4.23+\$16.82+\$3.51+\$20.12\approx\$8+\$4+\$17+\$4+\$20=\$53$

19. $95\text{ lb}+127\text{ lb}+210\text{ lb}\approx 100+100+200=400\text{ lb}$

21. 15% of \$26.32 ≈ 15% of \$26 = $0.15\times\$26=\$3.90$

23. $\$595+\$289+\$120+\$110+230\approx\$600+\$300+\$100+\$100+\$200=\$1300$

25. $11\times 8\times\$1.50\approx 10\times 8\times\$1.50=10\times\$12=\$120$

27. $\frac{599 \text{ Mexican pesos}}{20.14 \text{ Mexican pesos / U.S. dollars}}\approx\frac{600}{20}=\$30$

29. 90 miles

31. a) 23% of 700 ≈ 25% of 700 = $0.25\times 700=175$
    b) 12% of 700 ≈ 10% of 700 = $0.10\times 700=70$
    c) 21% of 700 ≈ 20% of 700 = $0.20\times 700=140$

33. a) 5 million
    b) 98 million
    c) 98 million − 33 million = 65 million
    d) 19 million + 79 million + 84 million + 65 million + 33 million = 280 million

35. a) 85%
    b) 68% − 53% = 15%
    c) 85% of 70 million acres = 59,500,000 acres
    d) No, since we are not given the area of each state.

37. 20

39. 120 bananas

41. 150°

43. 10%

45. 9 square units

47. 160 feet

49. – 57. Answers will vary.

59. There are 118 ridges around the edge.

Section 1.3: Problem-Solving Procedures

1. $\frac{1 \text{ in.}}{12 \text{ mi}} = \frac{4.25 \text{ in.}}{x \text{ mi}}$

   $x = 12(4.25)$

   $x = 51 \text{ mi}$

3. $\frac{3 \text{ ft}}{1.2 \text{ ft}} = \frac{x \text{ ft}}{15.36 \text{ ft}}$

   $3(15.36) = 1.2x$

   $\frac{46.03}{1.2} = \frac{1.2x}{1.2}$

   $x = \frac{46.03}{1.2} = 38.4 \text{ ft}$

5. $4 \times \$113 = \$452$

   $9.25\% \text{ of } \$452 = 0.0925 \times \$452 = \$41.81$

   $\$452 + \$41.81 = \$493.81$

7. a) Entertainment/Miscellaneous: $19.1\% \text{ of } \$1950 = 0.191 \times \$1950 = \$372.45$

   Food: $12.7\% \text{ of } \$1950 = 0.127 \times \$1950 = \$247.65$

   $\$372.45 - \$247.65 = \$124.80$

   b) Housing: $34.4\% \text{ of } \$1950 = 0.344 \times \$1950 = \$670.80$

   Transportation: $16\% \text{ of } \$1950 = 0.16 \times \$1750 = \$312.00$

   $\$670.80 - \$312.00 = \$358.80$

9. $43 \text{ rides} \times \$2.00 \text{ per ride} = \$86.00$; In order for the cost of rides with the \$84.50 MetroCard to be less than the cost of the rides without the CharlieCard, Marcelo would have to take 43 bus rides per month.

   Note that $\frac{\$84.50}{43 \text{ rides}} \approx \$1.97 \text{ per ride.}$

11. $\$250 + \$130(18) = \$250 + \$2340 = \$2590$

    Savings: $\$2590 - \$2500 = \$90$

13. 15-year mortgage: $\$887.63(12)(15) = \$159,773.40$

    30-year mortgage: $\$572.90(12)(30) = \$206,244.00$

    Savings: $\$206,244.00 - \$159,773.40 = \$46,470.60$

15. a) $10 \times 10 \times 10 \times 10 = 10,000$

    b) 1 in 10,000

17. $38,687.0 \text{ mi} - 38,451.4 \text{ mi} = 235.6 \text{ mi}$; $\frac{235.6 \text{ mi}}{12.6 \text{ gal}} \approx 18.7 \text{ mpg}$

19. By mail: $(\$52.80 + \$5.60 + \$8.56) \times 4 = \$66.96 \times 4 = \$267.84$

    Tire store: $\$324 + 0.08 \times \$324 = \$324 + \$25.92 = \$349.92$

    Savings: $\$349.92 - \$267.84 = \$82.08$

21. 15,000 ft – 3000 ft = 12,000 ft decrease in elevation. Temperature increases $2.4°\text{F}$ for every 1000 ft decrease in elevation.

    $-6°\text{F} + 28.8°\text{F} = 22.8°\text{F}$ and $-6°\text{F} + 28.8°\text{F} = 22.8°\text{F}$; The precipitation at the airport will be snow.

23. a) Since \$876 is less than $10\% \times \$9525 = \$952.50$, Shenile's tax was 10% of her taxable income. Her taxable income was $\frac{\$876}{0.10} = \$8760$.

b) Since \$2017.50 is greater than \$952.50 and less than \$4453.50, Logan's taxable income was in the 12% bracket. Since $\$2017.50 - \$962.50 = \$1065$ was 12% of his taxable income above \$9525, he made an additional $\frac{\$1065}{0.12} = \$8875$. His taxable income was $\$9625 + \$8875 = \$18,400$.

25. a) $1 \times 60 \times 24 \times 365 = 525,600 \text{ oz}$ and $\frac{525,600 \text{ oz}}{128} = 4106.25 \text{ gal}$

b) $\frac{4106.25}{1000} \times \$11.20 = 4.10625 \times \$11.20 = \$45.99$

27. a) $\frac{20,000}{20.8} - \frac{20,000}{21.6} \approx 961.538 - 925.926 = 35.612 \approx 35.61 \text{ gal}$

b) $35.61 \times \$3.00 = \$106.83$

c) $140,000,000 \times 35.61 = 4,985,400,000 \text{ gal}$

29. Cost after 1 year: $\$999 + 0.02(\$999) = \$999 + \$19.98 = \$1018.98$

Cost after second year: $\$1018.98 + 0.02(\$1018.98) = \$1018.98 + \$20.38 = \$1039.36$

31. After paying the \$100 deductible, Yungchen must pay 20% of the cost of x-rays.
First x-ray: $\$100 + 0.20(\$620) = \$100 + \$124 = \$224$

Second x-ray: $0.20(\$980) = \$196$

Total: $\$224 + \$196 = \$420$

33. a) water/milk: $3(1) = 3$ cups salt: $3\left(\frac{1}{8}\right) = \frac{3}{8}$ tsp

Cream of wheat: $3(3) = 9$ tbsp $= \frac{9}{16}$ cup (because 16 tbsp = 1 cup)

b) water/milk: $\frac{2 + 3\frac{3}{4}}{2} = \frac{\frac{23}{4}}{2} = \frac{23}{8} = 2\frac{7}{8}$ cups salt: $\frac{\frac{1}{4} + \frac{1}{2}}{2} = \frac{\frac{3}{4}}{2} = \frac{3}{8}$ tsp

cream of wheat: $\frac{\frac{1}{2} + \frac{3}{4}}{2} = \frac{\frac{5}{4}}{2} = \frac{5}{8}$ cups =5 tbsp

c) water/milk: $3\frac{3}{4} - 1 = \frac{15}{4} - \frac{4}{4} = \frac{11}{4} = 2\frac{3}{4}$ cups salt: $\frac{1}{2} - \frac{1}{8} = \frac{4}{8} - \frac{1}{8} = \frac{3}{8}$ tsp

cream of wheat: $\frac{3}{4} - \frac{3}{16} = \frac{12}{16} - \frac{3}{16} = \frac{9}{16}$ cup = 9 tbsp

d) Differences exist in water/milk because the amount for 4 servings is not twice that for 2 servings.

Differences also exist in Cream of Wheat because $\frac{1}{2}$ cup is not twice 3 tbsp.

35. a)

20 DVDs	12 DVDs	Total Number of DVDs
1	1	$20+12=32$
0	2	$12+12=24$

b) $\$240+\$180=\$420$

37. $1 \text{ ft}^2 = 12 \text{ in.} \times 12 \text{ in.} = 144$ square inches

39. Area of original rectangle: lw

Area of new rectangle: $(2l)(2w) = 4(lw)$

Thus, if the length and width of a rectangle are doubled, the area is 4 times as large.

41. Volume of original cube: lwh

Volume of new cube: $(2l)(2w)(2h) = 8(lwh)$

Thus, if the length, width, and height of a cube are doubled, the volume is 8 times as large.

43. $\dfrac{10 \text{ pieces}}{x} = \dfrac{1000 \text{ pieces}}{\$10}$

$1000x = 10(10)$

$\dfrac{1000x}{1000} = \dfrac{100}{1000}$

$x = \dfrac{100}{1000} = \$0.10 = 10¢$

45. You have three ties, each a different color.

47. a) refresh

b) workout

49.

		④		
	③		②	
⑤		①		⑥

51.

8	6	16
18	10	2
4	14	12

53. $8+6+2+4=20$; $3+7+5+1=16$; $10+14+12+8=44$

The sum of the four corner entries is 4 times the number in the center of the middle row.

55. 45, 36, 99; Multiply the number in the center of the middle row by 9.

57. $3\times2\times1=6$ ways

59. Other answers are possible, but 1 and 8 must appear in the center.

	7	
3	1	4
5	8	6
	2	

61. Other answers are possible.

1	2	3	4	5
2	3	4	5	1
3	4	5	1	2
4	5	1	2	3
5	1	2	3	4

63. Mark plays the drums.

65. The areas of the colored regions are 1×1, 1×1, 2×2, 3×3, 5×5, 8×8, 13×13, 21×21.

$1+1+4+9+25+64+169+441=714$ square units

67. Thomas would have opened the box labeled *grapes and cherries*. Because all the boxes are labeled incorrectly, whichever fruit he pulls from the box of grapes and cherries, will be the only fruit in that box. If he pulled a grape, he labeled the box *grape*. If he pulled a cherry, he labeled the box *cherries*. That left two boxes whose original labels were incorrect. Because all labels must be changed, there was only one way for Thomas to assign the two remaining labels.

Review Exercises

1. 23, 28, 33 (Add 5 to previous number.)
2. 16, 13, 10 (Subtract 3 from previous number)
3. 64, –128, 256 (Multiply previous number by –2.)
4. 25, 32, 40 $(19+6=25,\ \ 25+7=32,\ \ 32+8=40)$
5. 10, 4, –3 (Subtract 1, then 2, then 3, …)
6. $\frac{3}{8}, \frac{3}{16}, \frac{3}{32}$ (Multiply previous number by $\frac{1}{2}$.)
7. 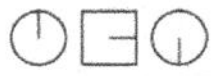 [figure]
8. 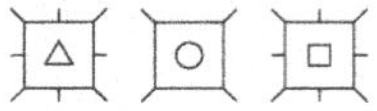 [figure]
9. c
10. a) The final number is twice the original number.
    b) The final number is twice the original number.
    c) Conjecture: The final number is twice the original number.
    d) $n \to 10n \to 10n+5 \to \frac{10n+5}{5} = \frac{10n}{5} + \frac{5}{5} = 2n+1 \to 2n+1-1 = 2n$
11. This process will always result in an answer of 3.
    $n \to n+5 \to 6(n+5) = 6n+30 \to 6n+30-12 = 6n+18$
    $\to \frac{6n+18}{2} = \frac{6n}{2} + \frac{18}{2} = 3n+9 \to \frac{3n+9}{3} = \frac{3n}{3} + \frac{9}{3} = n+3 \to,\ n+3-n=3$
12. $1^2 + 2^2 = 5$ and 5 is an odd number. Other answers are possible.

Answers for Exercises 13–25 will vary depending on how you round your numbers. The answers may differ from the answers in the back of the textbook. However, your answers should be something near the answers given. All answers are approximate.

13. $205{,}123 \times 4002 \approx 200{,}000 \times 4000$
    $= 800{,}000{,}000$
14. $215.9 + 128.752 + 3.6 + 861 + 792$
    $\approx 200 + 100 + 0 + 900 + 800 = 2000$
15. 21% of 2095 ≈ 20% of 2000
    $= 0.20 \times 2000 = 400$
16. Answers will vary.
17. 48 bricks × \$3.97 ≈ 50 × 4 = \$200
18. 8% of \$21,000 ≈ 8% of 20,000
    $= 0.08 \times 20{,}000 = \$1600$
19. $\frac{1.1\text{ mi}}{22\text{ min}} \approx \frac{1\text{ mi}}{20\text{ min}} = \frac{3\text{ mi}}{60\text{ min}} = 3\text{ mph}$
20. \$2.49 + \$0.79 + \$1.89 + \$0.10 + \$2.19 + \$6.75
    ≈ \$2 + \$1 + \$2 + \$0 + \$2 + \$7 = \$14.00
21. $5\text{ in.} = \frac{20}{4}\text{ in.} = 20\left(\frac{1}{4}\right)\text{ in.} = 20(0.1)\text{ mi}$
    $= 2\text{ mi}$
22. Approximately 70°F – 30°F = 40°F
23. Approximately 90°F – 80°F = 10°F
24. 13 square units
25. Length: 1.75 in., 1.75(1.25) = 21.875 ≈ 22 ft
    Height: 0.625 in., 0.625(12.5) = 7.8125 ≈ 8 ft
26. \$50 + \$40(12) = \$530
    Savings: \$530 – \$500 = \$30
27. 4(\$1.99) = \$7.96 for four six-packs
    Savings: \$7.96 – \$4.99 = \$2.97

28. Freemac: $\$15\times4\times2=\$120$

Sylvan: $\$25\times4$ hours $=\$100$

$\$120-\$100=\$20$

Sylvan Rental is less expensive by \$20.

29. Cost per person with 5 people: $\frac{\$445}{5}=\$89$

Cost per person with 5 people: $\frac{\$510}{6}=\$85$

Savings: $\$89-\$85=\$4$

30. a) $\frac{30\text{ lb}}{2500\text{ ft}^2}=\frac{x}{24{,}000\text{ ft}^2}$

$x=\frac{30\times24{,}000}{2500}$

$x=288$ lb

b) $\frac{150\text{ lb}}{30\text{ lb/bag}}=5$ bags; $5\times2500=12{,}500\text{ ft}^2$

31. 10% of $\$1030=0.10\times\$1030=\$103$

$7\times\$103=\$721$

Savings: $\$721-\$60=\$661$

32. $\frac{1.5\text{ mg}}{10\text{ lb}}=\frac{x\text{ mg}}{52\text{ lb}}$

$10x=52(1.5)$

$\frac{10x}{10}=\frac{78}{10}$

$x=7.8$ mg

33. $\$5500-0.30(\$5500)=\$5500-\$1650=\$3850$ take-home; 28% of $\$3850=0.28\times\$3850=\$1078$

34. 9 A.M. Eastern is 6 A.M. Pacific, from 6 A.M. Pacific to 1:35 P.M. Pacific is 7 hr 35 min, 7 hr 35 min − 50 min stop = 6 hr 45 min.

35. 3 P.M. − 4 hr = 11 A.M.; July 26, 11:00 A.M.

36. a) $\frac{65\text{ mi}}{1\text{ hr}}\times\frac{1.6\text{ km}}{\text{mi}}=\frac{104\text{ km}}{1\text{ hr}}\approx104\text{ km/hr}$

b) $\frac{90\text{ km}}{1\text{ hr}}\times\frac{1\text{ mi}}{1.6\text{ km}}=\frac{90\text{ mi}}{1.6\text{ km}}\approx56.25\text{ mi/hr}$

37. Each figure has an additional two dots. To get the hundredth figure, 97 more figures must be drawn, $97(2)=194$ dots added to the third figure. Thus, $194+7=201$.

38.

21	7	8	18
10	16	15	13
14	12	11	17
9	19	20	6

39.

23	25	15
13	21	29
27	17	19

40. 59 min 59 sec; Since it doubles every second, the jar was half full 1 second earlier than 1 hour.

41. 6

42. Nothing. Each friend paid \$9 for a total of \$27; \$25 to the hotel, \$2 to the clerk. \$25 for the room + \$3 for each friend + \$2 for the clerk = \$30

43. Let $x=$ the score on the fifth exam

$$\frac{93+88+81+86+x}{5}=80$$

$$\frac{348+x}{5}=80,$$

$$348+x=400,$$

$$x=52$$

44. Yes; 3 quarters and 4 dimes, or 1 half dollar, 1 quarter and 4 dimes, or 1 quarter and 9 dimes.

45. $6 \text{ cm} \times 6 \text{ cm} \times 6 \text{ cm} = 216 \text{ cm}^3$

46. Place six coins in each pan with one coin off to the side. If it balances, the heavier coin is the one on the side. If the pan does not balance, take the six coins on the heavier side and split them into two groups of three. Select the three heavier coins and weigh two coins. If the pan balances, it is the third coin. If the pan does not balance, you can identify the heavier coin.

47. $1+500=501,\ 2+499=501,\ \ldots$; There are 250 such pairs and $250(501)=125{,}250$.

48. 4 green weigh the same as 8 blue, 2 yellow weigh the same as 5 blue, 2 white weigh the same as 3 blue, $8+5+3=16$ blue

49. There are 90 such numbers: 101, 111, 121, 131, 141, 151, 161, 171, 181, 191, 202,h …, 292, 303, …, 393, 404, …, 494, 505, …, 595, 606, …, 696, 707, …, 797, 808, …, 898, 919, …, 999.

50. The fifth figure will be an octagon with sides of equal length. Inside the octagon will be a seven-sided figure with sides of equal length. The figure will have one antenna.

51. 61 orange tiles will be required. The sixth figure will have 6 rows of 6 tiles and 5 rows of 5 tiles $(6\times6+5\times5=36+25=61)$.

52. Some possible answers are given below. There are other possibilities.

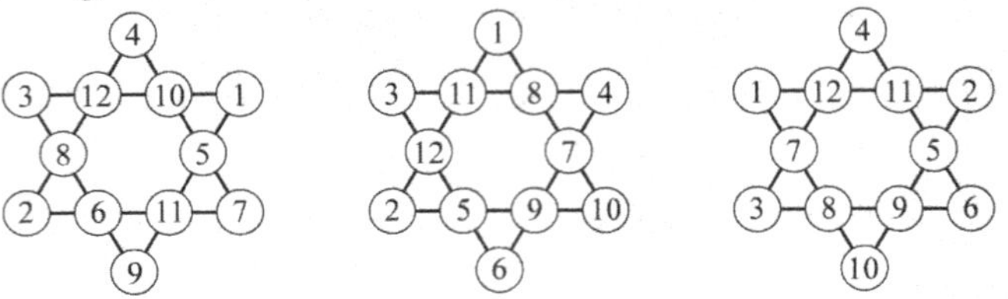

53. a) 2

b) There are 3 choices for the first spot. Once that person is standing, there are 2 choices for the second spot and 1 for the third. Thus, $3\times2\times1=6$.

c) $4\times3\times2\times1=24$

d) $5\times4\times3\times2\times1=120$

e) $n(n-1)(n-2)\cdots1$, (or $n!$), where $n=$ the number of people in line.

Chapter Test

1. 26, 32, 38 (Add 6 to previous number.)

2. $\frac{1}{5}, \frac{1}{6}, \frac{1}{7}$ (Add 1 to the denominator of the previous number.)

3. a) The result is the original number plus 1.

b) The result is the original number plus 1.

c) Conjecture: The result will always be the original number plus 1.

d) $n \rightarrow 5n \rightarrow 5n+10 \rightarrow \frac{5n+10}{5}=\frac{5n}{5}+\frac{10}{5}=n+2 \rightarrow n+2-1=n+1$

4. $0.51\times96{,}000 \approx 0.5\times100{,}000=50{,}000$

5. $\frac{188{,}000}{0.11} \approx \frac{200{,}000}{0.1} \approx 2{,}000{,}000$

6. 9 square units

7. a) $\frac{130 \text{ lb}}{63 \text{ in.}} \approx 2.0635;\ \frac{2.0635}{63 \text{ in.}}=0.032754;$ $0.032754\times703 \approx 23.03$

b) He is in the at-risk range.

8. a) 325,000 visitors

b) 100,000 visitors

9. $\$50.40-\$35.00=\$15.40;\ \dfrac{\$15.40}{\$0.20}=77$ miles

10. Since $\dfrac{\$15}{\$2.59}\approx 5.79$, a maximum of 5 six-packs can be purchased. There will be $\$15.00-(5\times\$2.59)$ $=\$15.00-\$12.95=\$2.05$ remaining, so $\dfrac{\$2.05}{\$0.80}=2.5625$, or two additional individual cans can be purchased. The following table was created using similar calculations for the other possible number of six-packs.

Six-packs	Individual cans	Number of cans
5	2	$6\times5+2=32$
4	5	$6\times4+5=29$
3	9	$6\times3+9=27$
2	12	$6\times2+12=24$
1	15	$6\times1+15=21$
0	18	$6\times0+18=18$

The maximum number of cans is 32.

11. $3\text{ cuts}\times\dfrac{2.5\text{ min}}{1\text{ cut}}=7\dfrac{1}{2}=7.5\text{ min}$

12. $10\times\dfrac{3}{4}=10\times0.75=7.5$ miles

13. $\$12.75\times40=\$510$

$\$12.75\times1.5\times10=\$191.25$

$\$510+\$191.25=\$701.25$

$\$652.25-\$701.25=-\$49.00$

She was underpaid by $49.00.

14.

40	15	20
5	25	45
30	35	10

15. Mary drove the first 15 miles at 60 mph which took $\dfrac{15}{60}=\dfrac{1}{4}$ hour and drove the second 15 miles at 30 mph took $\dfrac{15}{30}=\dfrac{1}{2}$ hour for a total time of $\dfrac{3}{4}$ hour, or 45 minutes. If she drove the entire 30 miles at 45 mph, the trip would take $\dfrac{30}{45}=\dfrac{2}{3}$ hour, or 40 minutes, which is less than $\dfrac{3}{4}$ hour.

16. $\dfrac{6\text{ lb}}{2\text{ lb}}=3;\ 3\times\dfrac{1}{2}\text{tsp}=\dfrac{3}{2}\text{tsp or }1\dfrac{1}{2}\text{tsp}=\dfrac{1}{2}\text{ tbsp}$

17. Area of lawn including walkway: $(10+2)\times(12+2)=12\times14=168\text{ m}^2$

Area of lawn only: $10\times12=120\text{ m}^2$

Area of walkway: $168-120=48\text{ m}^2$

18. 243 jelly beans; $260-17=243$, $234+9=243$, and $274-31=243$

19. a) $3\times\$3.99=\$11.97$

b) $9(\$1.75\times0.75)=11.8125\approx\$11.81$

c) $\$11.97-\$11.81=\$0.16$; Using the coupon is least expensive by $0.16.

20. $4\times3\times2\times1=24$; The first position can hold any of four letters, the second any of the three remaining letters, and so on.

Chapter Two: Sets

Section 2.1: Set Concepts

1. Set
3. Description, Roster form, Set-builder notation
5. Infinite
7. Equal
9. Empty or null
11. Not well defined, "best" is interpreted differently by different people.
13. Well defined, the contents can be clearly determined.
15. Well defined, the contents can be clearly determined.
17. Infinite, the number of elements in the set is not a natural number.
19. Infinite, the number of elements in the set is not a natural number.
21. Infinite, the number of elements in the set is not a natural number.
23. $\{\text{Hawaii}\}$
25. $\{11, 12, 13, 14, \ldots, 177\}$
27. $B = \{2, 4, 6, 8, \ldots\}$
29. $\{\ \}$ or $\varnothing$
31. $E = \{14, 15, 16, 17, \ldots, 84\}$
33. {Google Play Games, Pokémon Go}
35. {Solitaire by MobilityWare, Candy Crush Soda Saga, ApplsLoading}
37. $\{2015, 2016, 2017, 2018\}$
39. $\{\ \}$ or $\varnothing$
41. $B = \{x \mid x \in N \text{ and } 6 < x < 15\}$ or $B = \{x \mid x \in N \text{ and } 7 \le x \le 14\}$
43. $C = \{x \mid x \in N \text{ and } x \text{ is a multiple of } 3\}$
45. $E = \{x \mid x \in N \text{ and } x \text{ is odd}\}$
47. $C = \{x \mid x \text{ is February}\}$
49. Set A is the set of natural numbers less than or equal to 7.
51. Set V is the set of vowels in the English alphabet
53. Set T is the set of species of trees.
55. Set S is the set of seasons.
57. $\{\text{Facebook, Instagram, Facebook Messenger}\}$
59. $\{\text{Twitter, Pinterest, Snapchat}\}$
61. $\{2016, 2017, 2018, 2019\}$
63. $\{2013, 2014, 2015\}$
65. False; $\{e\}$ is a set, and not an element of the set.
67. False; h is not an element of the set.
69. False; 3 is an element of the set.
71. True; 9 is an odd natural number.
73. $n(A) = 4$
75. $n(C) = 0$
77. Both; A and B contain exactly the same elements.
79. Neither; the sets have a different number of elements.
81. Equivalent; both sets contain the same number of elements, 4.
83. a) Set A is the set of natural numbers greater than 2. Set B is the set of all numbers greater than 2.
    b) Set A contains only natural numbers. Set B contains other types of numbers, including fractions and decimal numbers.

83. (continued)

c) $A = \{3, 4, 5, 6, \ldots\}$

d) No; because there are an infinite number of elements between any two elements in set B, we cannot write set B in roster form.

85. Cardinal; 7 tells how many.

87. Ordinal; sixteenth tells Lincoln's relative position.

89. Answers will vary.

91. Answers will vary.

Section 2.2: Subsets

1. Subset

3. 2^n, where n is the number of elements in the set.

5. True; {table} is a subset of {sofa, chair, table}.

7. False; apple is not in the second set.

9. True; {AT&T, Verizon} is a proper subset of {Verizon, Sprint, T-Mobile, Verizon}.

11. False; no subset is a proper subset of itself.

13. True; book is an element of {book, magazine, newspaper}.

15. False; {cookie} is a set, not an element.

17. True; tiger is not an element of {zebra, giraffe, polar bear}.

19. True; {chair} is a proper subset of {sofa, table, chair}.

21. False; the set $\{\varnothing\}$ contains the element $\varnothing$.

23. False; the set $\{0\}$ contains the element 0.

25. False; 0 is a number and { } is a set.

27. $B \subseteq A, B \subset A$

29. $B \subseteq A, B \subset A$

31. None

33. $A = B, A \subseteq B, B \subseteq A$

35. { } is the only subset.

37. { }, {cow}, {horse}, {cow, horse}

39. a) $\{\ \}, \{a\}, \{b\}, \{c\}, \{d\}, \{a,b\}, \{a,c\}, \{a,d\}, \{b,c\}, \{b,d\}, \{c,d\}, \{a,b,c\}, \{a,b,d\}, \{a,c,d\}, \{b,c,d\}, \{a,b,c,d\}$

b) All the sets in part (a) are proper subsets of A except $\{a,b,c,d\}$.

41. False; A could be equal to B.

43. True; every set is a subset of itself.

45. True; $\varnothing$ is a proper subset of every set except itself.

47. True; every set is a subset of the universal set.

49. True; $\varnothing$ is a proper subset of every set except itself and $U \neq \varnothing$.

51. True; $\varnothing$ is a subset of every set.

53. The number of different variations is equal to the number of subsets of a set with 4 elements, which is $2^4 = 2 \times 2 \times 2 \times 2 = 16$.

55. The number of options is equal to the number of subsets of a set with 7 elements, which is $2^7 = 2\times2\times2\times2\times2\times2\times2 = 128$.

57. $E = F$ since they are both subsets of each other.

59. a) Yes.
    b) No, c is an element of set D.
    c) Yes, each element of $\{a,b\}$ is an element of set D.

61. A one element set has one proper subset, namely the empty set. A one element set has two subsets, namely itself and the empty set. One is one-half of two. Thus, the set must have one element.

63. Yes

Section 2.3: Venn Diagrams and Set Operations

1. Complement
3. Intersection
5. Cartesian
7. $m\times n$

9. U A B

11. U A B

13. U A B

15. 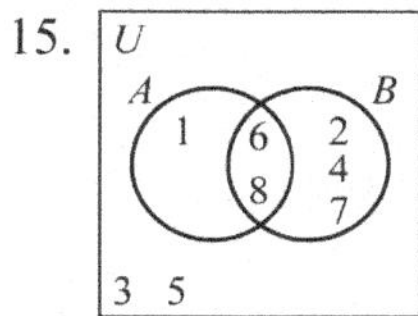

17. U A B Chick-fil-A Domino's Subway Panera Bread Burger King Chipotle Pizza Hut McDonald's

19. 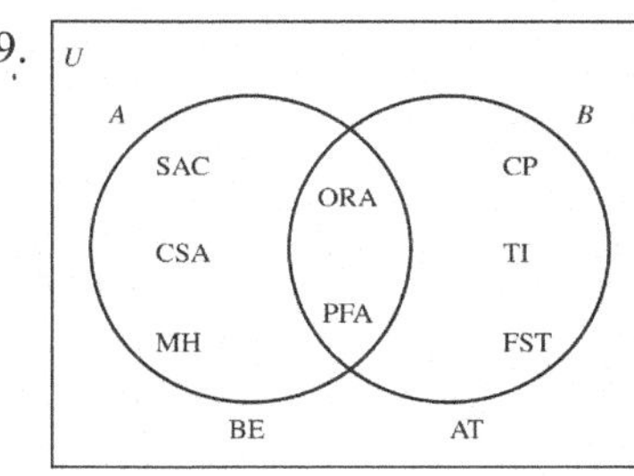

21. The set of retail stores in the United States that do not sell children's clothing

23. The set of cities in the United States that do not have a professional sports team

25. The set of cities in the United States that have a professional sports team or a symphony

27. The set of cities in the United States that have a professional sports team and do not have a symphony

29. The set of furniture stores in the U.S. that sell mattresses or leather furniture

31. The set of furniture stores in the U.S. that do not sell outdoor furniture and sell leather furniture

33. The set of furniture stores in the U.S. that sell mattresses or outdoor furniture or leather furniture

35. $A = \{a, b, e, g, h\}$

37. $A\cap B = \{a, b, e, g, h\}\cap\{b, c, e, f\} = \{b, e\}$

39. $A\cup B = \{a, b, e, g, h\}\cup\{b, c, e, f\}$
$= \{a, b, c, e, f, g, h\}$

41. $A' \cap B' = \{a, b, e, g, h\}' \cap \{b, c, e, f\}'$
$= \{c, d, f, i, j\} \cap \{a, d, g, h, i, j\}$
$= \{d, i, j\}$

43. $A - B = \{a, b, e, g, h\} - \{b, c, e, f\} = \{a, g, h\}$

45. $A = \{1, 2, 7, 8, 9\}$

47. $U = \{1, 2, 3, 4, 5, 6, 7, 8, 9, 10, 11\}$

49. $A' \cup B = \{3, 4, 5, 6, 10, 11\} \cup \{2, 3, 5, 6, 9\} = \{2, 3, 4, 5, 6, 9, 10, 11\}$

51. $A' \cap B = \{3, 4, 5, 6, 10, 11\} \cup \{2, 3, 5, 6, 9\} = \{3, 5, 6\}$

53. $A' - B = \{4, 10, 11\}$

55. $A \cup B = \{1, 2, 4, 5, 7\} \cup \{2, 3, 5, 6\}$
$= \{1, 2, 3, 4, 5, 6, 7\}$

57. $A \cup B = \{1, 2, 3, 4, 5, 6, 7\}$ (See Exercise 55)
$(A \cup B)' = \{1, 2, 3, 4, 5, 6, 7\}' = \{8\}$

59. $(A \cup B)' = \{8\}$ (See Exercise 57)
$(A \cup B)' \cap B = \{8\} \cap \{2, 3, 5, 6\} = \{\ \}$

61. $(B \cup A)' = (A \cup B)' = \{8\}$ (See Exercise 57)
$(B \cup A)' \cap (B' \cup A') = \{8\} \cap (\{2, 3, 5, 6\}' \cup \{1, 2, 4, 5, 7\}')$
$= \{8\} \cap (\{1, 4, 7, 8\} \cup \{3, 6, 8\})$
$= \{8\} \cap \{1, 3, 4, 6, 7, 8\} = \{8\}$

63. $(A - B)' = \{1, 4, 7\}' = \{2, 3, 5, 6, 8\}$

65. $B \cup C = \{b, c, d, f, g\} \cup \{a, b, f, i, j\} = \{a, b, c, d, f, g, i, j\}$

67. $A' \cup B' = \{b, e, h, j, k\} \cup \{a, e, h, i, j, k\} = \{a, b, e, h, i, j, k\}$

69. $(A \cap B) \cup C = (\{a, c, d, f, g, i\} \cap \{b, c, d, f, g\}) \cup \{a, b, f, i, j\} = \{c, d, f, g\} \cup \{a, b, f, i, j\}$
$= \{a, b, c, d, f, g, i, j\}$

71. $(A' \cup C) \cup (A \cap B) = (\{a, c, d, f, g, i\}' \cup \{a, b, f, i, j\}) \cup (\{a, c, d, f, g, i\} \cap \{b, c, d, f, g\})$
$= (\{b, e, h, j, k\} \cup \{a, b, f, i, j\}) \cup \{c, d, f, g\} = \{a, b, e, f, h, i, j, k\} \cup \{c, d, f, g\}$
$= \{a, b, c, d, e, f, g, h, i, j, k\}$, or U

73. $(A - B)' - C = \{a, i\}' - \{a, b, f, i, j\} = \{b, c, d, e, f, g, h, j, k\} - \{a, b, f, i, j\} = \{c, d, e, g, h, k\}$

75. $\{(1, a), (1, b), (2, a), (2, b), (3, a), (3, b)\}$

77. No; The ordered pairs are not the same.
For example: $(1, a) \neq (a, 1)$.

79. $n(\{(a, 1), (a, 2), (a, 3), (b, 1), (b, 2), (b, 3)\}) = 6$

81. $A \cap B = \{1, 3, 5, 7, 9\} \cap \{2, 4, 6, 8\} = \{\ \}$

83. $(B \cup C)' = (\{2, 4, 6, 8\} \cup \{1, 2, 3, 4, 5\})' = \{1, 2, 3, 4, 5, 6, 8\}' = \{7, 9\}$

85. $A \cap B' = \{1, 3, 5, 7, 9\} \cap \{2, 4, 6, 8\}' = \{1, 3, 5, 7, 9\} \cap \{1, 3, 5, 7, 9\} = \{1, 3, 5, 7, 9\}$, or A

87. $(A \cup C) \cap B = (\{1, 3, 5, 7, 9\} \cup \{1, 2, 3, 4, 5\}) \cap \{2, 4, 6, 8\} = \{1, 2, 3, 4, 5, 7, 9\} \cap \{2, 4, 6, 8\} = \{2, 4\}$

89. $(A'\cup B')\cap C = \left(\{1,3,5,7,9\}'\cup\{2,4,6,8\}'\right)\cap\{1,2,3,4,5\}$
$= (\{2,4,6,8\}\cup\{1,3,5,7,9\})\cap\{1,2,3,4,5\}$
$= \{1,2,3,4,5,6,7,8,9\}\cap\{1,2,3,4,5\}$
$= \{1,2,3,4,5\}$, or C

91. A set and its complement will always be disjoint since the complement of a set is all of the elements in the universal set that are not in the set. Therefore, a set and its complement will have no elements in common. For example, if $U=\{1,2,3\}$, $A=\{1,2\}$, and $A'=\{3\}$, so $A\cap A'=\{\ \}$.

93. Let $A=\{\text{customers who owned dogs}\}$ and $B=\{\text{customers who owned cats}\}$
$n(A\cup B)=n(A)+n(B)-n(A\cap B)=27+38-16=49$

95. a) $A\cup B=\{a,b,c,d\}\cup\{b,d,e,f,g,h\}=\{a,b,c,d,e,f,g,h\}$, $n(A\cup B)=8$
$A\cap B=\{a,b,c,d\}\cap\{b,d,e,f,g,h\}=\{b,d\}, n(A\cap B)=2.$
$n(A)+n(B)-n(A\cap B)=4+6-2=8$
Therefore, $n(A\cup B)=n(A)+n(B)-n(A\cap B)$.

b) Answers will vary.

c) Elements in the intersection of A and B are counted twice in $n(A)+n(B)$.

97. $A\cup B=\{1,2,3,4,\ldots\}\cup\{4,8,12,16,\ldots\}=\{1,2,3,4,\ldots\}$, or A

99. $B\cup C=\{4,8,12,16,\ldots\}\cup\{2,4,6,8,\ldots\}=\{2,4,6,8,\ldots\}$, or C

101. $A\cap C=\{1,2,3,4,\ldots\}\cap\{2,4,6,8,\ldots\}=\{2,4,6,8,\ldots\}$, or C

103. $B\cup C=C$ (See Exercise 99)
$(B\cup C)'\cup C=C'\cup C=\{2,4,6,8,\ldots\}'\cup\{2,4,6,8,\ldots\}=\{0,1,2,3,4,\ldots\}$, or U

105. $A\cap A'=\{\ \}$

107. $A\cup\varnothing=A$

109. $A'\cup U=U$

111. If $A\cap B=B$, then $B\subseteq A$.

113. If $A\cap B=\varnothing$, then A and B are disjoint sets.

Section 2.4: Venn Diagrams with Three Sets and Verification of Equality of Sets

1. 8

3. a) $(A\cup B)'=A'\cap B'$
b) $(A\cup B)'=A'\cup B'$

5. $A'\cap B'$ is represented by regions V and VI. If $B\cap C$ contains 12 elements and region V contains 4 elements, then region VI contains $12-4=8$ elements.

7. 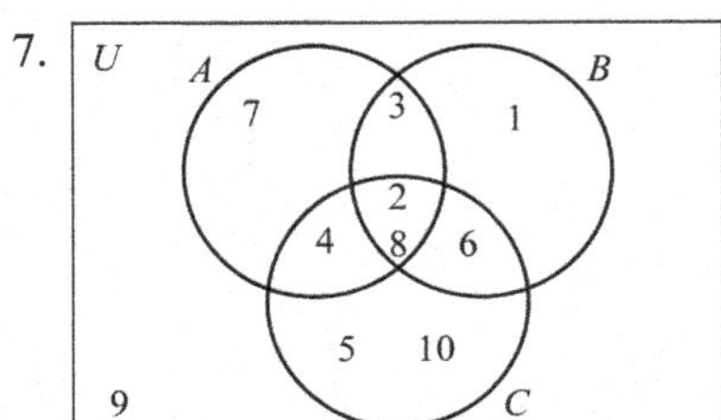

9. 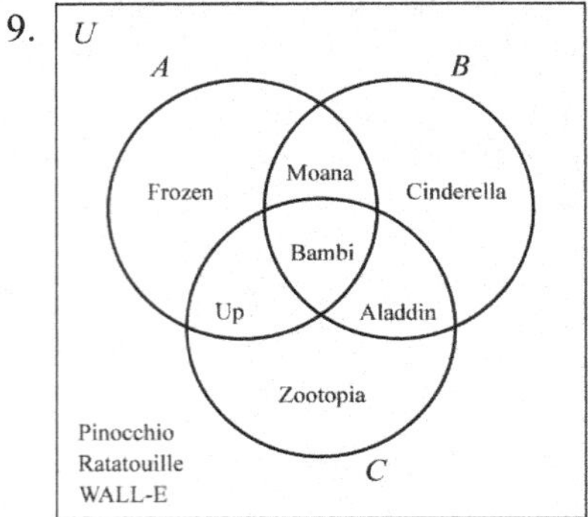

11. 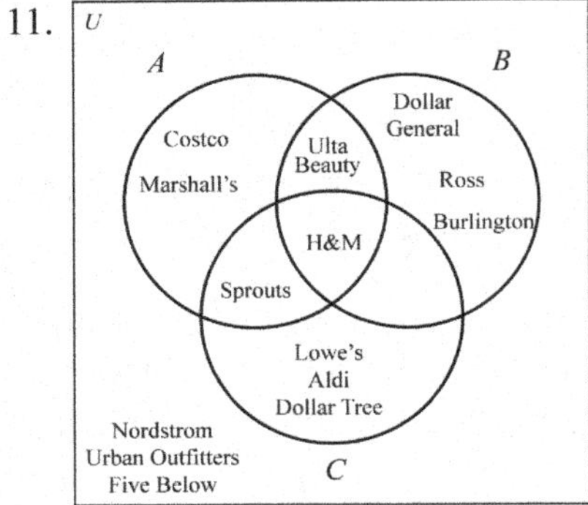

13. 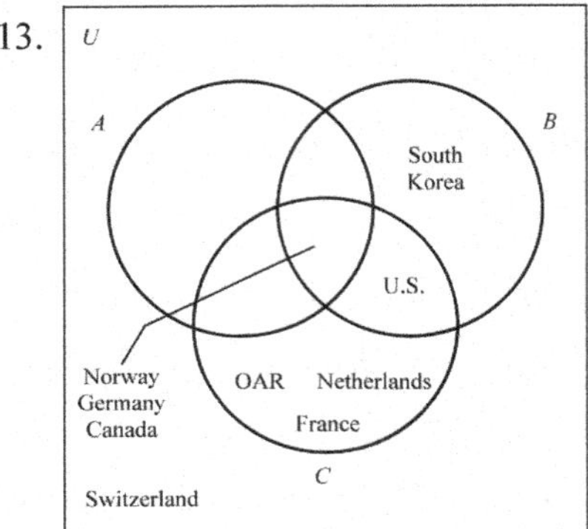

15. Canada, IV

17. China, V

19. Spain, VIII

21. VI

23. III

25. III

27. V

29. II

31. VII

33. I

35. VIII

37. VI

39. $A = \{1, 3, 4, 5, 9, 10\}$

41. $C = \{4, 5, 6, 8, 9, 11\}$

43. $A \cap B = \{3, 4, 5\}$

45. $(B \cap C)' = \{1, 2, 3, 7, 9, 10, 11, 12, 13, 14\}$

47. $(A \cup C)' = \{2, 7, 12, 13, 14\}$

49. $(A - B)' = \{2, 3, 4, 5, 6, 7, 8, 11, 12, 13, 14\}$

51.

$(A \cap B)'$

Set	Regions
A	I, II
B	II, III
$A \cap B$	II
$(A \cap B)'$	I, III, IV

$A' \cup B'$

Set	Regions
A	I, II
A'	III, IV
B	II, III
B'	I, IV
$A' \cup B'$	I, III, IV

Both statements are represented by the same regions, I, III, IV, of the Venn diagram. Therefore, $(A \cap B)' = A' \cup B'$ for all sets A and B.

53. $A' \cup B'$

Set	Regions
A	I, II
A'	III, IV
B	II, III
B'	I, IV
$A' \cup B'$	I, III, IV

$A \cap B$

Set	Regions
A	I, II
B	II, III
$A \cap B$	II

Since the two statements are not represented by the same regions, it is not true that $A' \cup B' = A \cap B$ for all sets A and B.

55. $A' \cap B'$

Set	Regions
A	I, II
A'	III, IV
B	II, III
B'	I, IV
$A' \cap B'$	IV

$A \cup B'$

Set	Regions
A	I, II
B	II, III
B'	I, IV
$A \cup B'$	I, II, IV

Since the two statements are not represented by the same regions, it is not true that $A' \cap B' = A \cup B'$ for all sets A and B.

57. $A \cap (B \cup C)$

Set	Regions
B	II, III, V, VI
C	IV, V, VI, VII
$B \cup C$	II, III, IV, V, VI, VII
A	I, II, IV, V
$A \cap (B \cup C)$	II, IV, V

$(A \cap B) \cup C$

Set	Regions
A	I, II, IV, V
B	II, III, V, VI
$A \cap B$	II, V
C	IV, V, VI, VII
$(A \cap B) \cup C$	II, IV, V, VI, VII

Since the two statements are not represented by the same regions, it is not true that $A \cap (B \cup C) = (A \cap B) \cup C$ for all sets A, B, and C.

59. $A \cap (B \cup C)$

Set	Regions
B	II, III, V, VI
C	IV, V, VI, VII
$B \cup C$	II, III, IV, V, VI, VII
A	I, II, IV, V
$A \cap (B \cup C)$	II, IV, V

$(B \cup C) \cap A$

Set	Regions
B	II, III, V, VI
C	IV, V, VI, VII
$B \cup C$	II, III, IV, V, VI, VII
A	I, II, IV, V
$(B \cup C) \cap A$	II, IV, V

Both statements are represented by the same regions, II, IV, V, of the Venn diagram. Therefore, $A \cap (B \cup C) = (B \cup C) \cap A$ for all sets A, B, and C.

61. $A \cap (B \cup C)$

Set	Regions
B	II, III, V, VI
C	IV, V, VI, VII
$B \cup C$	II, III, IV, V, VI, VII
A	I, II, IV, V
$A \cap (B \cup C)$	II, IV, V

$(A \cap B) \cup (A \cap C)$

Set	Regions
A	I, II, IV, V
B	II, III, V, VI
$A \cap B$	II, V
C	IV, V, VI, VII
$A \cap C$	IV, V
$(A \cap B) \cup (A \cap C)$	II, IV, V

Both statements are represented by the same regions, II, IV, V, of the Venn diagram. Therefore, $A \cap (B \cup C) = (A \cap B) \cup (A \cap C)$ for all sets A, B, and C.

63. $(A \cup B) \cap (B \cup C)$

Set	Regions
A	I, II, IV, V
B	II, III, V, VI
$A \cup B$	I, II, III, IV, V, VI
C	IV, V, VI, VII
$B \cup C$	II, III, IV, V, VI, VII
$(A \cup B) \cap (B \cup C)$	II, III, IV, V, VI

$B \cup (A \cap C)$

Set	Regions
A	I, II, IV, V
C	IV, V, VI, VII
$A \cap C$	IV, V
B	II, III, V, VI
$B \cup (A \cap C)$	II, III, IV, V, VI

Both statements are represented by the same regions, II, III, IV, V, VI, of the Venn diagram. Therefore, $(A \cup B) \cap (B \cup C) = B \cup (A \cap C)$ for all sets A, B, and C.

65. $(A \cup B)'$

67. $(A \cup B) \cap C'$

69. a) $A \cap B = \{3,6\}$, so $(A \cap B)' = \{1,2,4,5,7,8\}$
$A' = \{4,5,7,8\}$ and $B' = \{1,2,4,5,8\}$, so $A' \cup B' = \{1,2,4,5,7,8\}$
Both equal $\{1,2,4,5,7,8\}$.

b) Answers will vary.

71. 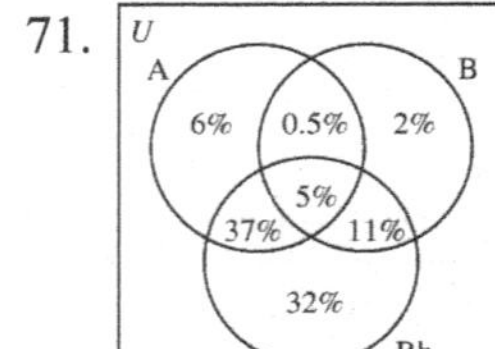

73. a) A: Office Building Construction Projects
B: Plumbing Projects
C: Budget Greater Than \$300,000

b) Region V; $A \cap B \cap C$

c) Region VI; $A' \cap B \cap C$

d) Region I; $A \cap B' \cap C'$

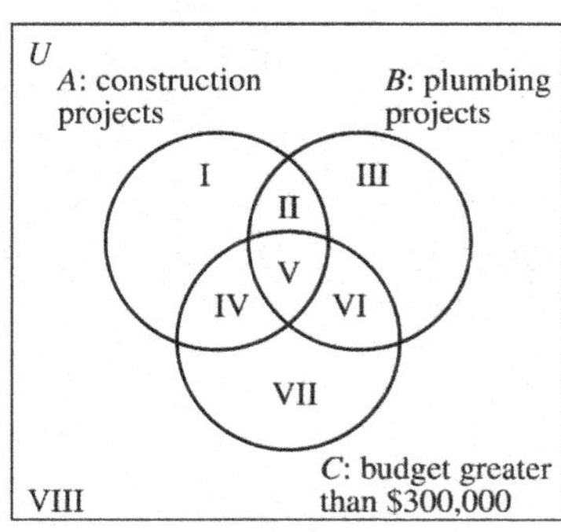

75. a)

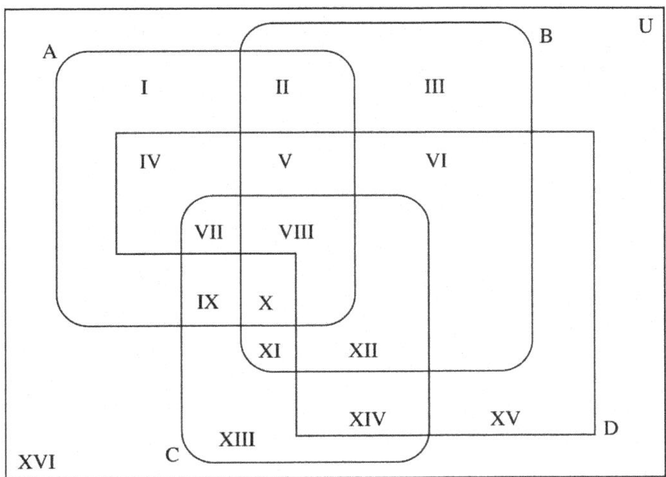

b)

Region	Set	Region	Set
I	$A \cap B' \cap C' \cap D'$	IX	$A \cap B' \cap C \cap D'$
II	$A \cap B \cap C' \cap D'$	X	$A \cap B \cap C \cap D'$
III	$A' \cap B \cap C' \cap D'$	XI	$A' \cap B \cap C \cap D'$
IV	$A \cap B' \cap C' \cap D$	XII	$A' \cap B \cap C \cap D$
V	$A \cap B \cap C' \cap D$	XIII	$A' \cap B' \cap C \cap D'$
VI	$A' \cap B \cap C' \cap D$	XIV	$A' \cap B' \cap C \cap D$
VII	$A \cap B' \cap C \cap D$	XV	$A' \cap B' \cap C' \cap D$
VIII	$A \cap B \cap C \cap D$	XVI	$A' \cap B' \cap C' \cap D'$

Section 2.5: Applications of Sets

1. a) 97
   b) 58
   c) $250-(97+80+58)=15$

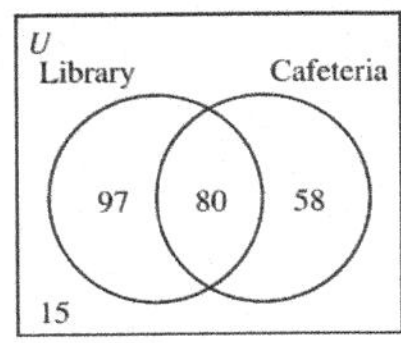

3. a) 12
   b) 17
   c) $12+35+17=64$

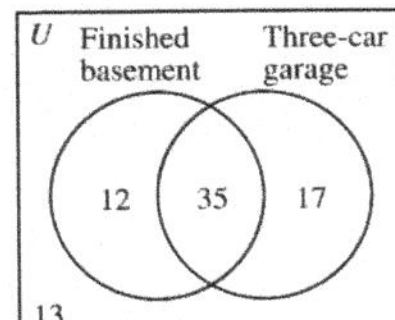

5. a) 11
   b) 12
   c) $11+12+8+23+22+16=92$
   d) $11+12+8=31$
   e) $12+16+23=51$

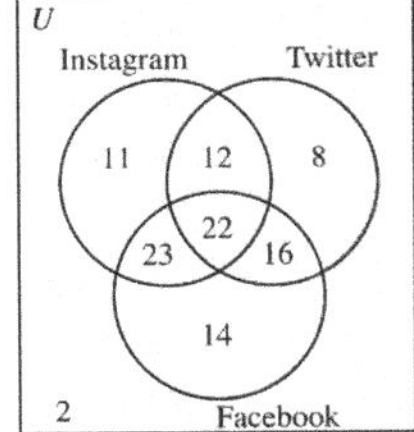

7. a) 20
   b) 121
   c) $121 + 83 + 40 = 244$
   d) $16 + 38 + 11 = 65$
   e) $350 - 20 - 40 = 290$

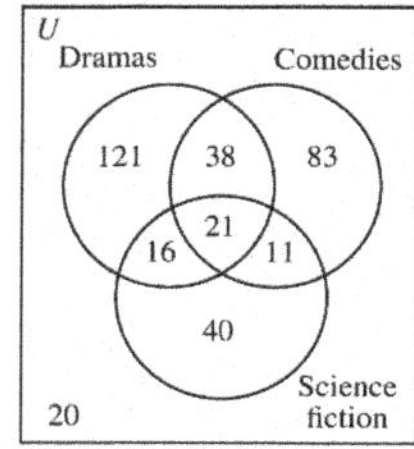

9. a) $128 + 40 + 51 + 27 + 58 + 32 + 85 + 29 = 450$
   b) 40
   c) 85
   d) $27 + 40 + 32 = 99$
   e) $450 - 29 = 421$

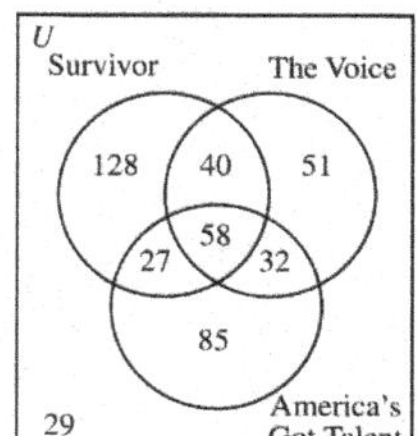

11. a) $30 + 37 = 67$
    b) $350 - 25 - 88 = 237$
    c) 37
    d) 25

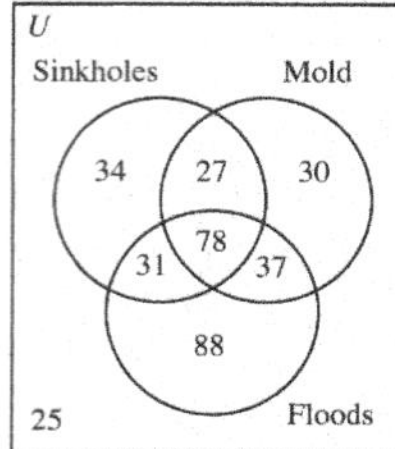

13. The Venn diagram shows the number of cars driven by women is 37, the sum of the numbers in Regions II, IV, V. This exceeds the 35 women the agent claims to have surveyed.

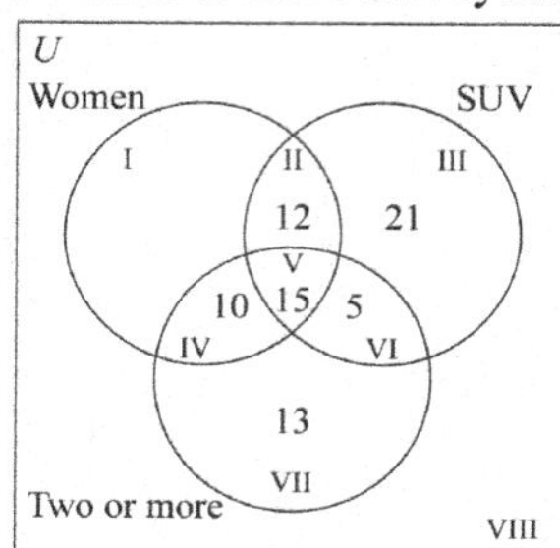

15. First fill in 125, 110, and 90 on the Venn diagram on the next page. Referring to the labels in the Venn diagram and the given information, we have the following.

    $a + c = 60$

    $b + c = 50$

    $a + b + c = 200 - 125 = 75$

    Adding the first two equations and subtracting the third from this sum gives $c = 60 + 50 - 75 = 35$. Then $a = 25$ and $b = 15$. Then $d = 180 - 110 - 25 - 35 = 10$. We now have labeled all the regions except the region outside the three circles, so the number of farmers growing at least one of the crops is $125 + 25 + 110 + 15 + 35 + 10 + 90$, or 410. Thus, the number growing none of the crops is $500 - 410$, or 90.

    a) 410
    b) 35
    c) 90
    d) $15 + 25 + 10 = 50$

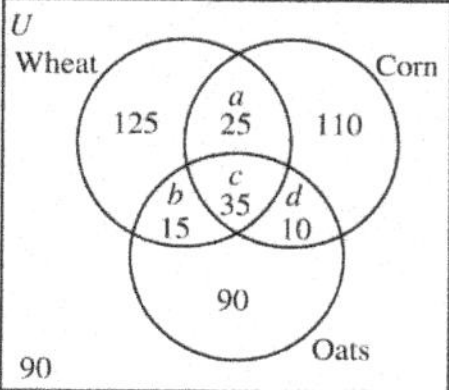

17. From the given information we can generate the Venn diagram. First fill in 4 for Region V. Then since the intersections in pairs all have 6 elements, we can fill in 2 for each of Regions II, IV, and VI. This already accounts for the 10 elements $A \cup B \cup C$, so the remaining 2 elements in U must be in Region VIII.

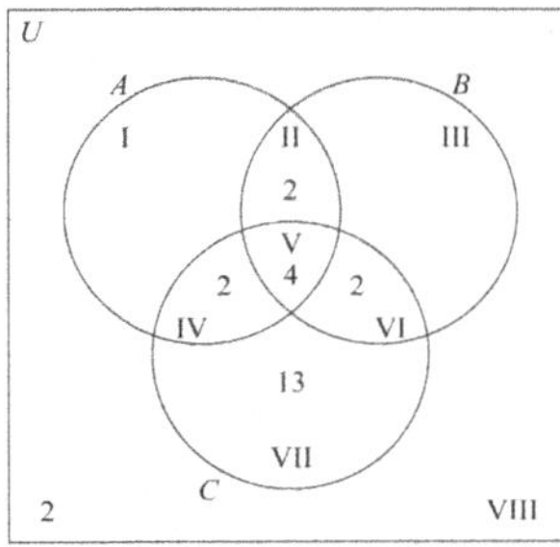

a) 10, the sum of the numbers in Regions I, II, III, IV, V, VI
b) 10, the sum of the numbers in Regions III, IV, V, VI, VIII
c) 6, the sum of the numbers in Regions I, III, IV, VI, VIII

Section 2.6: Infinite Sets

1. Infinite

3. $$\begin{array}{ccccccccc} \{5, & 6, & 7, & 8, & 9, & \ldots, & n+4, & \ldots\} \\ \downarrow & \downarrow & \downarrow & \downarrow & \downarrow & & \downarrow & \\ \{6, & 7, & 8, & 9, & 10, & \ldots, & n+5, & \ldots\} \end{array}$$

5. $$\begin{array}{cccccccc} \{6, & 8, & 10, & 12, & 14, & \ldots, & 2n+4, & \ldots\} \\ \downarrow & \downarrow & \downarrow & \downarrow & \downarrow & & \downarrow & \\ \{8, & 10, & 12, & 14, & 16, & \ldots, & 2n+6, & \ldots\} \end{array}$$

7. $$\begin{array}{cccccccc} \{5, & 7, & 9, & 11, & 13, & \ldots, & 2n+3, & \ldots\} \\ \downarrow & \downarrow & \downarrow & \downarrow & \downarrow & & \downarrow & \\ \{9, & 11, & 13, & 15, & 17, & \ldots, & 2n+5, & \ldots\} \end{array}$$

9. $$\begin{array}{cccccccc} \{\frac{1}{2}, & \frac{1}{4}, & \frac{1}{6}, & \frac{1}{8}, & \frac{1}{10}, & \ldots, & \frac{1}{2n}, & \ldots\} \\ \downarrow & \downarrow & \downarrow & \downarrow & \downarrow & & \downarrow & \\ \{\frac{1}{4}, & \frac{1}{6}, & \frac{1}{8}, & \frac{1}{10}, & \frac{1}{12}, & \ldots, & \frac{1}{2n+2}, & \ldots\} \end{array}$$

11. $$\begin{array}{cccccccc} \{\frac{4}{11}, & \frac{5}{11}, & \frac{6}{11}, & \frac{7}{11}, & \frac{8}{11}, & \ldots, & \frac{n+3}{11}, & \ldots\} \\ \downarrow & \downarrow & \downarrow & \downarrow & \downarrow & & \downarrow & \\ \{\frac{5}{11}, & \frac{6}{11}, & \frac{7}{11}, & \frac{8}{11}, & \frac{9}{11}, & \ldots, & \frac{n+4}{11}, & \ldots\} \end{array}$$

13. $$\begin{array}{cccccccc} \{1, & 2, & 3, & 4, & 5, & \ldots, & n, & \ldots\} \\ \downarrow & \downarrow & \downarrow & \downarrow & \downarrow & & \downarrow & \\ \{3, & 6, & 9, & 12, & 15, & \ldots, & 3n, & \ldots\} \end{array}$$

15. $$\begin{array}{cccccccc} \{1, & 2, & 3, & 4, & 5, & \ldots, & n, & \ldots\} \\ \downarrow & \downarrow & \downarrow & \downarrow & \downarrow & & \downarrow & \\ \{4, & 6, & 8, & 10, & 12, & \ldots, & 2n+2, & \ldots\} \end{array}$$

17. $$\begin{array}{cccccccc} \{1, & 2, & 3, & 4, & 5, & \ldots, & n & \ldots\} \\ \downarrow & \downarrow & \downarrow & \downarrow & \downarrow & & \downarrow & \\ \{2, & 5, & 8, & 11, & 14, & \ldots, & 3n-1, & \ldots\} \end{array}$$

19. $$\begin{array}{cccccccc} \{1, & 2, & 3, & 4, & 5, & \cdots, & n, & \cdots\} \\ \downarrow & \downarrow & \downarrow & \downarrow & \downarrow & & \downarrow & \\ \{\frac{1}{3}, & \frac{1}{6}, & \frac{1}{9}, & \frac{1}{12}, & \frac{1}{15}, & \cdots, & \frac{1}{3n}, & \cdots\} \end{array}$$

21. $$\begin{array}{cccccccc} \{1, & 2, & 3, & 4, & 5, & \cdots, & n, & \cdots\} \\ \downarrow & \downarrow & \downarrow & \downarrow & \downarrow & & \downarrow & \\ \{\frac{1}{3}, & \frac{1}{4}, & \frac{1}{5}, & \frac{1}{6}, & \frac{1}{7}, & \cdots, & \frac{1}{n+2}, & \cdots\} \end{array}$$

23. $$\begin{array}{cccccccc} \{1, & 2, & 3, & 4, & 5, & \ldots, & n, & \ldots\} \\ \downarrow & \downarrow & \downarrow & \downarrow & \downarrow & & \downarrow & \\ \{1, & 4, & 9, & 16, & 25, & \ldots, & n^2, & \ldots\} \end{array}$$

25. $$\begin{array}{cccccccc} \{1, & 2, & 3, & 4, & 5, & \ldots, & n, & \ldots\} \\ \downarrow & \downarrow & \downarrow & \downarrow & \downarrow & & \downarrow & \\ \{3, & 9, & 27, & 81, & 243, & \ldots, & 3^n, & \ldots\} \end{array}$$

27. =

29. =

31. =

32. a) Answers will vary.
    b) No

Review Exercises

1. True
2. False; the word best makes the statement not well-defined.
3. True
4. False; no set is a proper subset of itself.
5. False; the elements 6, 12, 18, 24, ... are members of both sets.
6. True
7. False; the two sets do not contain exactly the same elements.
8. True; both sets contain the same number of elements, 4.
9. True
10. True
11. True
12. True
13. True
14. True
15. $A = \{7, 9, 11, 13, 15\}$
16. $B = \{\text{Colorado, Nebraska, Missouri, Oklahoma}\}$
17. $C = \{1, 2, 3, 4, \ldots, 174\}$
18. $D = \{9, 10, 11, 12, \ldots, 80\}$
19. $A = \{x \mid x \in N \text{ and } 50 < x < 150\}$ or $A = \{x \mid x \text{ in } N \text{ and } 51 \le x \le 149\}$
20. $B = \{x \mid x \in N \text{ and } x > 42\}$
21. $C = \{x \mid x \in N \text{ and } x < 7\}$
22. $D = \{x \mid x \in N \text{ and } 27 \le x \le 51\}$
23. A is the set of capital letters in the English alphabet from E through M, inclusive.
24. B is the set of U.S. coins with a value of less than one dollar.
25. C is the set of the first three lowercase letters in the English alphabet.
26. D is the set of numbers greater than or equal to 3 and less than 9.
27. $A \cap B = \{1, 3, 5, 7\} \cap \{3, 7, 9, 10\} = \{3, 7\}$
28. $A \cup B' = \{1, 3, 5, 7\} \cup \{3, 7, 9, 10\}'$
$= \{1, 3, 5, 7\} \cup \{1, 2, 4, 5, 6, 8\}$
$= \{1, 2, 3, 4, 5, 6, 7, 8\}$
29. $A' \cap B = \{1, 3, 5, 7\}' \cap \{3, 7, 9, 10\}$
$= \{2, 4, 6, 8, 9, 10\} \cap \{5, 7, 9, 10\}$
$= \{9, 10\}$
30. $(A \cup B)' \cup C = (\{1, 3, 5, 7\} \cup \{3, 7, 9, 10\})' \cup \{1, 7, 10\}$
$= \{1, 3, 5, 7, 9, 10\}' \cup \{1, 7, 10\}$
$= \{2, 4, 6, 8\} \cup \{1, 7, 10\}$
$= \{1, 2, 4, 6, 7, 8, 10\}$
31. $A - B = \{1, 3, 5, 7\} - \{3, 7, 9, 10\} = \{1, 5\}$
32. $A - C' = \{1, 3, 5, 7\} - \{1, 7, 10\}'$
$= \{1, 3, 5, 7\} - \{2, 3, 4, 5, 6, 8, 9\}$
$= \{1, 7\}$
33. $\{(1, 1), (1, 7), (1, 10), (3, 1), (3, 7), (3, 10), (5, 1), (5, 7), (5, 10), (7, 1), (7, 7), (7, 10)\}$
34. $\{(3, 1), (3, 3), (3, 5), (3, 7), (7, 1), (7, 3), (7, 5), (7, 7), (9, 1), (9, 3), (9, 5), (9, 7), (10, 1), (10, 3), (10, 5), (10, 7)\}$

35. $2^4 = 16$

36. $2^4 - 1 = 16 - 1 = 15$

37.

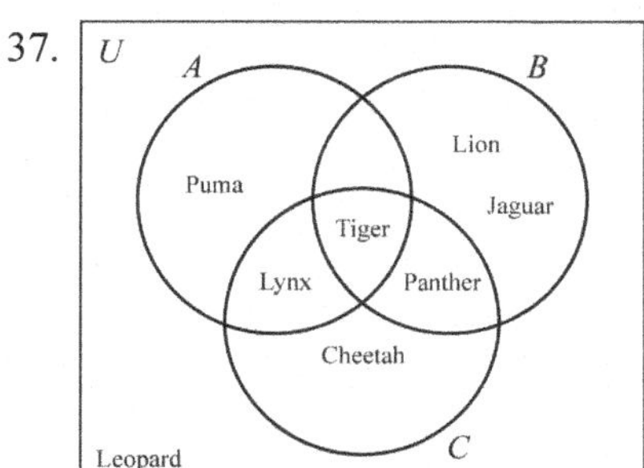

38. $A \cup C = \{b, c, d, e, f, h, k, l\}$

39. $A \cap B' = \{e, k\}$

40. $A \cup B \cup C = \{a, b, c, d, e, f, g, h, k, l\}$

41. $A \cap B \cap C = \{f\}$

42. $(A \cup B) \cap C = \{c, e, f\}$

43. $A - B' = \{d, f, l\}$

44. $(A' \cup B')'$

Set	Regions
A	I, II
A'	III, IV
B	II, III
B'	I, IV
$A' \cup B'$	I, III, IV
$(A' \cup B')'$	II

$A \cap B$

Set	Regions
A	I, II
B	II, III
$A \cap B$	II

Both statements are represented by the same region, II, of the Venn diagram. Therefore, $(A' \cup B')' = A \cap B$ for all sets A and B.

45. $(A \cup B') \cup (A \cup C')$

Set	Regions
A	I, II, IV, V
B	II, III, V, VI
B'	I, IV, VII, VIII
$A \cup B'$	I, II, IV, V, VII, VIII
C	IV, V, VI, VII
C'	I, II, III, VIII
$A \cup C'$	I, II, III, IV, V, VIII
$(A \cup B') \cup (A \cup C')$	I, II, III, IV, V, VII, VIII

$A \cup (B \cap C)'$

Set	Regions
B	II, III, V, VI
C	IV, V, VI, VII
$B \cap C$	V, VI
$(B \cap C)'$	I, II, III, IV, VII, VIII
A	I, II, IV, V
$A \cup (B \cap C)'$	I, II, III, IV, V, VII, VIII

Both statements are represented by the same regions, I, II, III, IV, V, VII, VIII, of the Venn diagram. Therefore, $(A \cup B') \cup (A \cup C') = A \cup (B \cap C)'$ for all sets A, B, and C.

46. II

47. III

48. I

49. IV

50. IV

51. II

52. II

53. The company paid $450 since the sum of the numbers in Regions I through IV is 450.

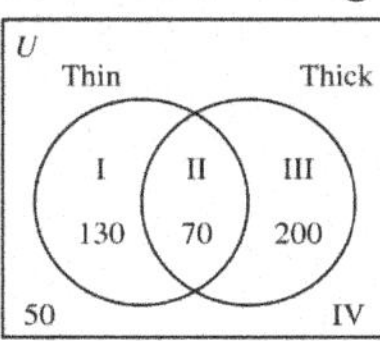

54. a) 131, the sum of the numbers in Regions I through VIII
    b) 32, Region I
    c) 10, Region II
    d) 65, the sum of the numbers in Regions I, IV, VII

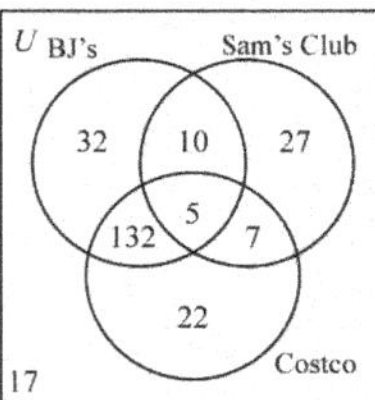

55. a) 31, Region I
    b) 74, the sum of the numbers in Regions I, III, VII
    c) 71, Region VI
    d) 185, the sum of the numbers in Regions I, II, III
    e) 328, the sum of the numbers in Regions II, IV, VI

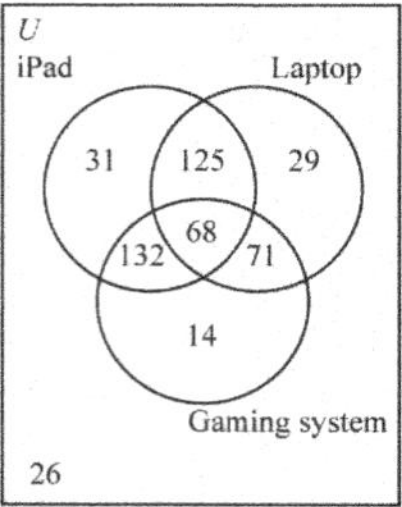

56. $\{2, 4, 6, 8, 10, \ldots, 2n, \ldots\}$
$\downarrow \downarrow \downarrow \downarrow \downarrow \quad \downarrow$
$\{4, 6, 8, 10, 12, \ldots, 2n+2, \ldots\}$

57. $\{3, 5, 7, 9, 11, \ldots, 2n+1, \ldots\}$
$\downarrow \downarrow \downarrow \downarrow \downarrow \quad \downarrow$
$\{5, 7, 9, 11, 13, \ldots, 2n+3, \ldots\}$

58. $\{1, 2, 3, 4, 5, \ldots, n, \ldots\}$
$\downarrow \downarrow \downarrow \downarrow \downarrow \quad \downarrow$
$\{5, 8, 11, 14, 17, \ldots, 2n+3, \ldots\}$

59. $\{1, 2, 3, 4, 5, \ldots, n, \ldots\}$
$\downarrow \downarrow \downarrow \downarrow \downarrow \quad \downarrow$
$\{4, 9, 14, 19, 24, \ldots, 5n-1, \ldots\}$

Chapter Test

1. True; both sets contain the same number of elements, 4.
2. False; the sets do not contain exactly the same elements.
3. True
4. False; the second set does not contain the element 7.
5. False; the set has $2^4 = 16$ subsets.
6. True
7. False; for any set A, $A \cup A' = U$, not { }.
8. True
9. $A = \{1, 2, 3, 4, 5, 6, 7, 8, 9, 10, 11\}$
10. Set A is the set of natural numbers less than 12.

For exercises 11–16, refer to the Venn Diagram to the right.

11. $A \cap B = \{2, 4, 6, 8\} \cap \{6, 8, 10, 12\} = \{6, 8\}$

12. $A \cup C' = \{2, 4, 6, 8\} \cup \{4, 6, 8, 12\} = \{2, 4, 6, 8, 12\}$

13. $A \cap \left(B \cap C'\right) = \{2, 4, 6, 8\} \cap \{6, 8, 12\} = \{6, 8\}$

14. $n(A \cap B') = n(\{2, 4\}) = 2$

15. $A - B = \{2, 4, 6, 8\} - \{6, 8, 10, 12\} = \{2, 4\}$

16. $A \times C = \{(2, 2), (2, 10), (2, 14), (4, 2), (4, 10), (4, 14), (6, 2), (6, 10), (6, 14), (8, 2), (8, 10), (8, 14)\}$

17. 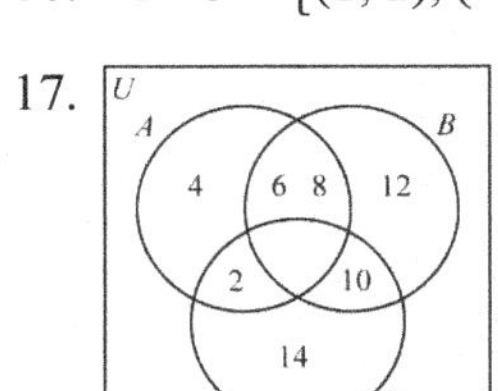

18.

$A \cap (B \cup C')$

Set	Regions
B	II, III, V, VI
C	IV, V, VI, VII
C'	I, II, III, VIII
$B \cup C'$	I, II, III, V, VI, VIII
A	I, II, IV, V
$A \cap (B \cup C')$	I, II, V

$(A \cap B) \cup (A \cap C')$

Set	Regions
A	I, II, IV, V
B	II, III, V, VI
$A \cap B$	II, V
C	IV, V, VI, VII
C'	I, II, III, VIII
$A \cap C'$	I, II
$(A \cap B) \cup (A \cap C')$	I, II, V

Both statements are represented by the same regions, I, II, V, of the Venn diagram. Therefore, $A \cap (B \cup C') = (A \cap B) \cup (A \cap C')$ for all sets A, B, and C.

19. a) 58, the sum of the numbers in Regions II, IV, VI
    b) 10, Region VIII
    c) 145, the sum of the numbers in all regions except VIII
    d) 22, Region II
    e) 69, the sum of the numbers in Regions I, II, III
    f) 16, Region III

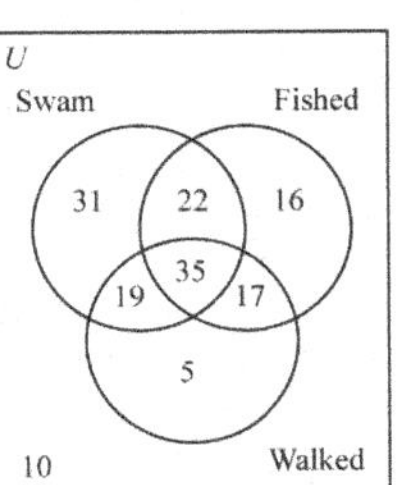

20. $\{7, 8, 9, 10, 11, \ldots, n+6, \ldots\}$

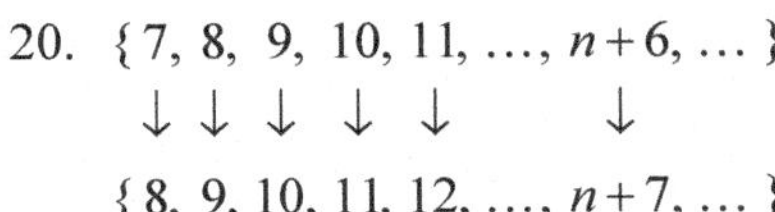

$\{8, 9, 10, 11, 12, \ldots, n+7, \ldots\}$

Chapter Three: Logic

Section 3.1: Statements and Logical Connectives

1. Statement
3. Compound
5. a) Not
   b) And
   c) Or
7. Simple statement
9. Compound; biconditional $\leftrightarrow$
11. Compound; conjunction, $\wedge$
13. Compound; conditional, $\rightarrow$
15. Compound; negation, $\sim$
17. Some Eco Sun scooters are not made by Amigo.
19. All turtles have claws.
21. Some bicycles have three wheels.
23. No pedestrians are in the crosswalk.
25. Some mountain climbers are teachers.
27. $\sim p$
29. $\sim q \vee \sim p$
31. $\sim p \rightarrow \sim q$
33. $p \wedge \sim q$
35. $\sim q \leftrightarrow p$
37. $\sim (p \vee q)$
39. Brie does not have a MacBook.
41. Joe has an iPad and Brie has a MacBook.
43. If Joe does not have an iPad then Brie has a MacBook.
45. Joe does not have an iPad or Brie does not have a MacBook.
47. It is false that Joe has an iPad and Brie has a MacBook.
49. $(p \wedge \sim q) \wedge r$
51. $(p \wedge q) \vee r$
53. $(r \wedge q) \rightarrow p$
55. $(r \leftrightarrow q) \wedge p$
57. The water is $70°$ or the sun is shining, and we do not go swimming.
59. The water is not $70°$, and the sun is shining or we go swimming.
61. If we do not go swimming, then the sun is shining and the water is $70°$.
63. If the sun is shining then we go swimming, and the water is $70°$.
65. The sun is shining if and only if the water is $70°$, and we go swimming.
67. Not permissible. In the list of choices, the connective "or" is the exclusive or, thus one can order either the soup or the salad, but not both items.
69. Not permissible. In the list of choices, the connective "or" is the exclusive or, thus one can order either potatoes or pasta, but not both items.
71. a) b: Johnny started the bonfire.
   m: Jaci forgot marshmallows.
   $b \wedge \sim m$
   b) Conjunction
73. a) w: work out
   g: gain weight
   $\sim (w \rightarrow \sim g)$
   b) Negation
75. a) f: Food has fiber.
   v: Food has vitamins.
   h: You will be healthy.
   $(f \vee v) \rightarrow h$
   b) Conditional
77. a) c: May take course.
   f: Failed previous exam.
   p: Passed placement test.
   $c \leftrightarrow (\sim f \vee p)$
   b) Biconditional

79. a) c: Classroom is empty
w: It is the weekend
s: It is 7:00 a.m.
$(c \leftrightarrow w) \vee s$

b) Disjunction

81. a) Answers will vary.
b) Answers will vary.

Section 3.2: Truth Tables for Negation, Conjunction, and Disjunction

1. Opposite

3. False

5.

p	p	$\wedge$	$\sim p$
T	T	F	F
F	F	F	T
	1	3	2

7.

p	q	$\sim p$	$\vee$	q
T	T	F	T	T
T	F	F	F	F
F	T	T	T	T
F	F	T	T	F
		1	3	2

9.

p	q	p	$\wedge$	$\sim q$
T	T	T	F	F
T	F	T	T	T
F	T	F	F	F
F	F	F	F	T
		1	3	2

11.

p	q	$\sim$	$(p$	$\wedge$	$\sim q)$
T	T	T	T	F	F
T	F	F	T	T	T
F	T	T	F	F	F
F	F	T	F	F	T
		4	1	3	2

13.

p	q	r	p	$\vee$	$(\sim q$	$\vee$	$r)$
T	T	T	T	T	F	T	T
T	T	F	T	T	F	F	F
T	F	T	T	T	T	T	T
T	F	F	T	T	T	T	F
F	T	T	F	T	F	T	T
F	T	F	F	F	F	F	F
F	F	T	F	T	T	T	T
F	F	F	F	T	T	T	F
			1	5	2	4	3

15.

p	q	r	$(p$	$\wedge$	$\sim q)$	$\vee$	r
T	T	T	T	F	F	T	T
T	T	F	T	F	F	F	F
T	F	T	T	T	T	T	T
T	F	F	T	T	T	T	F
F	T	T	F	F	F	T	T
F	T	F	F	F	F	F	F
F	F	T	F	F	T	T	T
F	F	F	F	F	T	F	F
			1	4	2	5	3

17.

p	q	r	$(\sim p$	$\wedge$	$q)$	$\vee$	$\sim r$
T	T	T	F	F	T	F	F
T	T	F	F	F	T	T	T
T	F	T	F	F	F	F	F
T	F	F	F	F	F	T	T
F	T	T	T	T	T	T	F
F	T	F	T	T	T	T	T
F	F	T	T	F	F	F	F
F	F	F	T	F	F	T	T
			2	4	3	5	1

19. p: Apples are a good source of fiber.
q: Oranges are a good source of vitamin C.
In symbolic form the statement is $p \wedge q$.

p	q	p	$\wedge$	q
T	T	T	T	T
T	F	T	F	F
F	T	F	F	T
F	F	F	F	F
		1	3	2

21. p: Joaquin will pitch.
q: Joaquin will play first base.
In symbolic form the statement is $p \vee \sim q$.

p	q	p	$\vee$	$\sim q$
T	T	T	T	F
T	F	T	T	T
F	T	F	F	F
F	F	F	T	T
		1	3	2

23. p: The car is a Chevrolet.
q: The car is a Corvette.
In symbolic form the statement is $\sim(\sim p \wedge q)$.

p	q	$\sim$	$(\sim p$	$\wedge$	$q)$
T	T	T	F	F	T
T	F	T	F	F	F
F	T	F	T	T	T
F	F	T	T	F	F
		4	1	3	2

27. a) $(p \wedge \sim q) \vee r$
$(\mathrm{T} \wedge \mathrm{T}) \wedge \mathrm{T}$
$\mathrm{T} \wedge \mathrm{T}$
T
Therefore, the statement is true.

b) $(p \wedge \sim q) \vee r$
$(\mathrm{F} \wedge \mathrm{F}) \wedge \mathrm{T}$
$\mathrm{F} \wedge \mathrm{T}$
T
Therefore, the statement is true.

29. a) $(\sim p \wedge \sim q) \vee \sim r$
$(\mathrm{F} \wedge \mathrm{T}) \wedge \mathrm{F}$
$\mathrm{F} \wedge \mathrm{F}$
F
Therefore, the statement is false.

b) $(\sim p \wedge \sim q) \vee \sim r$
$(\mathrm{T} \wedge \mathrm{F}) \wedge \mathrm{F}$
$\mathrm{F} \wedge \mathrm{F}$
F
Therefore, the statement is false.

25. p: School is in session.
q: The kids are home.
r: I am working.
In symbolic form the statement is $(\sim p \wedge q) \vee r$.

p	q	r	$(\sim p$	$\wedge$	$q)$	$\vee$	r
T	T	T	F	F	T	T	T
T	T	F	F	F	T	F	F
T	F	T	F	F	F	T	T
T	F	F	F	F	F	F	F
F	T	T	T	T	T	T	T
F	T	F	T	T	T	T	F
F	F	T	T	F	F	T	T
F	F	F	T	F	F	F	F
			1	3	2	5	4

31. a) $(p \vee \sim q) \wedge [\sim(p \wedge \sim r)]$
$(\mathrm{T} \vee \mathrm{T}) \wedge [\sim(\mathrm{T} \wedge \mathrm{F})]$
$\mathrm{T} \wedge \sim\mathrm{F}$
$\mathrm{T} \wedge \mathrm{T}$
T
Therefore, the statement is true.

b) $(p \vee \sim q) \wedge [\sim(p \wedge \sim r)]$
$(\mathrm{F} \vee \mathrm{F}) \wedge [\sim(\mathrm{F} \wedge \mathrm{F})]$
$\mathrm{F} \wedge \sim\mathrm{F}$
$\mathrm{F} \wedge \mathrm{T}$
F
Therefore, the statement is false.

33. a) $(\sim r \wedge p) \vee q$
$(\mathrm{F} \wedge \mathrm{T}) \vee \mathrm{F}$
$\mathrm{F} \vee \mathrm{F}$
F
Therefore, the statement is false.

b) $(\sim r \wedge p) \vee q$
$(\mathrm{F} \wedge \mathrm{F}) \vee \mathrm{T}$
$\mathrm{F} \vee \mathrm{T}$
T
Therefore, the statement is true.

35. a) $(\sim p \vee \sim q) \vee (\sim r \vee q)$

$(\text{F} \vee \text{T}) \vee (\text{F} \vee \text{F})$

$\text{T} \vee \text{F}$

T

Therefore, the statement is true.

b) $(\sim p \vee \sim q) \vee (\sim r \vee q)$

$(\text{T} \vee \text{F}) \vee (\text{F} \vee \text{T})$

$\text{T} \vee \text{F}$

T

Therefore, the statement is true.

37. $8+7=20+5$ and $63 \div 7 = 3 \cdot 3$

$\text{T} \wedge \text{T}$

T

Therefore, the statement is true.

39. C: Chevrolet makes trucks.
T: Toyota makes shoes.

$C \vee T$

$\text{T} \vee \text{F}$

T

Therefore, the statement is true.

41. G: George Washington was the first U.S. president.
A: Abraham Lincoln was the second U.S. president.
H: Harry Truman was the third U.S. president.

$(G \vee A) \wedge \sim H$

$(\text{T} \vee \text{F}) \wedge \sim\text{F}$

$\text{T} \wedge \text{T}$

T

Therefore, the statement is false.

43. C: Chicago is in Mexico.
L: Los Angeles is in California.
D: Dallas is in Canada.

$(C \vee L) \vee D$

$(\text{F} \vee \text{T}) \vee \text{F}$

$\text{T} \vee \text{F}$

T

Therefore, the statement is true.

45. p: Russia is larger than Canada.
q: China is larger than Brazil.

$p \wedge q$

$\text{T} \wedge \text{T}$

T

Therefore, the statement is true.

47. p: The area of Brazil and China combined is larger than the area of Russia.
q: The area of the United States and Canada combined is larger than the area of Russia.

$p \wedge \sim q$

$\text{T} \wedge \sim\text{T}$

$\text{T} \wedge \text{F}$

F

Therefore, the statement is false.

49. p: 30% of Americans get 6 hours of sleep.
q: 9% get 5 hours of sleep.

$\sim(p \wedge q)$

$\sim(\text{F} \wedge \text{T})$

$\sim\text{F}$

T

Therefore, the statement is true.

51. p: 13% of Americans get at most 5 hours of sleep.
q: 32% of Americans get at least 6 hours of sleep.
r: 30% of Americans get at least 8 hours of sleep.

$(p \vee q) \wedge r$

$(\text{T} \vee \text{F}) \wedge \text{F}$

$\text{T} \wedge \text{F}$

F

Therefore, the statement is false.

53.

p	q	p	$\wedge$	$\sim q$
T	T	T	F	F
T	F	T	T	T
F	T	F	F	F
F	F	F	F	T
		1	3	2

True in case 2.

55.

p	q	p	$\vee$	$\sim q$
T	T	T	T	F
T	F	T	T	T
F	T	F	F	F
F	F	F	T	T
		1	3	2

True in cases 1, 2, and 4.

57.

p	q	r	$(r$	$\vee$	$q)$	$\wedge$	p
T	T	T	T	T	T	T	T
T	T	F	F	T	T	T	T
T	F	T	T	T	F	T	T
T	F	F	F	F	F	F	T
F	T	T	T	T	T	F	F
F	T	F	F	T	T	F	F
F	F	T	T	T	F	F	F
F	F	F	F	F	F	F	F
			1	3	2	5	4

True in cases 1, 2, and 3.

59.

p	q	r	q	$\vee$	$(p$	$\wedge$	$\sim r)$
T	T	T	T	T	T	F	F
T	T	F	T	T	T	T	T
T	F	T	F	F	T	F	F
T	F	F	F	T	T	T	T
F	T	T	T	T	F	F	F
F	T	F	T	T	F	F	T
F	F	T	F	F	F	F	F
F	F	F	F	F	F	F	T
			4	5	1	2	3

True in cases 1, 2, 4, 5, and 6.

61. a) Both Mr. Duncan and Mrs. Tuttle qualify for the loan.
    b) Mrs. Rusinek does not qualify, since their combined income is less than $46,000.

63. a) Xavier qualifies for the special fare. The other four do not qualify.
    b) Gina is returning on April 3. Kara is returning on a Monday. Christos is not staying over on a Saturday. Chang is returning on a Monday.

65.

p	q	r	$[(q$	$\wedge$	$\sim r)$	$\wedge$	$(\sim p$	$\vee$	$\sim q)]$	$\vee$	$(p$	$\vee$	$\sim r)$
T	T	T	T	F	F	F	F	F	F	T	T	T	F
T	T	F	T	T	T	T	F	T	F	T	T	T	T
T	F	T	F	F	F	F	F	T	T	T	T	T	F
T	F	F	F	F	T	F	F	T	T	T	T	T	T
F	T	T	T	F	F	F	T	T	F	F	F	F	F
F	T	F	T	T	T	T	T	T	F	T	F	T	T
F	F	T	F	F	F	F	T	T	T	F	F	F	F
F	F	F	F	F	T	F	T	T	T	T	F	T	T
			1	3	2	7	4	6	5	11	8	10	9

67. Yes. Since only the locations of the variables change between statements, the combinations of true and false values will be the same in both.

p	q	r	$(p$	$\wedge$	$\sim q)$	$\vee$	r	$(q$	$\wedge$	$\sim r)$	$\vee$	p
T	T	T	T	F	F	T	T	T	F	F	T	T
T	T	F	T	F	F	F	F	T	T	T	T	T
T	F	T	T	T	T	T	T	F	F	F	T	T
T	F	F	T	T	T	T	F	F	F	T	T	T
F	T	T	F	F	F	T	T	T	F	F	F	F
F	T	F	F	F	F	F	F	T	T	T	T	F
F	F	T	F	F	T	T	T	F	F	F	F	F
F	F	F	F	F	T	F	F	F	F	T	F	F
			1	3	2	5	4	1	3	2	5	4

Section 3.3: Truth Tables for the Conditional and Biconditional

1. False

3. True

5. Self-contradiction

7.

p	q	$\sim p$	$\rightarrow$	q
T	T	F	T	T
T	F	F	T	F
F	T	T	T	T
F	F	T	F	F
		1	3	2

9.

p	q	$\sim p$	$\leftrightarrow$	$\sim q$
T	T	F	T	F
T	F	F	F	T
F	T	T	F	F
F	F	T	T	T
		1	3	2

11.

p	q	$(p$	$\rightarrow$	$q)$	$\leftrightarrow$	p
T	T	T	T	T	T	T
T	F	T	F	F	F	T
F	T	F	T	T	F	F
F	F	F	T	F	F	F
		1	3	2	5	4

13.

p	q	p	$\leftrightarrow$	$(q$	$\vee$	$p)$
T	T	T	T	T	T	T
T	F	T	T	F	T	T
F	T	F	F	T	T	F
F	F	F	T	F	F	F
		1	5	2	4	3

15.

p	q	$(\sim p$	$\rightarrow$	$q)$	$\leftrightarrow$	$(p$	$\rightarrow$	$\sim q)$
T	T	F	T	T	F	T	F	F
T	F	F	T	F	T	T	T	T
F	T	T	T	T	T	F	T	F
F	F	T	F	F	F	F	T	T
		1	3	2	7	4	6	5

17.

p	q	r	$\sim p$	$\rightarrow$	$(q$	$\wedge$	$r)$
T	T	T	F	T	T	T	T
T	T	F	F	T	T	F	F
T	F	T	F	T	F	F	T
T	F	F	F	T	F	F	F
F	T	T	T	T	T	T	T
F	T	F	T	F	T	F	F
F	F	T	T	F	F	F	T
F	F	F	T	F	F	F	F
			1	5	2	4	3

19.

p	q	r	$\sim p$	$\leftrightarrow$	$(q$	$\vee$	$\sim r)$
T	T	T	F	F	T	T	F
T	T	F	F	F	T	T	T
T	F	T	F	T	F	F	F
T	F	F	F	F	F	T	T
F	T	T	T	T	T	T	F
F	T	F	T	T	T	T	T
F	F	T	T	F	F	F	F
F	F	F	T	T	F	T	T
			1	5	2	4	3

21.

p	q	r	$(\sim p$	$\wedge$	$q)$	$\rightarrow$	$\sim r$
T	T	T	F	F	T	T	F
T	T	F	F	F	T	T	T
T	F	T	F	F	F	T	F
T	F	F	F	T	F	T	T
F	T	T	T	T	T	F	F
F	T	F	T	T	T	T	T
F	F	T	T	F	F	T	F
F	F	F	T	F	F	T	T
			1	3	2	5	4

23.

p	q	r	$(p$	$\rightarrow$	$q)$	$\leftrightarrow$	$(\sim q$	$\rightarrow$	$\sim r)$
T	T	T	T	T	T	T	F	T	F
T	T	F	T	T	T	T	F	T	T
T	F	T	T	F	F	T	T	F	F
T	F	F	T	F	F	F	T	T	T
F	T	T	F	T	T	T	F	T	F
F	T	F	F	T	T	T	F	T	T
F	F	T	F	T	F	F	T	F	F
F	F	F	F	T	F	T	T	T	T
			1	3	2	7	4	6	5

25. p: Today is Monday
q: The library is open;
r: We can study together.

p	q	r	p	$\rightarrow$	$(q$	$\wedge$	$r)$
T	T	T	T	T	T	T	T
T	T	F	T	F	T	F	F
T	F	T	T	F	F	F	T
T	F	F	T	F	F	F	F
F	T	T	F	T	T	T	T
F	T	F	F	T	T	F	F
F	F	T	F	T	F	F	T
F	F	F	F	T	F	F	F
			1	5	2	4	3

27. p: Blackberries are high in vitamin K.
q: Mangos are high in B vitamins.
r: Cherries are high in vitamin C.

p	q	r	$(p$	$\leftrightarrow$	$q)$	$\vee$	r
T	T	T	T	T	T	T	T
T	T	F	T	T	T	T	F
T	F	T	T	F	F	T	T
T	F	F	T	F	F	F	F
F	T	T	F	F	T	T	T
F	T	F	F	F	T	F	F
F	F	T	F	T	F	T	T
F	F	F	F	T	F	T	F
			1	3	2	5	4

29. p: It is too cold.
 q: We can take a walk.
 r: We can go to the gym.

p	q	r	$(\sim p$	$\to$	$q)$	$\vee$	r
T	T	T	F	T	T	T	T
T	T	F	F	T	T	T	F
T	F	T	F	T	F	T	T
T	F	F	F	T	F	T	F
F	T	T	T	T	T	T	T
F	T	F	T	T	T	T	F
F	F	T	T	F	F	T	T
F	F	F	T	F	F	F	F
			1	3	2	5	4

31. Tautology

p	q	$\sim p$	$\vee$	$(p$	$\vee$	$q)$
T	T	F	T	T	T	T
T	F	F	T	T	T	F
F	T	T	T	F	T	T
F	F	T	T	F	F	F
		1	5	2	4	3

33. Self-contradiction

p	q	$(\sim p$	$\wedge$	$q)$	$\wedge$	$(p$	$\vee$	$\sim q)$
T	T	F	F	T	F	T	T	F
T	F	F	F	F	F	T	T	T
F	T	T	T	T	F	F	F	F
F	F	T	F	F	F	F	T	T
		1	3	2	7	4	6	5

35. Neither

p	q	$\sim[$	$(p$	$\vee$	$q)$	$\leftrightarrow$	$q]$
T	T	F	T	T	T	T	T
T	F	T	T	T	F	F	F
F	T	F	F	T	T	T	T
F	F	F	F	F	F	T	F
		6	1	3	2	5	4

37. Not an implication

p	$\sim p$	$\to$	p
T	F	T	T
F	T	F	F
	1	3	2

39. Implication

p	q	$\sim p$	$\rightarrow$	$\sim$	$(p$	$\wedge$	$q)$
T	T	F	T	F	T	T	T
T	F	F	T	T	T	F	F
F	T	T	T	T	F	F	T
F	F	T	T	T	F	F	F
		1	6	5	2	4	3

41. Implication

p	q	$[(p$	$\rightarrow$	$q)$	$\wedge$	$(q$	$\rightarrow$	$p)]$	$\rightarrow$	$(p$	$\leftrightarrow$	$q)$
T	T	T	T	T	T	T	T	T	T	T	T	T
T	F	T	F	F	F	F	T	T	T	T	F	F
F	T	F	T	T	F	T	F	F	T	F	F	T
F	F	F	T	F	T	F	T	F	T	F	T	F
		1	3	2	7	4	6	5	11	8	10	9

43. $p \rightarrow (q \rightarrow r)$

$\text{T} \rightarrow (\text{F} \rightarrow \text{F})$

$\text{T} \rightarrow \quad \text{T}$

T

45. $(p \wedge q) \leftrightarrow (q \vee \sim r)$

$(\text{T} \wedge \text{F}) \leftrightarrow (\text{F} \vee \text{T})$

$\text{F} \leftrightarrow \text{F}$

F

47. $(\sim p \wedge \sim q) \vee \sim r$

$(\text{F} \wedge \text{T}) \vee \text{T}$

$\text{F} \vee \text{T}$

T

49. $(\sim p \leftrightarrow r) \vee (\sim q \leftrightarrow r)$

$(\text{F} \leftrightarrow \text{T}) \vee (\text{T} \leftrightarrow \text{F})$

$\text{F} \vee \text{T}$

T

51. If $1+1=3$, then $5-2=3$.

$\text{T} \rightarrow \text{T}$

T

53. If the United States has 60 states or the United States' capital is Washington D.C., then the capital of Mexico is Philadelphia.

$$(\text{F} \vee \text{T}) \rightarrow \text{F}$$
$$\text{T} \rightarrow \text{F}$$
$$\text{F}$$

55. Snickers is a brand of watches and Hershey's is a brand of cell phones, if and only if Reece's is a brand of trucks.

$$(\text{F} \wedge \text{F}) \leftrightarrow \text{F}$$
$$\text{F} \leftrightarrow \text{F}$$
$$\text{T}$$

57. Independence Day is in July and Labor Day is in September, if and only if Thanksgiving is in April.

$$(\text{T} \wedge \text{T}) \leftrightarrow \text{F}$$
$$\text{T} \leftrightarrow \text{F}$$
$$\text{F}$$

59. Io has a diameter of 1000–3161 miles, or Thebe may have water, and Io may have atmosphere.

$$(T \vee F) \wedge T$$
$$T \wedge T$$
$$T$$

61. Phoebe has a larger diameter than Rhea if and only if Callisto may have water ice, and Calypso has a diameter of 6–49 miles.

$$(F \leftrightarrow T) \wedge T$$
$$F \wedge T$$
$$F$$

63. The number of communications credits needed is more than the number of mathematics credits needed and the number of cultural issues credits needed is equal to the number of humanities credits needed, if and only if the number of social sciences credits needed is more than the number of natural sciences credits needed.

$$(T \wedge T) \leftrightarrow F$$
$$T \leftrightarrow F$$
$$F$$

65. $q \rightarrow p$
$F \rightarrow F$
T

67. $\sim q \leftrightarrow p$
$T \leftrightarrow F$
F

69. No, the statement only states what will occur if your sister gets straight A's. If your sister does not get straight A's, your parents may still get her a computer.

71.

p	q	r	$[p$	$\vee$	$(q$	$\rightarrow$	$\sim r)]$	$\leftrightarrow$	$(p$	$\wedge$	$\sim q)$
T	T	T	T	T	T	F	F	F	T	F	F
T	T	F	T	T	T	T	T	F	T	F	F
T	F	T	T	T	F	T	F	T	T	T	T
T	F	F	T	T	F	T	T	T	T	T	T
F	T	T	F	F	T	F	F	T	F	F	F
F	T	F	F	T	T	T	T	F	F	F	F
F	F	T	F	T	F	T	F	F	F	F	T
F	F	F	F	T	F	T	T	F	F	F	T
			1	5	2	4	3	9	6	8	7

73. a)

p	q	r	s	$(p$	$\vee$	$q)$	$\rightarrow$	$(r$	$\wedge$	$s)$
T	T	T	T	T	T	T	T	T	T	T
T	T	T	F	T	T	T	F	T	F	F
T	T	F	T	T	T	T	F	F	F	T
T	T	F	F	T	T	T	F	F	F	F
T	F	T	T	T	T	F	T	T	T	T
T	F	T	F	T	T	F	F	T	F	F
T	F	F	T	T	T	F	F	F	F	T
T	F	F	F	T	T	F	F	F	F	F
F	T	T	T	F	T	T	T	T	T	T
F	T	T	F	F	T	T	F	T	F	F
F	T	F	T	F	T	T	F	F	F	T
F	T	F	F	F	T	T	F	F	F	F
F	F	T	T	F	F	F	T	T	T	T
F	F	T	F	F	F	F	T	T	F	F
F	F	F	T	F	F	F	T	F	F	T
F	F	F	F	F	F	F	T	F	F	F
				1	3	2	7	4	6	5

b)

p	q	r	s	$(q$	$\rightarrow$	$\sim p)$	$\vee$	$(r$	$\leftrightarrow$	$s)$
T	T	T	T	T	F	F	T	T	T	T
T	T	T	F	T	F	F	F	T	F	F
T	T	F	T	T	F	F	F	F	F	T
T	T	F	F	T	F	F	T	F	T	F
T	F	T	T	F	T	F	T	T	T	T
T	F	T	F	F	T	F	T	T	F	F
T	F	F	T	F	T	F	T	F	F	T
T	F	F	F	F	T	F	T	F	T	F
F	T	T	T	T	T	T	T	T	T	T
F	T	T	F	T	T	T	T	T	F	F
F	T	F	T	T	T	T	T	F	F	T
F	T	F	F	T	T	T	T	F	T	F
F	F	T	T	F	T	T	T	T	T	T
F	F	T	F	F	T	T	T	T	F	F
F	F	F	T	F	T	T	T	F	F	T
F	F	F	F	F	T	T	T	F	T	F
				1	3	2	7	4	6	5

75.

Tiger	Boots	Sam	Sue
Blue	Yellow	Pink	Green
Nine Lives	Whiskas	Friskies	Meow Mix

Section 3.4: Equivalent Statements

1. Equivalent

3. $p \rightarrow q \Leftrightarrow \sim p \vee q$

5. $q \rightarrow p$

7. $\sim q \rightarrow \sim p$

9. The statements are equivalent.

p	q	p	$\to$	q	$\sim p$	$\vee$	q
T	T	T	T	T	F	T	T
T	F	T	F	F	F	F	F
F	T	F	T	T	T	T	T
F	F	F	T	F	T	T	F
		1	3	2	1	3	2

11. The statements are equivalent.

p	q	$\sim q$	$\to$	$\sim p$	p	$\to$	q
T	T	F	T	F	T	T	T
T	F	T	F	F	T	F	F
F	T	F	T	T	F	T	T
F	F	T	T	T	F	T	F
		1	3	2	1	3	2

13. The statements are not equivalent.

p	q	r	$(p$	$\vee$	$q)$	$\wedge$	r	p	$\vee$	$(q$	$\wedge$	$r)$
T	T	T	T	T	T	T	T	T	T	T	T	T
T	T	F	T	T	T	F	F	T	T	T	F	F
T	F	T	T	T	F	T	T	T	T	F	F	T
T	F	F	T	T	F	F	F	T	T	F	F	F
F	T	T	F	T	T	T	T	F	T	T	T	T
F	T	F	F	T	T	F	F	F	F	T	F	F
F	F	T	F	F	F	F	T	F	F	F	F	T
F	F	F	F	F	F	F	F	F	F	F	F	F
			1	3	2	5	4	1	5	2	4	3

15. The statements are equivalent.

p	q	$(p$	$\to$	$q)$	$\wedge$	$(q$	$\to$	$p)$	p	$\leftrightarrow$	q
T	T	T	T	T	T	T	T	T	T	T	T
T	F	T	F	F	F	F	T	T	T	F	F
F	T	F	T	T	F	T	F	F	F	F	T
F	F	F	T	F	T	F	T	F	F	T	F
		1	3	2	7	4	6	5	1	3	2

17. $\sim(p \wedge q) \Leftrightarrow \sim(\sim p) \vee q \Leftrightarrow \sim p \vee q$, which is not equivalent to $\sim p \wedge \sim q$.

19. Equivalent by law 2.

21. $\sim(\sim p \wedge q) \Leftrightarrow \sim(\sim p) \vee \sim q \Leftrightarrow p \vee \sim q$, which is not equivalent to $p \wedge \sim q$.

23. The statements are equivalent by law 2.

25. $\sim(p \wedge q) \Leftrightarrow \sim p \vee \sim q$; Jay-Z does not sing opera or Beyoncé does not sing country.

27. $\sim p \wedge \sim q \Leftrightarrow \sim(p \vee q)$; It is false that the dog was a bulldog or the dog was a boxer.

29. p: I am late for class.
q: I will get a bonus.
r: I will fail the test.

$p \to (\sim q \vee r) \Leftrightarrow p \to \sim(q \wedge \sim r)$

If I am late for class, then it is false that I will get bonus points and I will not fail the test.

31. I do not see a movie or I buy popcorn.

33. If Chase is hiding, then the pitcher is broken.

35. If Opal does not exercise daily, then she is not healthy.

37. We go to Chicago and we do not go to Navy Pier.

39. It is false that if I am cold then the heater is working.

41. Amazon has a sale and we will not buy $100 worth of books.

43. *Converse:* If you are on the president's list, then you got straight A's.
Inverse: If you did not get straight A's, then you are not on the president's list.
Contrapositive: If you are not on the president's list, then you did not get straight A's.

45. *Converse*: If I buy silver jewelry, then I go to Mexico.
Inverse: If I do not go to Mexico, then I do not buy silver jewelry.
Contrapositive: If I do not buy silver jewelry then I do not go to Mexico.

47. *Converse*: If I do not stay on my diet, then the menu includes calzones.
Inverse: If the menu does not include calzones, then I do stay on my diet.
Contrapositive: If I stay on my diet, then the menu does not include calzones.

49. If a natural number is not divisible by 7, then it is not divisible by 14. True

51. If a natural number is not divisible by 6, then it is not divisible by 3. False

53. If two lines are not parallel, then the two lines intersect in at least one point. True

55. p: The player gets a red card.
q: The player sits out.

			(a)			(b)			(c)	
p	q	p	$\rightarrow$	q	q	$\rightarrow$	p	$\sim q$	$\vee$	p
T	T	T	T	T	T	T	T	F	T	T
T	F	T	F	F	F	T	T	T	T	T
F	T	F	T	T	T	F	F	F	F	F
F	F	F	T	F	F	T	F	T	T	F
		1	3	2	1	3	2	1	3	2

Statements (b) and (c) are equivalent.

57. p: The shoes are on sale.
q: The purse is on sale.

			(a)			(b)			(c)		
p	q	$\sim p$	$\wedge$	q	$\sim p$	$\rightarrow$	$\sim q$	$\sim$	$(p$	$\vee$	$\sim q)$
T	T	F	F	T	F	T	F	F	T	T	F
T	F	F	F	F	F	T	T	F	T	T	T
F	T	T	T	T	T	F	F	T	F	F	F
F	F	T	F	F	T	T	T	F	F	T	T
		1	3	2	1	3	2	4	1	3	2

Statements (a) and (c) are equivalent.
Note: If we use DeMorgan's Laws on statement (a), we get statement (c).

59. p: We go hiking.
q: We go fishing.

		(a)			(b)			(c)		
p	q	$\sim p$	$\vee$	q	p	$\rightarrow$	$\sim q$	q	$\rightarrow$	$\sim p$
T	T	F	T	T	T	F	F	T	F	F
T	F	F	F	F	T	T	T	F	T	F
F	T	T	T	T	F	T	F	T	T	T
F	F	T	T	F	F	T	T	F	T	T
		1	3	2	1	3	2	1	3	2

Statements (b) and (c) are equivalent.

61. p: The grass grows.
q: The trees are blooming.

		(a)			(b)			(c)		
p	q	p	$\wedge$	q	q	$\rightarrow$	$\sim p$	$\sim q$	$\vee$	$\sim p$
T	T	T	T	T	T	F	F	F	F	F
T	F	T	F	F	F	T	F	T	T	F
F	T	F	F	T	T	T	T	F	T	T
F	F	F	F	F	F	T	T	T	T	T
		1	3	2	1	3	2	1	3	2

Statements (b) and (c) are equivalent.

Note: Since $p \rightarrow q \Leftrightarrow \sim p \vee q$, statement (b) is equivalent to statement (c).

63. p: The corn bag goes in the hole.
q: you are awarded three points.

		(a)			(b)				(c)		
p	q	p	$\rightarrow$	q	$\sim$	$(p$	$\wedge$	$q)$	$\sim p$	$\wedge$	$\sim q$
T	T	T	T	T	F	T	T	T	F	F	F
T	F	T	F	F	T	T	F	F	F	F	T
F	T	F	T	T	T	F	F	T	T	F	F
F	F	F	T	F	T	F	F	F	T	T	T
		1	3	2	4	1	3	2	1	3	2

None of the statements are equivalent.

65. p: The pay is good.
q: Today is Monday.
r: I will take the job.

			(a)					(b)						(c)				
p	q	r	$(p$	$\wedge$	$q)$	$\rightarrow$	r	$\sim r$	$\rightarrow$	$\sim$	$(p$	$\vee$	$q)$	$(p$	$\wedge$	$q)$	$\vee$	r
T	T	T	T	T	T	T	T	F	T	F	T	T	T	T	T	T	T	T
T	T	F	T	T	T	F	F	T	F	F	T	T	T	T	T	T	T	F
T	F	T	T	F	F	T	T	F	T	F	T	T	F	T	F	F	T	T
T	F	F	T	F	F	T	F	T	F	F	T	T	F	T	F	F	F	F
F	T	T	F	F	T	T	T	F	T	F	F	T	T	F	F	T	T	T
F	T	F	F	F	T	T	F	T	F	F	F	T	T	F	F	T	F	F
F	F	T	F	F	F	T	T	F	T	T	F	F	F	F	F	F	T	T
F	F	F	F	F	F	T	F	T	T	T	F	F	F	F	F	F	F	F
			1	3	2	5	4	1	6	5	2	4	3	1	3	2	5	4

None of the statements are equivalent.

67. p: The spacecraft was made by SpaceX.
q: The spacecraft was made by Orbital.
r: The spacecraft was made by Boeing.

					(a)					(b)					(c)		
p	q	r	p	$\vee$	$(\sim q$	$\wedge$	$r)$	r	$\leftrightarrow$	$(p$	$\vee$	$\sim q)$	$\sim p$	$\rightarrow$	$(\sim q$	$\wedge$	$r)$
T	T	T	T	T	F	F	T	T	T	T	T	F	F	T	F	F	T
T	T	F	T	T	F	F	F	F	F	T	T	F	F	T	F	F	F
T	F	T	T	T	T	T	T	T	T	T	T	T	F	T	T	T	T
T	F	F	T	T	T	F	F	F	F	T	T	T	F	T	T	F	F
F	T	T	F	F	F	F	T	T	F	F	F	F	T	F	F	F	T
F	T	F	F	F	F	F	F	F	T	F	F	F	T	F	F	F	F
F	F	T	F	T	T	T	T	T	T	F	T	T	T	T	T	T	T
F	F	F	F	F	T	F	F	F	F	F	T	T	T	F	T	F	F
			1	5	2	4	3	1	5	2	4	3	1	5	2	4	3

Statements (a) and (c) are equivalent.

Note: Since $p \rightarrow q \Leftrightarrow \sim p \vee q$, statement (c) is equivalent to $p \vee (\sim q \wedge r)$.

69. True. If $p \rightarrow q$ is false, it must be of the form $\text{T} \rightarrow \text{F}$. Therefore, the converse must be of the form $\text{F} \rightarrow \text{T}$, which is true.

71. False. A conditional statement and its contrapositive always have the same truth values.

73. Answers will vary.

75. a) $\sim p = 1 - p = 1 - 0.25 = 0.75$

b) $\sim q = 1 - q = 1 - 0.20 = 0.80$

c) $p \rightarrow q$ has a truth value equal to the lesser of $p = 0.75$ and $q = 0.20$, thus $p \rightarrow q = 0.20$.

d) $p \leftrightarrow q$ has truth value equal to the greater of $p = 0.25$ and $q = 0.20$, thus $p \rightarrow q = 0.25$.

e) $p \rightarrow q$ has truth value equal to the lesser of 1 and $1 - p + q = 1 - 0.25 + 0.20 = 0.95$, thus $p \rightarrow q = 0.95$.

f) $p \leftrightarrow q$ has a truth value equal to $1 - |p - q| = 1 - |0.25 - 0.20| = 1 - 0.05 = 0.95$, thus $p \leftrightarrow q = 0.95$.

Section 3.5: Symbolic Arguments

1. Valid
3. Fallacy
5. Valid
7. Syllogism
9. Inverse
11. Syllogism
13. This argument is the fallacy of the inverse and therefore it is invalid.
15. This is the law of detachment, so it is a valid argument.
17. This argument is a disjunctive syllogism and therefore is valid.
19. This argument is the fallacy of the converse. Therefore, it is invalid.
21. This argument is the law of contraposition and therefore it is valid.
23. This argument is the law of syllogism and therefore it is valid.

25. The argument is valid.

p	q	r	$[(p \leftrightarrow q)$	$\wedge$	$(q \wedge r)]$	$\rightarrow$	$(p \vee r)$
T	T	T	T	T	T	T	T
T	T	F	T	F	F	T	T
T	F	T	F	F	F	T	T
T	F	F	F	F	F	T	T
F	T	T	F	F	T	T	T
F	T	F	F	F	F	T	F
F	F	T	T	F	F	T	T
F	F	F	T	F	F	T	F
			1	3	2	5	4

27. The argument is invalid.

p	q	r	$[(p \wedge q)$	$\wedge$	$(q \vee r)]$	$\rightarrow$	r
T	T	T	T	T	T	T	T
T	T	F	T	T	T	F	F
T	F	T	F	F	T	T	T
T	F	F	F	F	F	T	F
F	T	T	F	F	T	T	T
F	T	F	F	F	T	T	F
F	F	T	F	F	T	T	T
F	F	F	F	F	F	T	F
			1	3	2	5	4

29. The argument is invalid.

p	q	r	$[(p \rightarrow q)$	$\wedge$	$(q \vee r)$	$\wedge$	$(r \vee p)]$	$\rightarrow$	p
T	T	T	T	T	T	T	T	T	T
T	T	F	T	T	T	T	T	T	T
T	F	T	F	F	T	F	T	T	T
T	F	F	F	F	F	F	T	T	T
F	T	T	T	T	T	T	T	F	F
F	T	F	T	T	T	F	F	T	F
F	F	T	T	T	T	T	T	F	F
F	F	F	T	F	F	F	F	T	F
			1	4	2	5	3	7	6

31. The argument is valid.

p	q	r	$[(p \rightarrow q)$	$\wedge$	$(r \rightarrow \sim p)$	$\wedge$	$(p \vee r)]$	$\rightarrow$	$(q \vee \sim p)$
T	T	T	T	F	F	F	T	T	T
T	T	F	T	T	T	T	T	T	T
T	F	T	F	F	F	F	T	T	F
T	F	F	F	F	T	F	T	T	F
F	T	T	T	T	T	T	T	T	T
F	T	F	T	T	T	F	F	T	T
F	F	T	T	T	T	T	T	T	T
F	F	F	T	T	T	F	F	T	T
			1	4	2	5	3	7	6

33. a) p: The Rays win the division.
q: The Rays win go to the playoffs.

$$\begin{array}{l} p \rightarrow q \\ \underline{\sim p} \\ \therefore \sim q \end{array}$$

b) The argument is invalid by the fallacy of the inverse.

35. a) p: We visit the zoo.
q: We see a zebra.

$$\begin{array}{l} p \rightarrow q \\ \underline{p} \\ \therefore q \end{array}$$

b) The argument is valid by the law of detachment.

37. a) p: The guitar is a Les Paul model.
q: The guitar is made by Gibson.

$$\begin{array}{l} p \rightarrow q \\ \underline{\sim q} \\ \therefore \sim p \end{array}$$

b) This argument is valid by the law of contraposition.

39. a) p: We take kayaks on the river.
q: We see alligators.

$$\begin{array}{l} p \rightarrow q \\ \underline{q} \\ \therefore p \end{array}$$

b) The argument is invalid by the fallacy of the converse.

41. a) p: Lucious will give the company to Hakeem.
q: Lucious will give the company to Cookie.

$$\begin{array}{l} p \vee q \\ \underline{\sim p} \\ \therefore q \end{array}$$

b) This argument is valid by disjunctive syllogism.

43. a) p: It is cold.
q: The graduation will be held indoors.
r: The fireworks will be postponed.

$$\begin{array}{l} p \rightarrow q \\ \underline{q \rightarrow r} \\ \therefore p \rightarrow r \end{array}$$

b) This argument is valid by the law of syllogism.

45. a) p: Jevon is a bowler.
q: Michael is a golfer.
r: Alisha is a curler.

$$\begin{array}{l} p \wedge q \\ \underline{q \rightarrow r} \\ \therefore r \rightarrow p \end{array}$$

p	q	r	$[(p \wedge q)$	$\wedge$	$(q \rightarrow r)]$	$\rightarrow$	$(r \rightarrow p)$
T	T	T	T	T	T	T	T
T	T	F	T	F	F	T	T
T	F	T	F	F	T	T	T
T	F	F	F	F	T	T	T
F	T	T	F	F	T	T	F
F	T	F	F	F	F	T	T
F	F	T	F	F	T	T	F
F	F	F	F	F	T	T	T
			1	3	2	5	4

b) The argument is valid.

47. a) p: Javier is a police officer.
q: Javier is a little league coach.
r: Javier is a community leader.

$$\begin{array}{l} p \vee q \\ \underline{p \rightarrow r} \\ \therefore\ q \leftrightarrow r \end{array}$$

p	q	r	$[(p \vee q)$	$\wedge$	$(p \rightarrow r)]$	$\rightarrow$	$(q \leftrightarrow r)$
T	T	T	T	T	T	T	T
T	T	F	T	F	F	T	F
T	F	T	T	T	T	T	T
T	F	F	T	F	F	T	F
F	T	T	T	T	T	T	T
F	T	F	T	T	T	F	F
F	F	T	F	F	T	T	F
F	F	F	F	F	T	T	T
			1	3	2	5	4

b) The argument is invalid.

49. p: You read Riptide Ultra-Glide.
q: You can understand *Tiger Shrimp Tango*.

$$\begin{array}{l} p \rightarrow q \\ \underline{\sim q} \\ \therefore\ \sim p \end{array}$$

The argument is valid by the law of contraposition.

51. p: It rains on Monday.
q: We will go shopping.

$$\begin{array}{l} p \rightarrow q \\ \underline{\sim p} \\ \therefore\ \sim q \end{array}$$

This argument is invalid by the fallacy of the inverse.

53. p: You submit your application.
q: You will be accepted to school.

$$\begin{array}{l} \sim p \rightarrow \sim q \\ \underline{\sim p} \\ \therefore\ \sim q \end{array}$$

The argument is valid by the law of detachment.

55. p: The baby is crying.
q: The baby is hungry.

$$\begin{array}{l} p \wedge \sim q \\ \underline{q \rightarrow p} \\ \therefore\ q \end{array}$$

p	q	$[(p \wedge \sim q)$	$\wedge$	$(q \rightarrow p)]$	$\rightarrow$	q
T	T	F	F	T	T	T
T	F	T	T	T	F	F
F	T	F	F	F	T	T
F	F	F	F	T	T	F
		1	3	2	5	4

The argument is invalid.

57. p: You liked *This Is Spinal Tap*.
q: You liked *Best in Show*
r: You liked *A Mighty Wind.*

$$\begin{array}{l} p \rightarrow q \\ \underline{q \rightarrow \sim r} \\ \therefore\ p \rightarrow r \end{array}$$

Using the law of syllogism, the conclusion should be $p \rightarrow \sim r$, so this argument is invalid.

59. Therefore, the radio is made by RCA. (law of detachment)

61. Therefore, I am stressed out. (disjunctive syllogism)

63. Therefore, you did not close the deal. (law of contraposition)

65. Yes, if the conclusion necessarily follows from the premises, the argument is valid, even if the conclusion is false.

67. Yes, if the conclusion does not necessarily follow from the premises, the argument is invalid, even if the premises are true.

69. p: Lynn wins the contest
q: Lynn strikes oil.
r: Lynn will be rich.
s: Lynn will stop working.

$$\begin{array}{l} (p \vee q) \rightarrow r \\ r \rightarrow s \\ \hline \therefore\ {\sim}s \rightarrow {\sim}p \end{array}$$

p	q	r	s	$[((p \vee q)$	$\rightarrow$	$r)$	$\wedge$	$(r \rightarrow s)]$	$\rightarrow$	$({\sim}s$	$\rightarrow$	${\sim}p)$
T	T	T	T	T	T	T	T	T	T	F	T	F
T	T	T	F	T	T	T	F	F	T	T	F	F
T	T	F	T	T	F	F	F	T	T	F	T	F
T	T	F	F	T	F	F	F	T	T	T	F	F
T	F	T	T	T	T	T	T	T	T	F	T	F
T	F	T	F	T	T	T	F	F	T	T	F	F
T	F	F	T	T	F	F	F	T	T	F	T	F
T	F	F	F	T	F	F	F	T	T	T	F	F
F	T	T	T	T	T	T	T	T	T	F	T	T
F	T	T	F	T	T	T	F	F	T	T	T	T
F	T	F	T	T	F	F	F	T	T	F	T	T
F	T	F	F	T	F	F	F	T	T	T	T	T
F	F	T	T	F	T	T	T	T	T	F	T	T
F	F	T	F	F	T	T	F	F	T	T	T	T
F	F	F	T	F	T	F	T	T	T	F	T	T
F	F	F	F	F	T	F	T	T	T	T	T	T
				1	3	2	5	4	9	6	8	7

The argument is valid.

71. a) p: I think.
q: I am.

$$\begin{array}{l} p \rightarrow q \\ {\sim}p \\ \hline \therefore\ {\sim}q \end{array}$$

b) No, the argument is invalid.

c) This argument is the fallacy of the inverse.

Section 3.6: Euler Diagrams and Syllogistic Arguments

1. Euler

3. Invalid

5. No

7. Valid

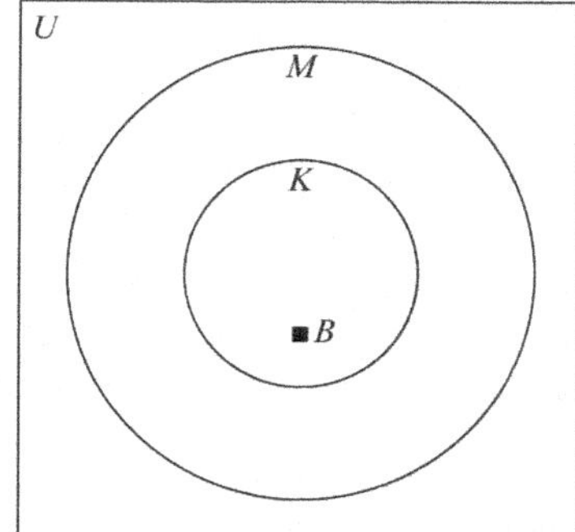

9. Invalid

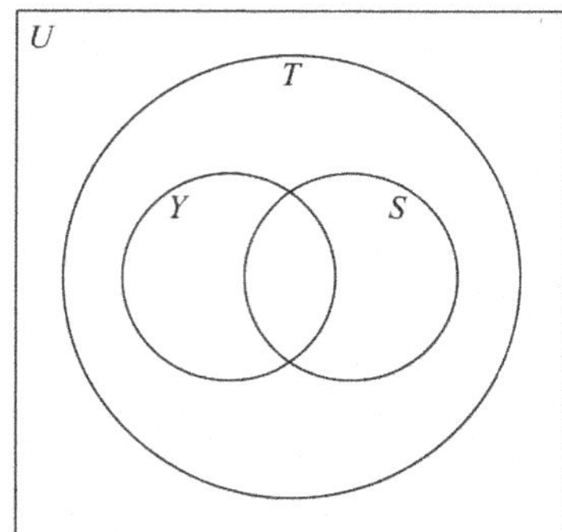

11. Valid

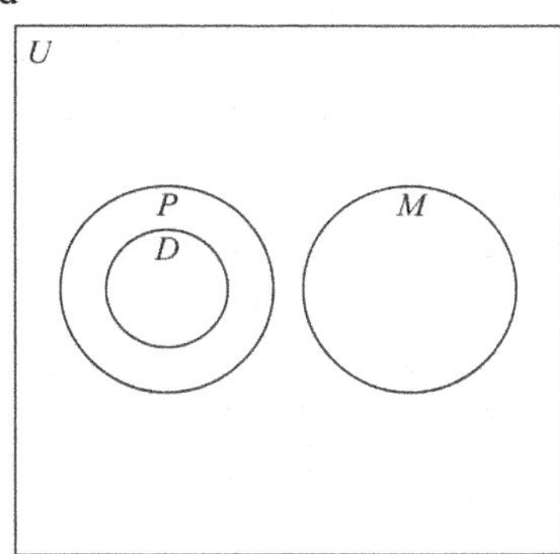

13. Invalid

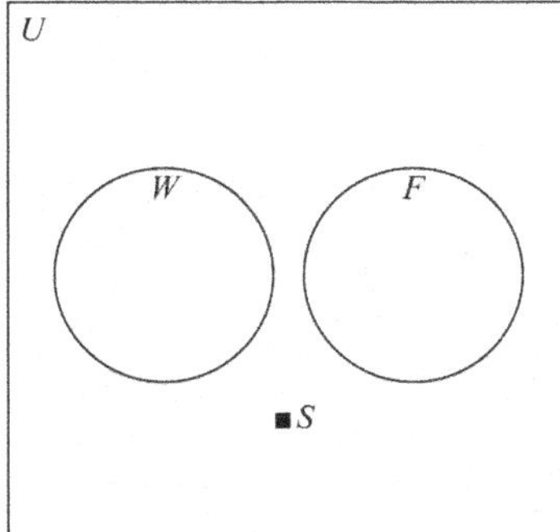

15. Invalid

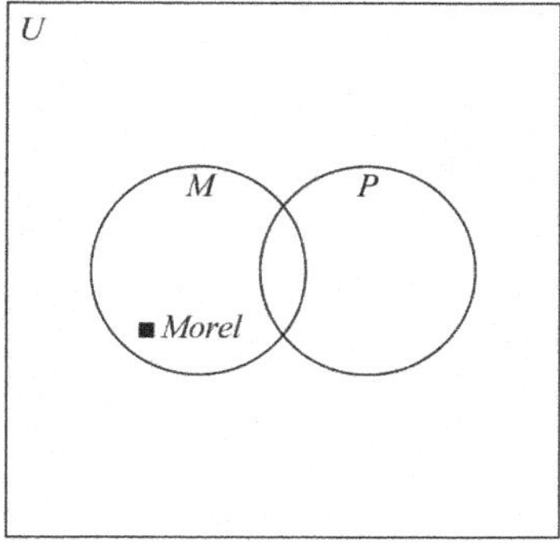

17. Invalid

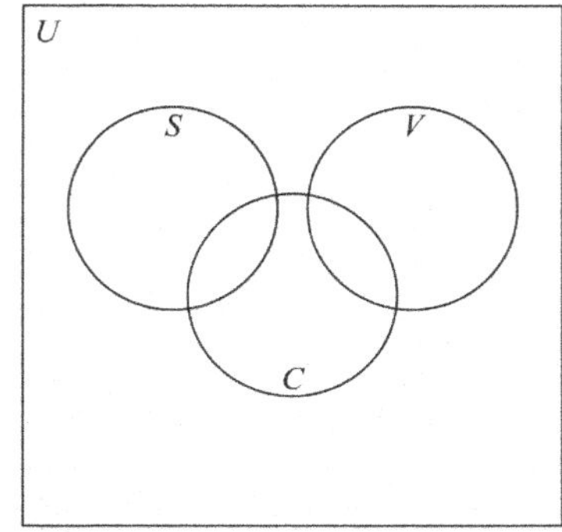

19. Valid

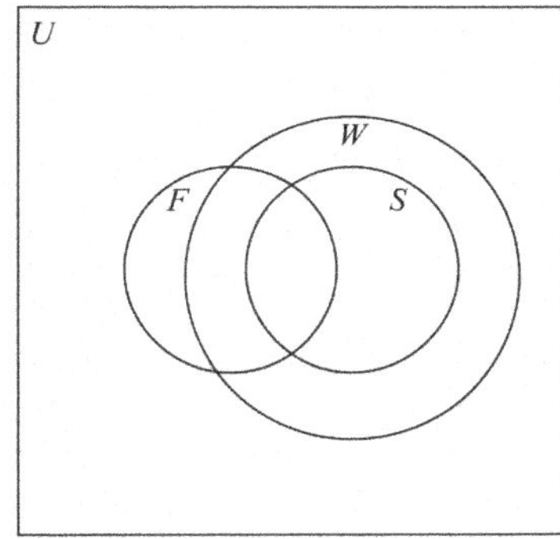

21. Invalid

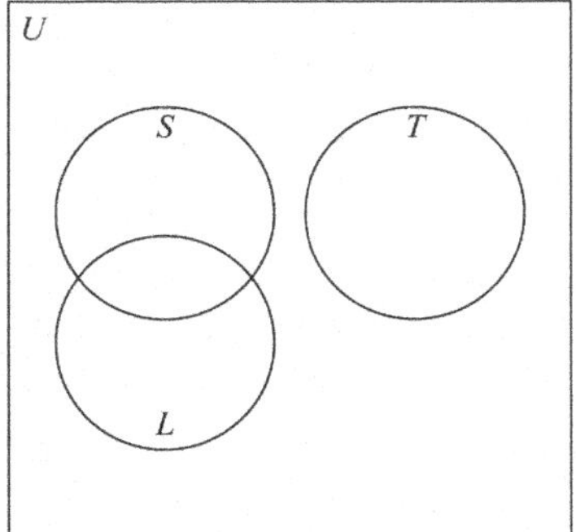

23. Invalid

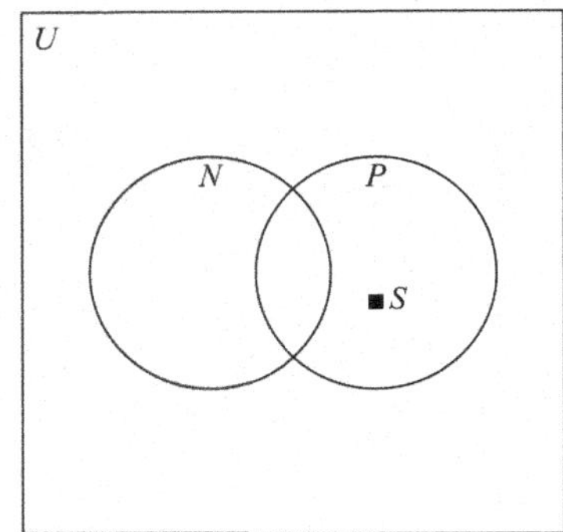

25. Invalid

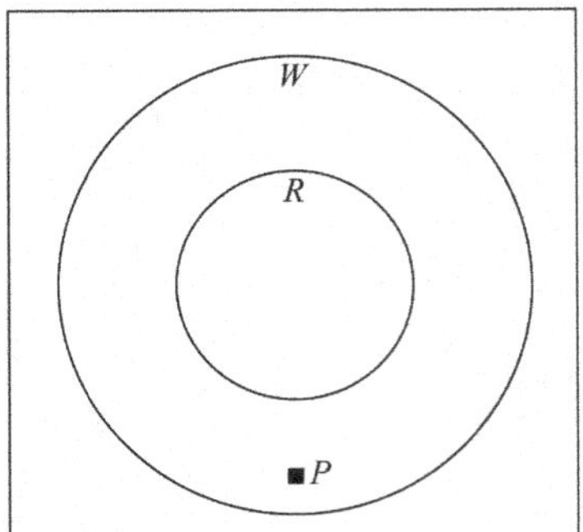

27. Valid

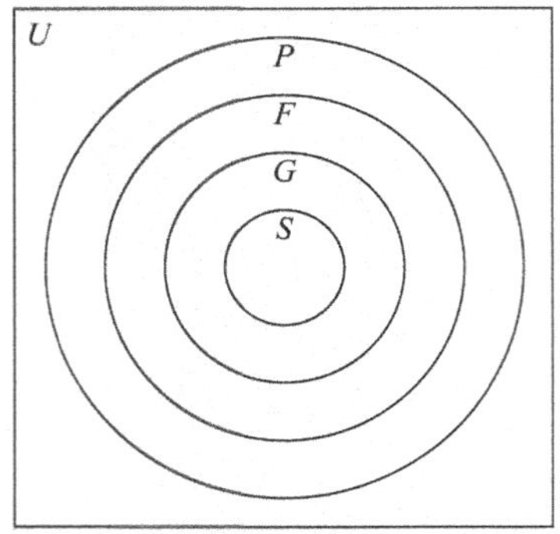

29. Yes, if the conclusion necessarily follows from the premises, the argument is valid.

31. $[(p \to q) \wedge (p \vee q)] \to \sim p$ can be expressed as a set statement by $[(P' \cup Q) \cap (P \cup Q)] \subseteq P'$. If this statement is true, then the argument is valid; otherwise, the argument is invalid. Since $(P' \cup Q) \cap (P \cup Q)$ is not a subset of P', the argument is invalid.

Set	Regions
$P' \cup Q$	II, III, IV
$P \cup Q$	I, II, III
$(P' \cup Q) \cap (P \cup Q)$	II, III

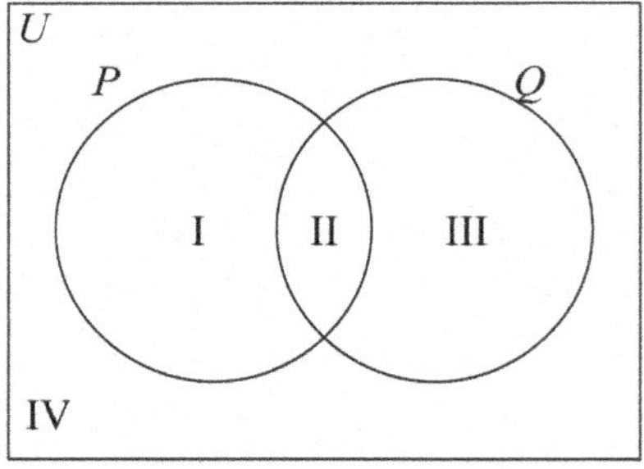

Section 3.7: Switching Circuits

1. Series

3. Closed

5. a) $p \vee q$

   b) The lightbulb will be on in all cases except when p is open and q is open.

p	q	$p \vee q$
T	T	T
T	F	T
F	T	T
F	F	F

7. a) $(p \vee q) \wedge \sim q$

   b) The lightbulb will be on only when p is closed and q is open.

p	q	$(p \vee q)$	$\wedge$	$\sim q$
T	T	T	F	F
T	F	T	T	T
F	T	T	F	F
F	F	F	F	T
		1	3	2

9. a) $(p \wedge q) \wedge [(p \wedge \sim q) \vee r]$

b) The lightbulb will be on only when p, q, and r are all closed.

p	q	r	$(p \wedge q)$	$\wedge$	$[(p \wedge \sim q)$	$\vee$	$r]$
T	T	T	T	T	F	T	T
T	T	F	T	F	F	F	F
T	F	T	F	F	T	T	T
T	F	F	F	F	T	F	F
F	T	T	F	F	F	F	T
F	T	F	F	F	F	F	F
F	F	T	F	F	F	F	T
F	F	F	F	F	F	F	F
			1	5	2	4	3

11. a) $p \vee q \vee (r \wedge \sim p)$

b) The lightbulb will be on in all cases except when p, q, and r are all open.

p	q	r	p	$\vee$	q	$\vee$	$(r \wedge \sim p)$
T	T	T	T	T	T	T	F
T	T	F	T	T	T	T	F
T	F	T	T	T	F	T	F
T	F	F	T	T	F	T	T
F	T	T	F	T	T	T	T
F	T	F	F	T	T	T	F
F	F	T	F	F	F	T	T
F	F	F	F	F	F	F	F
			1	3	2	5	4

13.

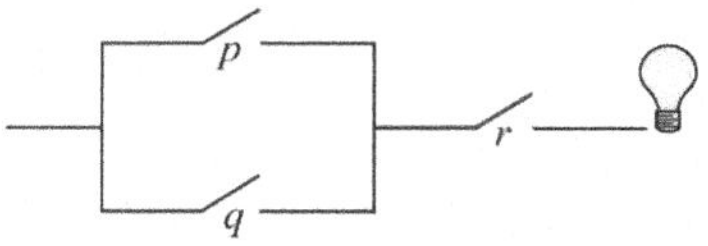

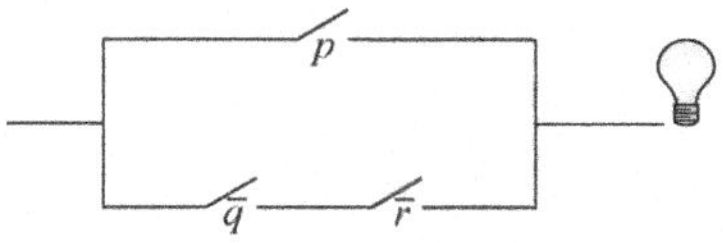

17.

15.

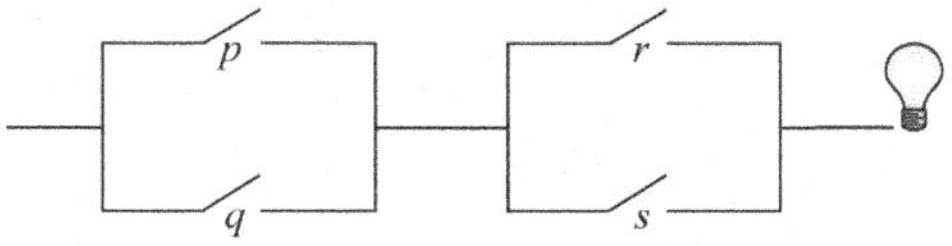

19.

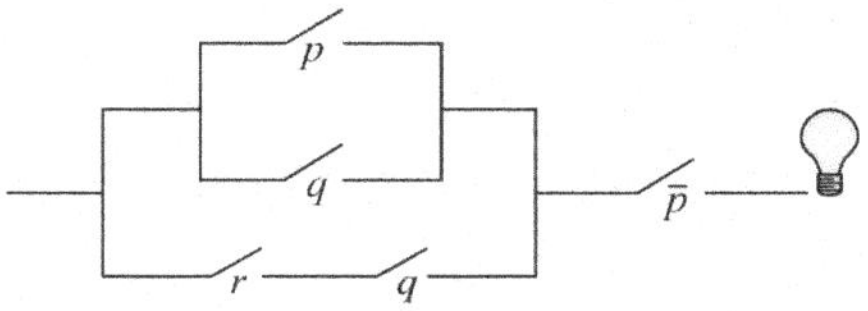

21. $p \vee \sim q$ and $\sim p \wedge q$; The circuits are not equivalent.

p	q	p	$\vee$	$\sim q$	$\sim p$	$\wedge$	q
T	T	T	T	F	F	F	T
T	F	T	T	T	F	F	F
F	T	F	F	F	T	T	T
F	F	F	F	T	T	F	F
		1	3	2	1	3	2

23. $[(p \wedge q) \vee r] \wedge p$ and $(q \vee r) \wedge p$; The circuits are equivalent.

p	q	r	$[(p$	$\wedge$	$q)$	$\vee$	$r]$	$\wedge$	p	$(q$	$\vee$	$r)$	$\wedge$	p
T	T	T	T	T	T	T	T	T	T	T	T	T	T	T
T	T	F	T	T	T	T	F	T	T	T	T	F	T	T
T	F	T	T	F	F	T	T	T	T	F	T	T	T	T
T	F	F	T	F	F	F	F	F	T	F	F	F	F	T
F	T	T	F	F	T	T	T	F	F	T	T	T	F	F
F	T	F	F	F	T	F	F	F	F	T	T	F	F	F
F	F	T	F	F	F	T	T	F	F	F	T	T	F	F
F	F	F	F	F	F	F	F	F	F	F	F	F	F	F
			1	3	2	5	4	7	6	1	3	2	5	4

25. $(p \vee \sim p) \wedge q \wedge r$ and $p \wedge q \wedge r$; The circuits are not equivalent.

p	q	r	$(p$	$\wedge$	$\sim p)$	$\wedge$	q	$\wedge$	r	p	$\wedge$	q	$\wedge$	r
T	T	T	T	F	F	F	T	F	T	T	T	T	F	T
T	T	F	T	F	F	F	T	F	F	T	T	T	T	F
T	F	T	T	F	F	F	F	F	T	T	F	F	T	T
T	F	F	T	F	F	F	F	F	F	T	F	F	T	F
F	T	T	F	F	T	F	T	F	T	F	F	T	T	T
F	T	F	F	F	T	F	T	F	F	F	F	T	T	F
F	F	T	F	F	T	F	F	F	T	F	F	F	T	T
F	F	F	F	F	T	F	F	F	F	F	F	F	T	F
			1	3	2	5	4	7	6	1	3	2	5	4

27. One of the two switches will always be open.

29. a) $p \rightarrow q \Leftrightarrow \sim p \vee q$

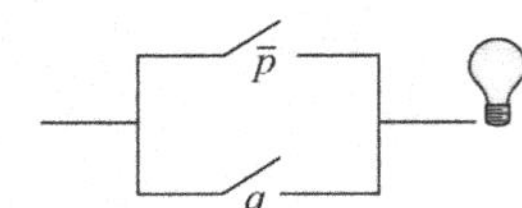

b) $\sim(p \rightarrow q) \Leftrightarrow \sim p \wedge q$

Review Exercises

1. Some diamonds are not made of carbon.
2. Some pets are allowed in this park.
3. No women are presidents.
4. All pine trees are green.
5. The coffee is hot or the coffee is strong.
6. The coffee is not hot and the coffee is strong.
7. If the coffee is hot, then the coffee is strong and it is not Maxwell House.
8. The coffee is Maxwell House if and only if the coffee is not strong.
9. The coffee is not Maxwell House, if and only if the coffee is strong and the coffee is not hot.
10. The coffee is Maxwell House or the coffee is not hot, and the coffee is not strong.
11. $q \wedge \sim r$
12. $r \rightarrow \sim p$
13. $(r \rightarrow q) \vee \sim p$
14. $(q \leftrightarrow p) \wedge \sim r$
15. $(r \wedge q) \vee \sim p$
16. $\sim(r \wedge q)$
17.

p	q	$(p \vee q)$	$\wedge$	$\sim p$
T	T	T	F	F
T	F	T	F	F
F	T	T	T	T
F	F	F	F	T
		1	3	2

18.

p	q	q	$\leftrightarrow$	$(p$	$\vee$	$\sim q)$
T	T	T	T	T	T	F
T	F	F	F	T	T	T
F	T	T	F	F	F	F
F	F	F	F	F	T	T
		1	5	2	4	3

19.

p	q	r	$(p \vee q)$	$\leftrightarrow$	$(p \vee r)$
T	T	T	T	T	T
T	T	F	T	T	T
T	F	T	T	T	T
T	F	F	T	T	T
F	T	T	T	T	T
F	T	F	T	F	F
F	F	T	F	F	T
F	F	F	F	T	F
			1	5	5

20.

p	q	r	p	$\wedge$	$(\sim q$	$\vee$	$r)$
T	T	T	T	T	F	T	T
T	T	F	T	F	F	F	F
T	F	T	T	T	T	T	T
T	F	F	T	T	T	T	F
F	T	T	F	F	F	T	T
F	T	F	F	F	F	F	F
F	F	T	F	F	T	T	T
F	F	F	F	F	T	T	F
			1	5	2	4	3

21.

p	q	r	p	$\rightarrow$	$(q$	$\wedge$	$\sim r)$
T	T	T	T	F	T	F	F
T	T	F	T	T	T	T	T
T	F	T	T	F	F	F	F
T	F	F	T	F	F	F	T
F	T	T	F	T	T	F	F
F	T	F	F	T	T	T	T
F	F	T	F	T	F	F	F
F	F	F	F	T	F	F	T
			1	5	2	4	3

22.

p	q	r	$(p \wedge q)$	$\rightarrow$	$\sim r$
T	T	T	T	F	F
T	T	F	T	T	T
T	F	T	F	T	F
T	F	F	F	T	T
F	T	T	F	T	F
F	T	F	F	T	T
F	F	T	F	T	F
F	F	F	F	T	T
			1	3	2

23. p: Apple makes iPhones.
q: Dell makes canoes.
r: Hewlett Packard makes laser printers.

$(p \wedge q) \vee r$

$(\text{T} \wedge \text{F}) \vee \text{T}$

$\text{F} \quad \vee \text{T}$

T

The statement is true.

24. p: A minute has 60 seconds.
q: An hour has 60 minutes.
r: A day has 24 hours.

$p \rightarrow (q \leftrightarrow r)$

$\text{T} \rightarrow (\text{T} \leftrightarrow \text{F})$

$\text{T} \rightarrow \quad \text{F}$

F

The statement is false.

25. p: Oregon borders the Pacific Ocean.
q: California borders the Atlantic Ocean.
r: Minnesota is south of Texas.

$(p \vee q) \rightarrow r$

$(\text{T} \vee \text{F}) \rightarrow \text{F}$

$\text{T} \rightarrow \text{F}$

F

The statement is false.

26. p: President's Day is in February.
q: Memorial Day is in May.
r: Labor Day is in December.

$p \vee (q \wedge r)$

$\text{T} \vee (\text{T} \wedge \text{F})$

$\text{T} \vee \text{F}$

T

The statement is true.

27. $(\sim p \vee q) \rightarrow \sim(p \wedge \sim q)$

$(\text{F} \vee \text{F}) \rightarrow \sim(\text{T} \vee \text{T})$

$\text{F} \rightarrow \sim\text{T}$

$\text{F} \rightarrow \text{F}$

T

28. $(p \leftrightarrow q) \rightarrow (\sim p \vee r)$

$(\text{T} \leftrightarrow \text{F}) \rightarrow (\text{F} \vee \text{F})$

$\text{F} \rightarrow \text{F}$

T

29. $\sim r \leftrightarrow [(p \vee q) \leftrightarrow \sim p]$

$\text{T} \leftrightarrow [(\text{T} \vee \text{F}) \leftrightarrow \text{F}]$

$\text{T} \leftrightarrow [\text{T} \leftrightarrow \text{F}]$

$\text{T} \leftrightarrow \text{F}$

F

30. $\sim[(q \wedge r) \rightarrow (\sim p \vee r)]$

$\sim[(\text{F} \wedge \text{F}) \rightarrow (\text{F} \vee \text{F})]$

$\sim[\text{F} \rightarrow \text{F}]$

$\sim\text{T}$

F

31. The statements are not equivalent.

p	q	$\sim$	$(p$	$\wedge$	$\sim q)$	$\sim p$	$\wedge$	q
T	T	T	T	F	F	F	F	T
T	F	F	T	T	T	F	F	F
F	T	T	F	F	F	T	T	T
F	F	T	F	F	T	F	T	T
		4	1	3	2	1	3	2

32. The statements are equivalent.

p	q	p	$\vee$	q	$\sim p$	$\rightarrow$	q
T	T	T	T	T	F	T	T
T	F	T	T	F	F	T	F
F	T	F	T	T	T	T	T
F	F	F	F	F	T	F	F
		1	3	2	1	3	2

33. The statements are equivalent.

p	q	r	$\sim p$	$\vee$	$(q$	$\wedge$	$r)$	$(\sim p$	$\vee$	$q)$	$\wedge$	$(\sim p$	$\vee$	$r)$
T	T	T	F	T	T	T	T	F	T	T	T	F	T	T
T	T	F	F	F	T	F	F	F	T	T	F	F	F	F
T	F	T	F	F	F	F	T	F	F	F	F	F	T	T
T	F	F	F	F	F	F	F	F	F	F	F	F	F	F
F	T	T	T	T	T	T	T	T	T	T	T	T	T	T
F	T	F	T	T	T	F	F	T	T	T	T	T	T	F
F	F	T	F	F	F	F	T	T	T	F	F	T	T	T
F	F	F	F	F	F	F	F	T	T	F	F	T	T	F
			1	5	2	4	3	1	3	2	7	4	6	5

34. The statements are not equivalent.

p	q	$(\sim q$	$\to$	$p)$	$\wedge$	p	$\sim$	$(\sim p$	$\leftrightarrow$	$q)$	$\vee$	p
T	T	F	T	T	T	T	T	F	F	T	T	T
T	F	T	T	T	T	T	F	F	T	F	T	T
F	T	F	T	F	F	F	F	T	T	T	F	F
F	F	T	F	F	F	F	T	T	F	F	T	F
		1	3	2	5	4	4	1	3	2	6	5

35. It is false that if a grasshopper is an insect then a spider is an insect.

36. If Lynn Swann did not play for the Steelers, then Jack Tatum played for the Raiders.

37. Altec Lansing does not produce only speakers and Harman Kardon does not produce only stereo receivers.

38. It is false that we did go to the beach or we did find sharks' teeth.

39. The temperature is above 32° or we will go ice fishing at O'Leary's Lake.

40. a) If Maya sits on the bench, then she plays basketball.
    b) If Maya does not play basketball, then she does not sit on the bench.
    c) If Maya does not sit on the bench, then she does not play basketball.

41. a) If we will learn the table's value, then we take the table to *Antiques Roadshow*.
    b) If we do not take the table to *Antiques Roadshow*, then we will not learn the table's value.
    c) If we will not learn the table's value, then we do not take the table to Antiques Roadshow.

42. a) If you do not sell more doughnuts, then you do not advertise.
    b) If you advertise, then you sell more doughnuts.
    c) If you sell more doughnuts, then you advertise.

43. a) If you fail the course, then you do not study for the math test.
    b) If you study for the math test, then you do not fail the course.
    c) If you do not fail the course, then you study for the math test.

44. a) If I let you attend the prom, then you will get straight A's on your report card.
    b) If you do not get straight A's on your report card, then I will not let you attend the prom.
    c) If I will not let you attend the prom, then you did not get straight A's on your report card.

45. p: Cheap Trick plays at the White House.
q: Jacque is the president.

			(a)			(b)			(c)		
p	q	p	$\to$	q	$\sim p$	$\vee$	q	$\sim$	$(p$	$\wedge$	$\sim q)$
T	T	T	T	T	F	T	T	T	T	F	F
T	F	T	F	F	F	F	F	F	T	T	T
F	T	F	T	T	T	T	T	T	F	F	F
F	F	F	T	F	T	T	F	T	F	F	T
		1	3	2	1	3	2	4	1	3	2

All three statements are equivalent.
Note: Since $p \to q \Leftrightarrow \sim p \vee q$, statement (a) is equivalent to statement (b)
Statement (b) is equivalent to statement (c) by DeMorgan's Laws.

46. p: The screwdriver is on the workbench.
q: The screwdriver is on the counter.

		(a)			(b)			(c)			
p	q	p	$\leftrightarrow$	$\sim q$	$\sim q$	$\rightarrow$	$\sim p$	$\sim$	$(q$	$\wedge$	$\sim p)$
T	T	T	F	F	F	T	F	T	T	F	F
T	F	T	T	T	T	F	F	T	F	F	F
F	T	F	T	F	F	T	T	F	T	T	T
F	F	F	F	T	T	T	T	T	F	F	T
		1	3	2	1	3	2	4	1	3	2

None of the statements are equivalent.

47. p: $2+3=6$
q: $3+1=5$

		(a)			(b)			(c)		
p	q	p	$\rightarrow$	q	p	$\leftrightarrow$	$\sim q$	$\sim q$	$\rightarrow$	$\sim p$
T	T	T	T	T	T	F	F	F	T	F
T	F	T	F	F	T	T	T	T	F	F
F	T	F	T	T	F	T	F	F	T	T
F	F	F	T	F	F	F	T	T	T	T
		1	3	2	1	3	2	1	3	2

Statements (a) and (c) are equivalent.
Note: Statement (c) is the contrapositive of statement (c), so they are equivalent.

48. p: The sale is on Tuesday.
q: I have money.
r: I will go to the sale.

p	q	r	$(p \wedge q)$	$\rightarrow$	r	r	$\rightarrow$	$(p \wedge q)$	r	$\vee$	$(p \wedge q)$
T	T	T	T	T	T	T	T	T	T	T	T
T	T	F	T	F	F	F	T	T	F	T	T
T	F	T	F	T	T	T	F	F	T	T	F
T	F	F	F	T	F	F	T	F	F	F	F
F	T	T	F	T	T	T	F	F	T	T	F
F	T	F	F	T	F	F	T	F	F	F	F
F	F	T	F	T	T	T	F	F	T	T	F
F	F	F	F	T	F	F	T	F	F	F	F
			1	3	2	1	3	2	1	3	2

None of the statements are equivalent.

49. This is the fallacy of the inverse. The argument is invalid.

50. The argument is valid.

p	q	r	$[(p \wedge q)$	$\wedge$	$(q \to r)]$	$\to$	$(p \to r)$
T	T	T	T	T	T	T	T
T	T	F	T	F	F	T	F
T	F	T	F	F	T	T	T
T	F	F	F	F	T	T	F
F	T	T	F	F	T	T	T
F	T	F	F	F	F	T	T
F	F	T	F	F	T	T	T
F	F	F	F	F	T	T	T
			1	3	2	5	4

51. p: Jose is the manager.
q: Kevin is the coach.
r: Tim is the umpire

$$\begin{array}{l} p \to q \\ q \to r \\ \hline \therefore\ p \to r \end{array}$$

This argument is valid by the law of syllogism.

52. p: The truck is a diesel.
q: The truck is too cold to start.
r: The car has a flat tire.

$$\begin{array}{l} p \to q \\ q \vee r \\ \hline \therefore\ \sim p \end{array}$$

p	q	r	$[(p \to q)$	$\wedge$	$(q \vee r)]$	$\to$	$\sim p$
T	T	T	T	T	T	F	F
T	T	F	T	T	T	F	F
T	F	T	F	F	T	T	F
T	F	F	F	F	F	T	F
F	T	T	T	T	T	T	T
F	T	F	T	T	T	T	T
F	F	T	T	T	T	T	T
F	F	F	T	F	F	T	T
			1	3	2	5	4

The argument is invalid.

53. Invalid

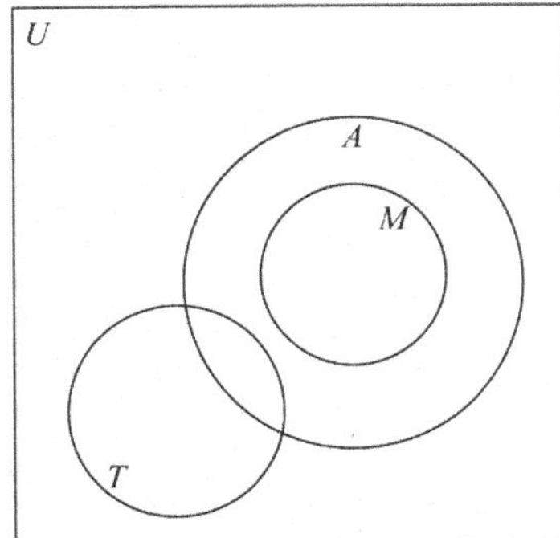

54. Valid

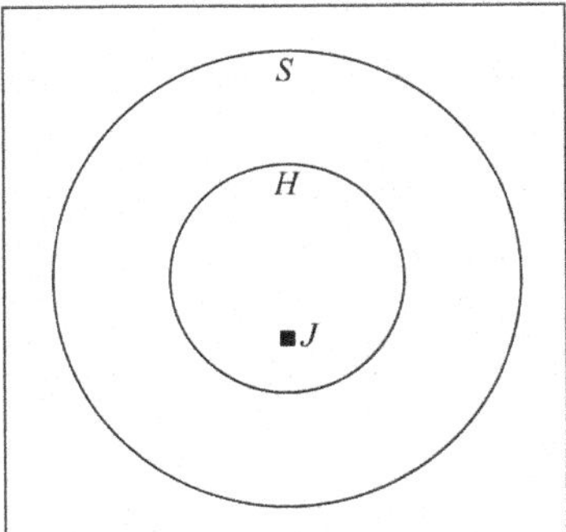

55. Invalid

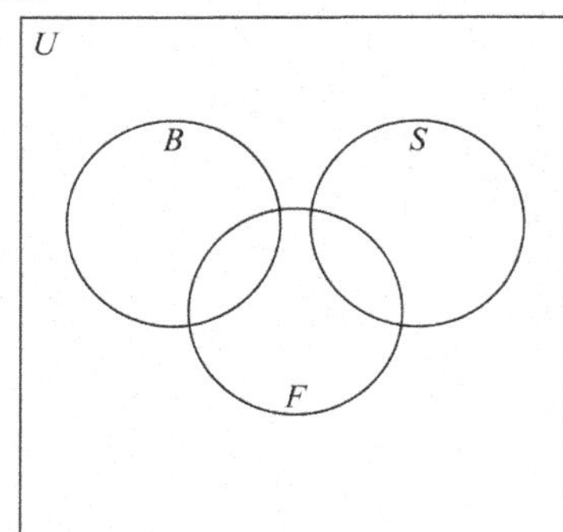

56. Invalid

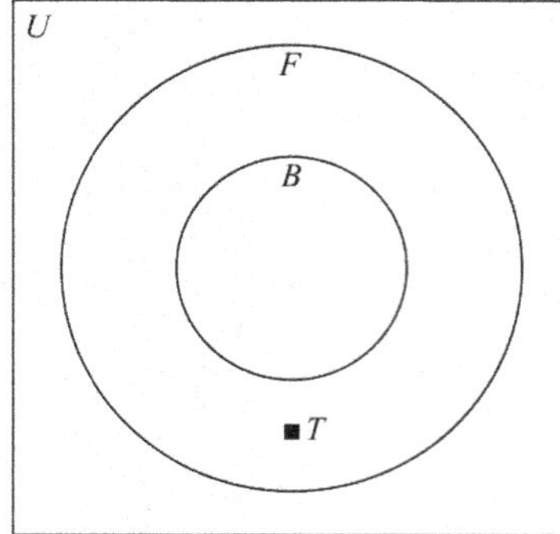

57. a) $p \wedge [(q \wedge r) \vee \sim p]$

b) For the bulb to be on, the first switch on the left must be closed. Thus, the switch on the bottom branch of the parallel portion will be open. Therefore, both switches on the top branch must be closed. Thus, the bulb lights only when p, q, and r are all closed.

58.

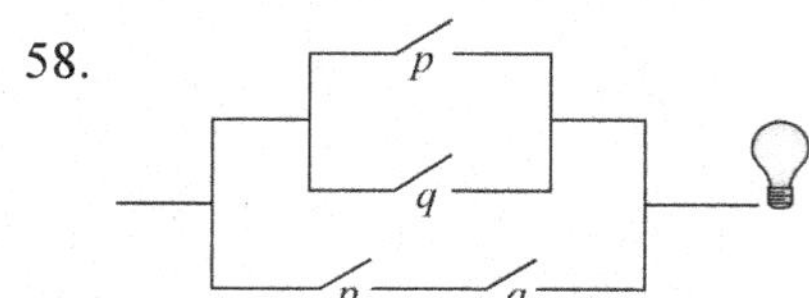

59. Symbolically the two circuits are $(p \vee q) \wedge (\sim q \vee \sim p)$ and $(p \wedge \sim q) \vee (q \wedge \sim p)$.

p	q	$\sim p$	$\sim q$	$(p \vee q)$	$\wedge$	$(\sim q \vee \sim p)$	$(p \wedge \sim q)$	$\wedge$	$(q \wedge \sim p)$
T	T	F	F	T	F	F	F	F	F
T	F	F	T	T	T	T	T	T	F
F	T	T	F	T	T	T	F	T	T
F	F	T	T	F	F	T	F	F	F
				1	3	2	1	3	2

The circuits are equivalent.

Chapter Test

1. $(\sim p \wedge q) \wedge \sim r$

2. $(r \rightarrow q) \vee \sim p$

3. $\sim (r \leftrightarrow \sim q)$

4. Phobos is not a moon of Mars and Rosalind is a moon of Uranus, if and only if Callisto is not a moon of Jupiter.

5. If Phobos is a moon of Mars or Callisto is not a Moon of Jupiter, then Rosalind is a moon of Uranus.

6.

p	q	r	$[\sim$	$(p \rightarrow r)]$	$\wedge$	q
T	T	T	F	T	F	T
T	T	F	T	F	T	T
T	F	T	F	T	F	F
T	F	F	T	F	F	F
F	T	T	F	T	F	T
F	T	F	F	T	F	T
F	F	T	F	T	F	F
F	F	F	F	T	F	F
			2	1	4	3

7.

p	q	r	$(q$	$\leftrightarrow$	$\sim r)$	$\vee$	p
T	T	T	T	F	F	T	T
T	T	F	T	T	T	T	T
T	F	T	F	T	F	T	T
T	F	F	F	F	T	T	T
F	T	T	T	F	F	F	F
F	T	F	T	T	T	T	F
F	F	T	F	T	F	T	F
F	F	F	F	F	T	F	F
			1	3	2	5	4

8. p: $2+6=8$
q: $7-12=5$

$p \vee q$
$\text{T} \vee \text{F}$
T

The statement is true.

9. p: A leap year has 366 days.
q: A week has 8 days.
r: An hour has 24 minutes.

$(p \wedge q) \leftrightarrow r$
$(\text{T} \wedge \text{F}) \leftrightarrow \text{F}$
$\text{F} \leftrightarrow \text{F}$
T

The statement is true.

10. $(\sim p \wedge q) \leftrightarrow (q \vee \sim r)$
$(\text{F} \wedge \text{F}) \leftrightarrow (\text{F} \vee \text{F})$
$\text{F} \leftrightarrow \text{F}$
T

11. $[\sim(r \rightarrow \sim p)] \wedge (q \rightarrow p)$
$[\sim(\text{T} \rightarrow \text{F})] \wedge (\text{F} \rightarrow \text{T})$
$\sim(\text{F}) \wedge \text{T}$
$\text{T} \wedge \text{T}$
T

12. The statements are equivalent by DeMorgan's Laws: $\sim p \vee q \Leftrightarrow \sim(\sim p) \wedge \sim q \Leftrightarrow p \wedge \sim q$.

13. p: The bird is red.
q: It is a cardinal.

			(a)			(b)			(c)	
p	q	p	$\rightarrow$	q	$\sim p$	$\vee$	q	$\sim p$	$\rightarrow$	$\sim q$
T	T	T	T	T	F	T	T	F	T	F
T	F	T	F	F	F	F	F	F	T	T
F	T	F	T	T	T	T	T	T	F	F
F	F	F	T	F	T	T	F	T	T	T
		1	3	2	1	3	2	1	3	2

Statements (a) and (b) are equivalent.

Note: Since $p \rightarrow q \Leftrightarrow \sim p \vee q$, statement (a) is equivalent to statement (b).

Statement (c) is the inverse of statement (a), so they are not equivalent.

14. p: The test is today.
q: The concert is tonight.

			(a)				(b)			(c)	
p	q	$\sim$	$(p$	$\vee$	$q)$	$\sim p$	$\wedge$	$\sim q$	$\sim p$	$\rightarrow$	$\sim q$
T	T	F	T	T	T	F	T	F	F	T	F
T	F	T	T	F	F	F	T	T	F	T	T
F	T	T	F	F	T	T	F	F	T	F	F
F	F	T	F	F	F	T	T	T	T	T	T
		4	1	3	2	1	3	2	1	3	2

Statements (a) and (b) are equivalent.

Note: Statement (a) is equivalent to statement (b) by DeMorgan's Laws.

15. s: The soccer team won the game.
f: Sue played fullback.
p: The team is in second place.

$s \rightarrow f$
$f \rightarrow p$
$\therefore s \rightarrow p$

This argument is valid by the law of syllogism.

16. Invalid

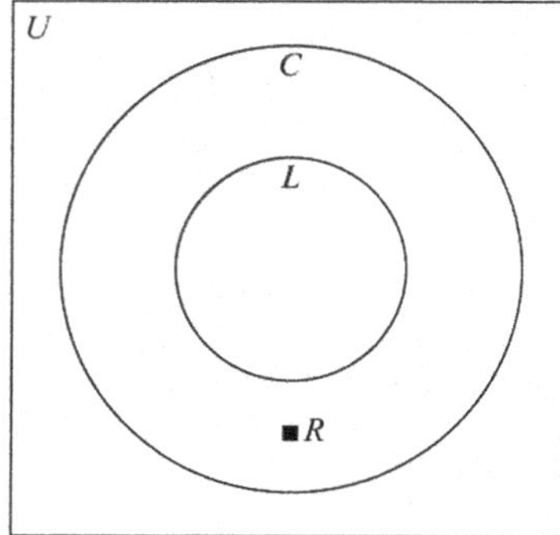

17. Some coffee beans do not contain caffeine.

18. Nick did not play football or Max did not play baseball.

19. *Converse*: If today is Saturday, then the garbage truck comes.
Inverse: If the garbage truck does not come today, then today is not Saturday.
Contrapositive: If today is not Saturday, then the garbage truck does not come.

20. 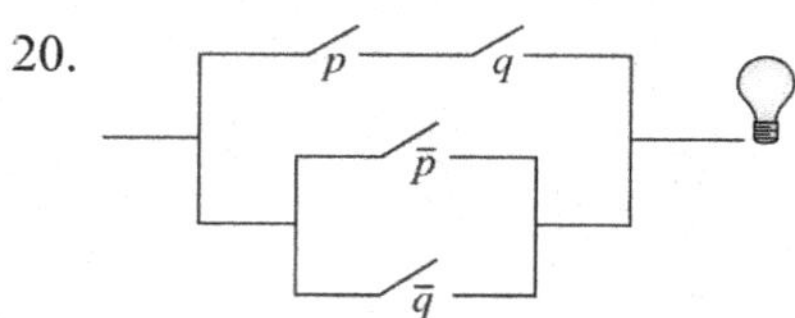

Chapter Four: Systems of Numeration

Section 4.1: Additive, Multiplicative, and Ciphered Systems of Numeration

1. How many
3. Hindu-Arabic
5. Subtract
7. Multiplicative
9. $100+100+10+10+1+1+1+1+1=225$
11. $10,000+1000+1000+100+100+100+10+10+10+10+1=12,341$
13. $100,000+100,000+100,000+10,000+10,000+10,000+1000+1000+1000+1000$
    $+100+100+10+1+1+1+1=334,214$
15. ∩∩||
17. 𓆼𓆼∩∩∩∩|||||
19. 𓆐𓂭𓂭𓂭𓂭𓂭𓆼𓆼𓆼99999999∩∩∩∩|||||
21. $10+10+10+5+1=36$
23. $100+(100-10)+(5-1)=194$
25. $1000+1000+500+100+(50-10)+1+1=2642$
27. $1000+1000+(1000-100)+(50-10)+5+1=2946$
29. $4(1000)+(500-100)+(100-10)+(10-1)=4499$
31. $10(1000)+1000+1000+500+100+50+10+5+1=12,666$
33. LXIII
35. CCCXLIX
37. MCMXIV
39. $\overline{\text{IV}}$DCCXCIII
41. $\overline{\text{IX}}$CMXCIX
43. $\overline{\text{XLVI}}$CCLXXXI
45. $7(10)+4=74$
47. $4(1000)+8(10)+1=4081$
49. $8(1000)+5(100)+5(10)=8550$
51. $4(1000)+3=4003$
53. 五十三
55. 三百七十八
57. 四千二百六十
59. 七千零五十六
61. $40+5=45$
63. $800+70+8=878$
65. $2\times1000+800+80+3=2883$
67. $\sigma\mu\gamma$
69. $\psi\,\pi\,\delta$
71. $^{\iota}\varepsilon\,\varepsilon$

73. Hindu-Arabic: $1000+20+20+1=1021$

Roman: MXXI $(100+10+10+1)$

Chinese: 一
千
零
二
十
一

Greek: ͵$\alpha\kappa\alpha$

75. Hindu-Arabic: $5(100)+2(10)+7=537$

Egyptian: 𓍢𓍢𓍢𓍢𓍢𓎆𓎆𓏺𓏺𓏺𓏺𓏺𓏺𓏺

Roman: DXXVII $(500+10+10+5+1+1)$

Greek: $\phi\kappa\xi$

77. A number is a quantity, and it answers the question "How many?" A numeral is a symbol used to represent the number.

79. $\overline{\text{CMXCIX}}$CMXCIX

81. Turn the book upside down.

83. 1888, MDCCCLXXXVIII

Section 4.2: Place-Value or Positional-Value Numeration Systems

1. 10

3. Hundreds

5. Units

7. a) 1
   b) 10
   c) Subtraction

9. a) 0
   b) 1
   c) 5

11. $(2\times10)+(5\times1)$

13. $(7\times100)+(1\times10)+(2\times1)$

15. $(4\times1000)+(3\times100)+(8\times10)+(7\times1)$

17. $(1\times10{,}000)+(6\times1000)+(4\times100)+(0\times10)+(2\times1)$

19. $(3\times100{,}000)+(4\times10{,}000)+(6\times1000)+(8\times100)+(6\times10)+(1\times1)$

21. $10+1+1+1+1=14$

23. $(10+1+1+1)(60)+(1+1+1+1)(1)=13(60)+4(1)=780+4=784$

25. $1(60^2)+(10+10+1)(60)+(10-(1+1))(1)=3600+21(60)+(10-2)(1)=3600+1260+8=4868$

27. 23 is 23 units. 𒌋𒌋𒁹𒁹𒁹

29. 471 is 7 groups of 60 and 51 units remaining. 𒌋⊤𒁹𒁹𒁹 𒌋𒌋𒌋𒌋𒌋𒁹

31. 3605 is 1 group of 3600 and 5 units remaining. 𒁹 𒁹𒁹𒁹𒁹𒁹

33. $2(20)+17(1)=40+17=57$

35. $7(18\times20)+9(20)+7(1)=2520+180+7$
$=2707$

37. $11(18\times20)+2(20)+0(1)=3960+40+0$
$=4000$

39. •
≡

41. $20\overline{)297}$ quotient 1
280
17

••••
═
••
≡

$297=14(20)+17(1)$

43. $360\overline{)2163}$ quotient 6
2160
3

$20\overline{)3}$ quotient 0
0
3

•
—
⬭
•••

$2163=6(360)+0(20)+3(1)$

45. Hindu-Arabic: $5(18\times 20)+7(20)+4(1)=1800+140+4=1944$

Babylonian: $1944 = 32(60)+24(1) =$ <<<▼▼ <<▼▼▼▼

47. $(\triangle \times ⊕^2) + (\square \times ⊕) + (\phi \times 1)$

49. The Mayan system has a different base and the numbers are written vertically.

51. a) $999{,}999 = 6\left(18\times 20^3\right)+18\left(18\times 20^2\right)+17\left(18\times 20\right)+13\left(20\right)+19\left(1\right)$

•
—

•••
≡

••
≡

•••
≡

••••
≡

b) $999{,}999 = 4(60)^3+37(60)^2+46(60)+39(1)$

▼▼▼▼ <<<<▼▼▼▼ <<<<▼▼▼▼▼▼ <<<<▼▼

53. $3(60)+33(1)=180+33=213$

32

$213-32=181$

$181=3(60)+1(1)$ ▼▼▼ ▼

55. $7(18\times 20)+6(20)+15(1)=2520+120+15=2655$

$6(18\times 20)+7(20)+13(1)=2160+140+13=2313$

$2655-2313=342$

$342=17(20)+2(1)$

••
≡

••

Section 4.3: Other Bases

1. Base

3. 5

5. $12_3 = (1\times 3)+(2\times 1)=5$

7. $412_6 = (4\times 36)+(1\times 6)+(2\times 1)=152$

9. $241_8 = (2\times 64)+(4\times 8)+(1\times 1)=161$

11. $309_{12} = (3\times 144)+(0\times 12)+(9\times 1)=441$

13. $573_{16} = (5\times 256)+(7\times 16)+(3\times 1)=1395$

15. $110101_2 = 32+16+4+1=53$

17. $7654_8 = (7\times 512)+(6\times 64)+(5\times 8)+(4\times 1)$
$= 4012$

19. $A91_{12} = (10\times 144)+(9\times 12)+1=1549$

21. $C679_{16} = (12\times 4096)+(6\times 256)+(7\times 16)+9$
$= 50{,}809$

23. $2^2=4$ and $2^3=8$

$$\begin{array}{r} 1 \\ 4\overline{)7} \\ \underline{4} \\ 3 \end{array} \qquad \begin{array}{r} 1 \\ 2\overline{)3} \\ \underline{2} \\ 1 \end{array}$$

$7 = 111_2$

25. $3^1=3,\ 3^2=9,\ 3^3=27$

$$\begin{array}{r} 2 \\ 9\overline{)25} \\ \underline{18} \\ 7 \end{array} \qquad \begin{array}{r} 2 \\ 3\overline{)7} \\ \underline{6} \\ 1 \end{array}$$

$25 = 221_3$

27. $4^1 = 4,\ \ 4^2 = 16,\ \ 4^3 = 64,\ \ 4^4 = 256$

$$\begin{array}{r}2\\64\overline{)190}\\\underline{128}\\62\end{array}\qquad\begin{array}{r}3\\16\overline{)62}\\\underline{48}\\14\end{array}\qquad\begin{array}{r}3\\4\overline{)14}\\\underline{12}\\2\end{array}$$

$190 = 2332_4$

29. $5^1 = 5,\ \ 5^2 = 25,\ \ 5^3 = 125$

$$\begin{array}{r}4\\25\overline{)102}\\\underline{100}\\2\end{array}\qquad\begin{array}{r}0\\5\overline{)2}\\\underline{0}\\2\end{array}$$

$102 = 402_5$

31. $8^3 = 512$ and $8^4 = 4096$

$$\begin{array}{r}2\\512\overline{)1098}\\\underline{1024}\\74\end{array}\qquad\begin{array}{r}1\\64\overline{)74}\\\underline{64}\\10\end{array}\qquad\begin{array}{r}1\\8\overline{)10}\\\underline{8}\\2\end{array}$$

$1098 = 2112_8$

33. $12^3 = 1728$ and $12^4 = 20,736$

$$\begin{array}{r}5\\1728\overline{)9004}\\\underline{8640}\\364\end{array}\qquad\begin{array}{r}2\\144\overline{)364}\\\underline{288}\\76\end{array}\qquad\begin{array}{r}6\\12\overline{)76}\\\underline{72}\\4\end{array}$$

$9004 = 5264_{12}$

35. $16^2 = 256$ and $16^3 = 4096$

$$\begin{array}{r}2\\4096\overline{)9455}\\\underline{8192}\\1263\end{array}\qquad\begin{array}{r}4\\256\overline{)1263}\\\underline{1024}\\239\end{array}\qquad\begin{array}{r}\text{(E)}\\14\\16\overline{)239}\\\underline{224}\\15\\\text{(F)}\end{array}$$

$9455 = 24EF_{16}$

37. $3^6 = 729$ and $3^7 = 2187$

$$\begin{array}{r}2\\729\overline{)2021}\\\underline{1458}\\563\end{array}\qquad\begin{array}{r}2\\243\overline{)563}\\\underline{486}\\77\end{array}\qquad\begin{array}{r}0\\81\overline{)77}\\\underline{0}\\77\end{array}\qquad\begin{array}{r}2\\27\overline{)77}\\\underline{54}\\23\end{array}\qquad\begin{array}{r}2\\9\overline{)23}\\\underline{18}\\5\end{array}\qquad\begin{array}{r}1\\3\overline{)5}\\\underline{3}\\2\end{array}$$

$2021 = 2202212_3$

39. $5^4 = 625$ and $5^5 = 3125$

$$\begin{array}{r}3\\625\overline{)2021}\\\underline{1875}\\146\end{array}\qquad\begin{array}{r}1\\125\overline{)146}\\\underline{125}\\21\end{array}\qquad\begin{array}{r}0\\25\overline{)21}\\\underline{0}\\21\end{array}\qquad\begin{array}{r}4\\5\overline{)21}\\\underline{20}\\1\end{array}$$

$2021 = 31041_5$

41. $7^3 = 343$ and $7^4 = 2401$

$$\begin{array}{r}5\\343\overline{)2021}\\\underline{1715}\\306\end{array}\qquad\begin{array}{r}6\\49\overline{)306}\\\underline{294}\\12\end{array}\qquad\begin{array}{r}1\\7\overline{)12}\\\underline{7}\\5\end{array}$$

$2021 = 5615_7$

43. $12^3 = 1728$ and $12^4 = 20,736$

$$\begin{array}{r}1\\1728\overline{)2021}\\\underline{1728}\\293\end{array}\qquad\begin{array}{r}2\\144\overline{)293}\\\underline{288}\\5\end{array}\qquad\begin{array}{r}0\\7\overline{)5}\\\underline{0}\\5\end{array}$$

$2021 = 1205_{12}$

45. $2(5)+3(1) = 10+3 = 13$

47. $$\begin{aligned}3(5^2)+0(5)+3(1) &= 3(25)+0+3\\&= 75+0+3\\&= 78\end{aligned}$$

49. $5^1 = 5$ and $5^2 = 25$

$$\begin{array}{r} 3 \\ 5\overline{)19} \\ \underline{15} \\ 4 \end{array}$$

$19 = 34_5$ or

51. $5^2 = 25$ and $5^3 = 125$

$$\begin{array}{r} 2 \\ 25\overline{)74} \\ \underline{50} \\ 24 \end{array} \qquad \begin{array}{r} 4 \\ 5\overline{)24} \\ \underline{20} \\ 4 \end{array}$$

$74 = 244_5$ or

53. $1(4) + 3(1) = 4 + 3 = 7$

55. $2(4^2) + 1(4) + 0(1) = 2(16) + 4 + 0$
$= 32 + 4 + 0 = 36$

For Exercises 57 and 59, (b) = blue = 0, (r) = red = 1, (go) = gold = 2, (gr) = green = 3

57. $4^1 = 4$ and $4^2 = 16$

$$\begin{array}{r} 2 \\ 4\overline{)10} \\ \underline{8} \\ 2 \end{array}$$

$10 = 22_4$ or (go)(go)$_4$

59. $4^2 = 16$ and $4^3 = 64$

$$\begin{array}{r} 3 \\ 16\overline{)60} \\ \underline{48} \\ 12 \end{array} \qquad \begin{array}{r} 3 \\ 4\overline{)12} \\ \underline{12} \\ 0 \end{array}$$

$60 = 330_4$ or (gr)(gr)(b)$_4$

61. a) $683 = 10213_5$

		Remainder
5	683	
5	136	3 ↑
5	27	1
5	5	2
5	1	0
	0	1

b) $763 = 1373_8$

		Remainder
8	763	
8	95	3 ↑
8	11	7
8	1	3
	0	1

63. a) $10_2 = (1 \times 2) + (0 \times 1) = 2$

b) $10_8 = (1 \times 8) + (0 \times 1) = 8$

c) $10_{16} = (1 \times 16) + (0 \times 1) = 16$

d) $10_{32} = (1 \times 32) + (0 \times 1) = 32$

e) In general, for any base b,
$10_b = (1 \times b) + (0 \times 1) = b$

65. $1(b^2) + 1(b) + 1 = 43$

$b^2 + b + 1 = 43$

$b^2 + b - 42 = 0$

$(b + 7)(b - 6) = 0$

$b + 7 = 0$ or $b - 6 = 0$

$b = -7$ or $b = 6$

Since the base cannot be negative, $b = 6$.

67. a) 1_3, 2_3, 10_3, 11_3, 12_3, 20_3, 21_3, 22_3, 100_3, 101_3, 102_3, 110_3, 111_3, 112_3, 120_3, 121_3, 122_3, 200_3, 201_3, 202_3

b) 1000_3

69. a) Answers will vary.

b) Answers will vary.

71. a) $3(4^4)+1(4^3)+2(4^2)+3(4)+0(1)=3(256)+64+2(16)+12+0=768+64+32+12+0=876$

b) ⓑ = blue = 0, ⓡ = red = 1, (go) = gold = 2, (gr) = green = 3

$4^1=4,\ 4^2=16,\ 4^3=64,\ 4^4=256$

$$\begin{array}{r} 2 \\ 64\overline{)177} \\ \underline{128} \\ 49 \end{array} \quad \begin{array}{r} 3 \\ 16\overline{)49} \\ \underline{48} \\ 1 \end{array} \quad \begin{array}{r} 0 \\ 4\overline{)1} \\ \underline{0} \\ 1 \end{array}$$

$177 = 2301_4$ or (go)(gr)(b)(r)$_4$

Section 4.4: Perform Computations in Other Bases

1. 12_5

3. 5

5. 22_5

7. $$\begin{array}{r} 21_3 \\ \underline{+20_3} \\ 111_3 \end{array}$$

9. $$\begin{array}{r} 132_5 \\ \underline{+34_5} \\ 221_5 \end{array}$$

11. $$\begin{array}{r} 654_7 \\ \underline{+436_7} \\ 1450_7 \end{array}$$

13. $$\begin{array}{r} 1011_2 \\ \underline{+1110_2} \\ 11001_2 \end{array}$$

15. $$\begin{array}{r} A734_{12} \\ \underline{+128B_{12}} \\ BA03_{12} \end{array}$$

17. $$\begin{array}{r} A734_{16} \\ \underline{+128B_{16}} \\ B9BF_{16} \end{array}$$

19. $$\begin{array}{r} 201_5 \\ \underline{-120_5} \\ 11_5 \end{array}$$

21. $$\begin{array}{r} 338_9 \\ \underline{-274_9} \\ 54_9 \end{array}$$

23. $$\begin{array}{r} 1101_2 \\ \underline{-111_2} \\ 110_2 \end{array}$$

25. $$\begin{array}{r} 4223_7 \\ \underline{-304_7} \\ 3616_7 \end{array}$$

27. $$\begin{array}{r} 4232_5 \\ \underline{-2341_5} \\ 1341_5 \end{array}$$

29. $$\begin{array}{r} A3B3_{12} \\ \underline{-21B4_{12}} \\ 81BB_{12} \end{array}$$

31. $$\begin{array}{r} 22_3 \\ \underline{\times 2_3} \\ 121_5 \end{array}$$

33. $$\begin{array}{r} 647_8 \\ \underline{\times 5_8} \\ 4103_8 \end{array}$$

35. $$\begin{array}{r} 37_8 \\ \underline{\times 21_8} \\ 37 \\ \underline{76\ \ } \\ 1017_8 \end{array}$$

37. $$\begin{array}{r} B12_{12} \\ \underline{\times 83_{12}} \\ 2936 \\ \underline{7494\ \ } \\ 77676_{12} \end{array}$$

39.
$$\begin{array}{r} 110_2 \\ \times\ 11_2 \\ \hline 110 \\ 110 \\ \hline 10010_2 \end{array}$$

41.
$$\begin{array}{r} 316_7 \\ \times\ 16_7 \\ \hline 2541 \\ 316 \\ \hline 6031_7 \end{array}$$

43. $2_4 \times 1_4 = 2_4$
$2_4 \times 2_4 = 10_4$
$2_4 \times 3_4 = 12_4$

$$2_4\overline{)312_4} = 123_4$$
$$\begin{array}{r} 2 \\ \hline 11 \\ 10 \\ \hline 12 \\ 12 \\ \hline 1 \end{array}$$

45. $3_7 \times 1_7 = 3_7$
$3_7 \times 2_7 = 6_7$
$3_7 \times 3_7 = 12_7$
$3_7 \times 4_7 = 15_7$
$3_7 \times 5_7 = 21_7$
$3_7 \times 6_7 = 24_7$

$$3_7\overline{)506_7} = 146_7$$
$$\begin{array}{r} 3 \\ \hline 20 \\ 15 \\ \hline 26 \\ 24 \\ \hline 2 \end{array}$$

146_7 R2_7

47. $3_8 \times 1_8 = 3_8$
$3_8 \times 2_8 = 6_8$
$3_8 \times 4_8 = 14_8$
$3_8 \times 5_8 = 17_8$

$$3_8\overline{)200_8} = 52_8$$
$$\begin{array}{r} 17 \\ \hline 10 \\ 6 \\ \hline 2 \end{array}$$

52_8 R2_8

49. $2_4 \times 1_4 = 2_4$
$2_4 \times 2_4 = 10_4$
$2_4 \times 3_4 = 12_4$

$$2_4\overline{)213_4} = 103_4$$
$$\begin{array}{r} 2 \\ \hline 01 \\ 00 \\ \hline 13 \\ 12 \\ \hline 1 \end{array}$$

103_4 R1_4

51. $3_5 \times 1_5 = 3_5$
$3_5 \times 2_5 = 11_5$
$3_5 \times 3_5 = 14_5$
$3_5 \times 4_5 = 22_5$

$$3_5\overline{)224_5} = 41_5$$
$$\begin{array}{r} 22 \\ \hline 4 \\ 3 \\ \hline 1 \end{array}$$

41_5 R1_5

53. $6_7 \times 1_7 = 6_7$
$6_7 \times 2_7 = 15_7$
$6_7 \times 3_7 = 24_7$
$6_7 \times 4_7 = 33_7$
$6_7 \times 5_7 = 42_7$
$6_7 \times 6_7 = 51_7$

$$\begin{array}{r} 45_7 \\ 6_7\overline{)404_7} \\ \underline{33} \\ 44 \\ \underline{42} \\ 2 \end{array}$$

45_7 R1_7

55. $$\begin{array}{r} 2_5 \\ +3_5 \\ \hline 10_5 \end{array}$$

57. $$\begin{array}{r} 23_5 \\ +13_5 \\ \hline 41_5 \end{array}$$

For Exercises 59–65, (b) = blue = 0, (r) = red = 1, (go) = gold = 2, (gr) = green = 3

59. $$\begin{array}{r} 3_4 \\ +\ 2_4 \\ \hline 11_4 \end{array}$$

(r)(r)$_4$

61. $$\begin{array}{r} 32_4 \\ +\ 11_4 \\ \hline 103_4 \end{array}$$

(r)(b)(gr)$_4$

63. $$\begin{array}{r} 31_4 \\ -\ 13_4 \\ \hline 12_4 \end{array}$$

(r)(go)$_4$

65. $$\begin{array}{r} 231_4 \\ -\ 103_4 \\ \hline 122_4 \end{array}$$

(r)(go)(go)$_4$

67. $$\begin{array}{r} \text{FAB}_{16} \\ \times\ 4_{16} \\ \hline \text{2C} \\ 28 \\ \text{3C} \\ \hline \text{3EAC}_{16} \end{array}$$

69. a) $$\begin{array}{r} 462_8 \\ \times\ 35_8 \\ \hline 2772 \\ 1626 \\ \hline 21252_8 \end{array}$$

b) $462_8 = 4(8^2) + 6(8) + 2(1) = 4(64) + 48 + 2 = 256 + 48 + 2 = 306$

$35_8 = 3(8) + 5(1) = 24 + 5 = 29$

69. (continued)

c)
$$\begin{array}{r} 306 \\ \times 29 \\ \hline 2754 \\ 612 \\ \hline 8874 \end{array}$$

d) $21252_8 = 2(8^4)+1(8^3)+2(8^2)+5(8)+2(1) = 2(4096)+512+2(64)+40+2$
$= 8192+512+128+40+2 = 8874$

e) Yes, in part (a), the numbers were multiplied in base 8 and then converted to base 10 in part (d). In part (b), the numbers were converted to base 10 first, then multiplied in part (c).

71. Gold = 0, Purple = 1, Blue = 2, Red =3

$$\begin{array}{r} 301_4 \\ +120_4 \\ \hline 1021_4 \end{array}$$

Section 4.5: Early Computational Methods

1. a) Divided
   b) Doubled

3. Three, two

5. $11 \times 19 = 19 + 38 + 152 = 209$

11 — 19
5 — 38
~~2 — 76~~
1 — 152

7. $29 \times 35 = 35 + 140 + 280 + 560 = 1015$

29 — 35
~~14 — 70~~
7 — 140
3 — 280
1 — 560

9. $35 \times 236 = 236 + 472 + 7552 = 8260$

35 — 236
17 — 472
~~8 — 944~~
~~4 — 1888~~
~~2 — 3776~~
1 — 7552

11. $93 \times 93 = 93 + 372 + 744 + 1488 + 5952 = 8649$

93 — 93
~~46 — 186~~
23 — 372
11 — 744
5 — 1488
~~2 — 2976~~
1 — 5952

13. $3 \times 229 = 687$

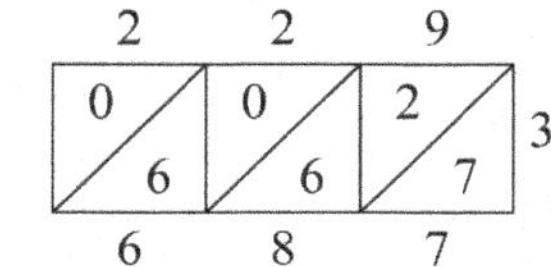

15. $6 \times 425 = 2550$

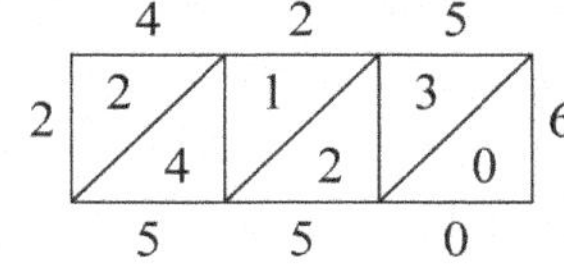

17. $75 \times 12 = 900$

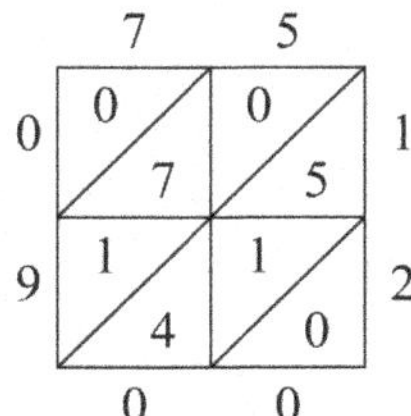

19. $314 \times 652 = 204,728$

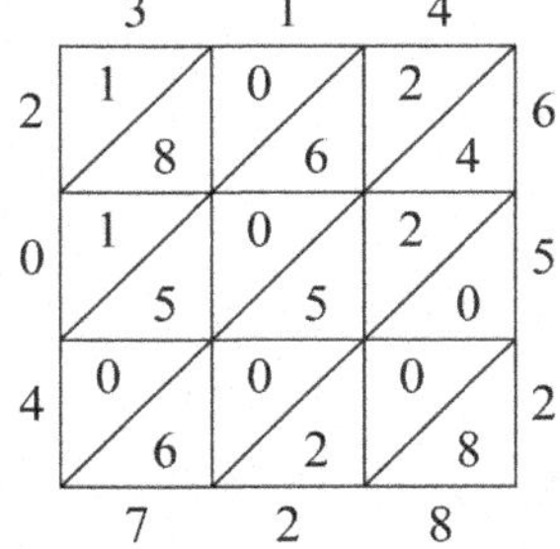

21. $2 \times 52 = 104$

	Index	5	2
2	1	1 / 0	0 / 4
		0	4

23. $5 \times 81 = 405$

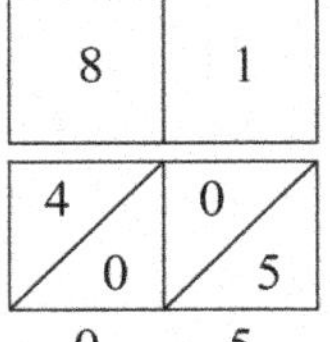

25. $5 \times 125 = 625$

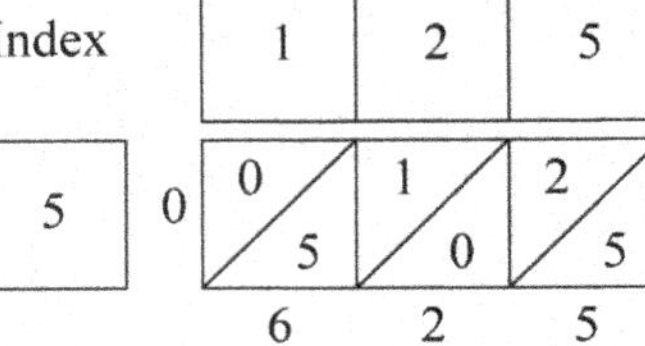

27. $9 \times 6742 = 60,678$

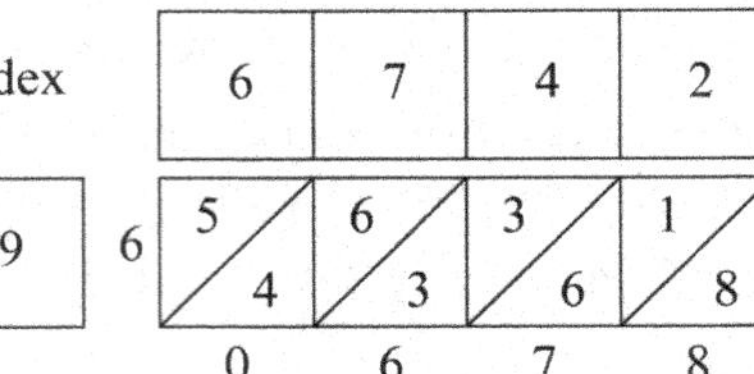

29. a) 253×46; Place the factors of 8 until the correct factors and placements are found so the rest of the rectangle can be completed.

b) $253 \times 46 = 11,638$

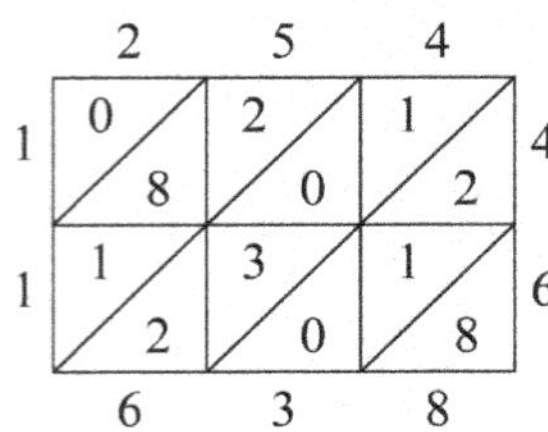

31. a) 4×382; Place the factors of 12 until the correct the factors and placements are found so the rest can be completed.

b) $4 \times 382 = 1528$

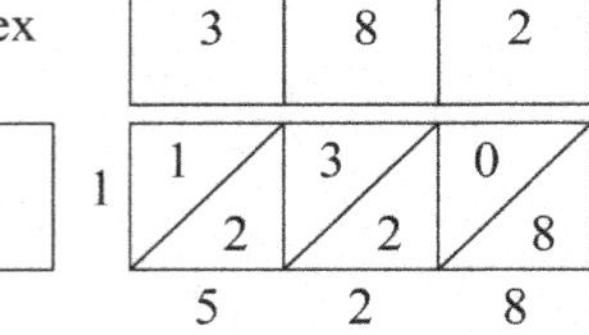

33. $13 \times 22 = 22 + 88 + 176$
$= 286 =$ 99∩∩∩∩∩∩∩∩||||||

13 — 22
~~6 — 44~~
3 — 88
1 — 176

35. $21_3 \times 21_3 = 1211_3$

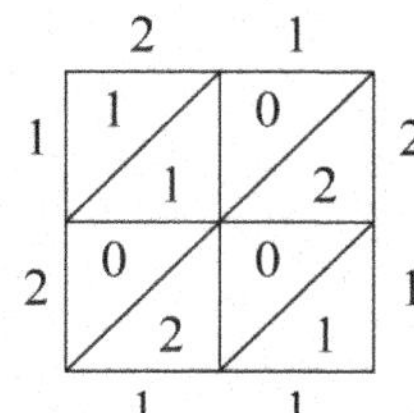

Review Exercises

1. $100 + 100 + 10 + 10 + 10 + 1 + 1 = 232$
2. $10,000 + 10,000 + 10,000 + 10,000 + 1000 + 1000 + 10 + 10 + 10 + 10 + 10 + 1 + 1 + 1 = 42,053$
3. $100,000 + 100,000 + 100 + 100 + 100 + 100 + 10 + 10 + 1 = 200,421$

4. $1,000,000+100,000+100,000+10,000+1000+1000+1000+1000+100+100+100+10+10+10$
$=1,214,330$

5. 𓆼𓆼𓆼𓍢𓍢𓎆𓏺𓏺𓏺𓏺

6. 𓂭𓍢𓍢

7. 𓆐𓂭𓂭𓆼𓆼𓆼𓆼𓍢𓍢𓍢𓎆𓎆𓏺

8. 𓁨𓆼𓆼𓆼𓆼𓎆𓎆𓎆𓎆𓏺𓏺

9. $10+(5-1)=14$

10. $100+(50-10)+1+1=142$

11. $1000+1000+(500-100)+10+10+10+5+1+1+1=2437$

12. $5(1000)+(500+200)+50+(10-1)=5759$

13. XXIV

14. DXLIII

15. MCMLXIV

16. $\overline{\text{VI}}$CDXCI

17. $3(10)+2=32$

18. $4(10)+5=45$

19. $2(100)+6(10)+7=267$

20. $3(1000)+4(100)+2(10)+9=3429$

21. 二十三

22. 五十四

23. 四百九十二

24. 二千六百五十二

25. $80+1=81$

26. $500+40+8=548$

27. $600+5=605$

28. $3000+300+30+4=3334$

29. $\kappa\beta$

30. $\psi o\gamma$

31. $\omega\xi\zeta$

32. ͵$\delta\sigma\iota\theta$

33. $(3\times 60)+(11\times 1)=191$

34. $(3\times 60)+(9\times 1)=189$

35. $(1\times 60^2)+(1\times 60)+(13\times 1)=3673$

36. $(2\times 60^2)++(3\times 60)+(8\times 1)=7388$

37. 𒁹𒁹𒁹 𒁹

38. 𒁹 𒁹 𒌋𒌋𒈨𒁹

39. 𒁹𒁹 𒁹 𒌋𒌋

40. 𒌋 𒁹𒁹𒁹 𒁹𒁹𒁹𒁹

41. $2(20)+8=48$

42. $3(18\times 20)+14(20)+7=1367$

43. $7(18\times 20)+0(20)+1=2521$

44. $2(18\times 20^2)+4(18\times 20)+4(20)+3=15,923$

45. $69=3(3\times 20)+9(1)$

$$\begin{array}{r} 3 \\ 20\overline{)69} \\ \underline{60} \\ 9 \end{array}$$

•••
••••
——

46. $812=2(18\times 20)+4(20)+12(1)$

$$\begin{array}{r} 2 \\ 360\overline{)812} \\ \underline{720} \\ 92 \end{array} \qquad \begin{array}{r} 4 \\ 20\overline{)92} \\ \underline{80} \\ 12 \end{array}$$

••
••••
••
——
——

47. $1571=4(18\times 20)+6(20)+11(1)$

$$\begin{array}{r} 4 \\ 360\overline{)1571} \\ \underline{1440} \\ 131 \end{array} \qquad \begin{array}{r} 6 \\ 20\overline{)131} \\ \underline{120} \\ 11 \end{array}$$

••••
•
——
•
——
——

48. $17{,}913 = 2\left(18\times 20^2\right)+9(18\times 20)+13(20)+13(1)$

$$\begin{array}{rrr} 2 & 9 & 13 \\ 7200\overline{)17{,}913} & 360\overline{)3513} & 20\overline{)273} \\ \underline{14{,}440} & \underline{3240} & \underline{260} \\ 3513 & 273 & 13 \end{array}$$

••
••••
___
•••
___
___
•••
___
___

49. $47_8 = 4(8)+7(1)=39$

50. $111_2 = 1\left(2^2\right)+1(2)+1(1)=4+2+1=7$

51. $130_4 = 1\left(4^2\right)+3(4)+0(1)=16+12+0=28$

52. $3425_7 = 3\left(7^3\right)+4\left(7^2\right)+2(7)+5(1)=3(343)+4(49)+14+5=1029+196+14+5=1244$

53. $A94_{12} = 10\left(12^2\right)+9(12)+4(1)=1440+108+4=1552$

54. $20220_3 = 2\left(3^4\right)+0\left(3^3\right)+2\left(3^2\right)+2(3)+0(1)=2(81)+0+2(9)+6+0=162+0+18+6+0=186$

55. $2^8=256$ and $2^9=512$

$$\begin{array}{rrrrrrrr} 1 & 1 & 1 & 0 & 0 & 1 & 1 & 1 \\ 256\overline{)463} & 128\overline{)207} & 64\overline{)79} & 32\overline{)15} & 16\overline{)15} & 8\overline{)15} & 4\overline{)7} & 2\overline{)3} \\ \underline{256} & \underline{128} & \underline{64} & \underline{0} & \underline{0} & \underline{8} & \underline{4} & \underline{2} \\ 207 & 79 & 15 & 15 & 15 & 7 & 3 & 1 \end{array}$$

$463 = 111001111_2$

56. $4^4=256$ and $4^5=1056$

$$\begin{array}{rrrr} 1 & 3 & 0 & 3 \\ 256\overline{)463} & 64\overline{)207} & 16\overline{)15} & 4\overline{)15} \\ \underline{256} & \underline{192} & \underline{0} & \underline{12} \\ 207 & 15 & 15 & 3 \end{array}$$

$463 = 13033_4$

57. $5^3=125$ and $5^4=625$

$$\begin{array}{rrr} 3 & 3 & 2 \\ 125\overline{)463} & 25\overline{)88} & 5\overline{)13} \\ \underline{375} & \underline{75} & \underline{10} \\ 88 & 13 & 3 \end{array}$$

$463 = 3323_5$

58. $8^2=64$ and $8^3=512$

$$\begin{array}{rr} 7 & 1 \\ 64\overline{)463} & 8\overline{)15} \\ \underline{448} & \underline{8} \\ 15 & 7 \end{array}$$

$463 = 717_8$

59. $12^2=144$ and $12^3=1728$

$$\begin{array}{rrr} 3 & 2 & 2 \\ 144\overline{)463} & 12\overline{)31} & 5\overline{)13} \\ \underline{432} & \underline{24} & \underline{10} \\ 31 & 7 & 3 \end{array}$$

$463 = 327_{12}$

60. $16^1=16$ and $16^2=256$

$$\begin{array}{rr} & \text{(C)} \\ 3 & 12 \\ 256\overline{)463} & 16\overline{)207} \\ \underline{256} & \underline{192} \\ 207 & 15 \\ & \text{(F)} \end{array}$$

$493 = 1CF_{16}$

61.
$$\begin{array}{r} 121_4 \\ \underline{+322_4} \\ 1103_4 \end{array}$$

62.
$$\begin{array}{r} 10110_2 \\ \underline{+11001_2} \\ 101111_2 \end{array}$$

63. $$\begin{array}{r} 9B_{12} \\ +87_{12} \\ \hline 166_{12} \end{array}$$

64. $$\begin{array}{r} 2B9_{16} \\ +456_{16} \\ \hline 70F_{16} \end{array}$$

65. $$\begin{array}{r} 3024_5 \\ +4023_5 \\ \hline 12102_5 \end{array}$$

66. $$\begin{array}{r} 3407_8 \\ +7014_8 \\ \hline 12423_8 \end{array}$$

67. $$\begin{array}{r} 321_4 \\ -133_4 \\ \hline 122_4 \end{array}$$

68. $$\begin{array}{r} 1001_2 \\ -101_2 \\ \hline 100_2 \end{array}$$

69. $$\begin{array}{r} A7B_{12} \\ -95_{12} \\ \hline 9A6_{12} \end{array}$$

70. $$\begin{array}{r} 4321_5 \\ -442_5 \\ \hline 3324_5 \end{array}$$

71. $$\begin{array}{r} 1713_8 \\ -1243_8 \\ \hline 450_8 \end{array}$$

72. $$\begin{array}{r} F64_{16} \\ -2A3_{16} \\ \hline CC1_{16} \end{array}$$

73. $$\begin{array}{r} 431_6 \\ \times 3_6 \\ \hline 2133_6 \end{array}$$

74. $$\begin{array}{r} 2321_4 \\ \times 3_4 \\ \hline 20223_4 \end{array}$$

75. $$\begin{array}{r} 34_5 \\ \times 21_5 \\ \hline 34 \\ 123 \\ \hline 1314_5 \end{array}$$

76. $$\begin{array}{r} 476_8 \\ \times 23_8 \\ \hline 1672 \\ 1174 \\ \hline 13632_8 \end{array}$$

77. $$\begin{array}{r} 126_{12} \\ \times 47_{12} \\ \hline 856 \\ 4A0 \\ \hline 5656_{12} \end{array}$$

78. $$\begin{array}{r} 1A3_{16} \\ \times 12_{16} \\ \hline 346 \\ 1A3 \\ \hline 1D76_{16} \end{array}$$

79. $2_3 \times 1_3 = 2_3$

$2_3 \times 2_3 = 11_3$

$$\begin{array}{r} 21_3 \\ 2_3 \overline{)120_3} \\ \underline{11} \\ 10 \\ \underline{2} \\ 1 \end{array}$$

21_3 R1_3

80. $2_4 \times 1_4 = 2_4$

$2_4 \times 2_4 = 10_4$

$2_4 \times 3_4 = 12_4$

$$\begin{array}{r} 130_4 \\ 2_4 \overline{)320_4} \\ \underline{21} \\ 120 \\ \underline{120} \\ 0 \\ \underline{0} \\ 1 \end{array}$$

81. $3_5 \times 1_5 = 3_5$
$3_5 \times 2_5 = 11_5$
$3_5 \times 3_5 = 14_5$
$3_5 \times 4_5 = 22_5$

$$\begin{array}{r} 23_5 \\ 3_5\overline{)130_5} \\ \underline{11} \\ 20 \\ \underline{14} \\ 1 \end{array}$$

23_5 R1_5

82. $4_6 \times 1_6 = 4_6$
$4_6 \times 2_6 = 12_6$
$4_6 \times 3_6 = 20_6$
$4_6 \times 4_6 = 24_6$
$4_6 \times 5_6 = 32_6$

$$\begin{array}{r} 433_6 \\ 4_6\overline{)3020_6} \\ \underline{24} \\ 22 \\ \underline{20} \\ 20 \\ \underline{20} \\ 0 \end{array}$$

83. $3_6 \times 1_6 = 3_6$
$3_6 \times 2_6 = 10_6$
$3_6 \times 3_6 = 13_6$
$3_6 \times 4_6 = 20_6$
$3_6 \times 5_6 = 23_6$

$$\begin{array}{r} 411_6 \\ 3_6\overline{)2034_6} \\ \underline{20} \\ 3 \\ \underline{3} \\ 4 \\ \underline{3} \\ 1 \end{array}$$

411_6 R1_6

84. $6_8 \times 1_8 = 6_8$
$6_8 \times 2_8 = 14_8$
$6_8 \times 3_8 = 22_8$
$6_8 \times 4_8 = 30_8$
$6_8 \times 5_8 = 36_8$
$6_8 \times 6_8 = 44_8$
$6_8 \times 7_8 = 52_8$

$$\begin{array}{r} 664_8 \\ 3_8\overline{)5072_8} \\ \underline{44} \\ 47 \\ \underline{44} \\ 32 \\ \underline{30} \\ 2 \end{array}$$

664_8 R2_8

85. $125 \times 23 = 23 + 92 + 184 + 368 + 736 + 1472$
$= 2875$

125 — 23
~~62 — 46~~
31 — 92
15 — 184
7 — 368
3 — 736
1 — 1472

86. $125 \times 23 = 2875$

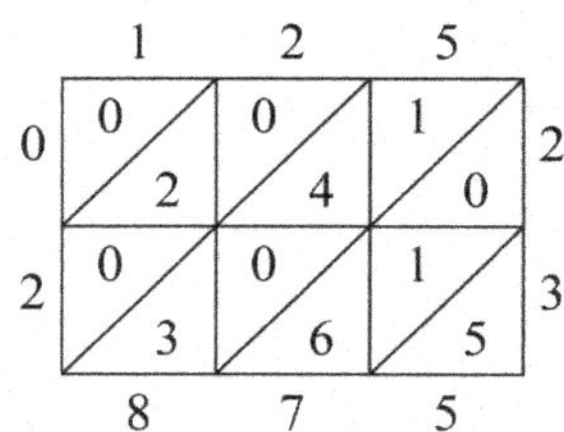

87. $125 \times 23 = 250 \times 10 + 375 = 2500 + 375 = 2875$

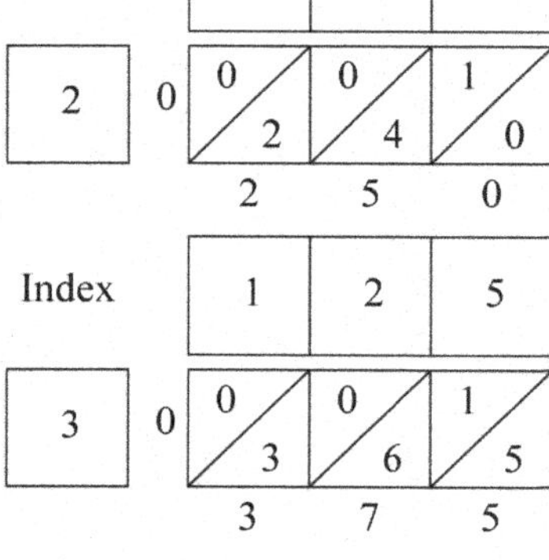

Chapter Test

1. MMCDLXXIX $= 1000 + 1000 + (500 - 100) + 50 + 20 + 9 = 2479$

2. $31(60) + 13(1) = 1860 + 13 = 1873$

3. $8(1000) + 0 + 9(10) = 8000 + 0 + 90 = 8090$

4. $2(18 \times 20) + 12(20) + 9(1) = 2(360) + 240 + 9 = 720 + 240 + 9 = 969$

5. $100,000 + 10,000 + 10,000 + 1000 + 1000 + 100 + 10 + 10 + 10 + 10 + 1 + 1 = 122,142$

6. $2(1000) + 700 + 40 + 5 = 2000 + 700 + 40 + 5 = 2745$

7. 𓆼𓆼𓍢𓎆𓎆𓏺𓏺𓏺𓏺

8. ʹβυοϛ

9. $1434 = 3(18 \times 20) + 17(20) + 14(1)$

$$\begin{array}{r} 3 \\ 360\overline{)1434} \\ \underline{1080} \\ 354 \end{array} \qquad \begin{array}{r} 17 \\ 20\overline{)354} \\ \underline{340} \\ 14 \end{array}$$

•••
••
═══
••••
═══

10. $1596 = 26(60) + 36(1) =$ <<YYYYYY <<<YYYYYY

$$\begin{array}{r} 26 \\ 60\overline{)1596} \\ \underline{1560} \\ 36 \end{array}$$

11. MMDCCXLIX

12. $23_4 = 2(4) + 3(1) = 8 + 3 = 11$

13. $B92_{12} = 11\left(12^2\right) + 9(12) + 2(1) = 11(144) + 108 + 2 = 1584 + 108 + 2 = 1694$

14. $2^5 = 32$ and $2^6 = 64$

$$\begin{array}{r} 1 \\ 32\overline{)36} \\ \underline{32} \\ 4 \end{array} \quad \begin{array}{r} 0 \\ 16\overline{)4} \\ \underline{0} \\ 4 \end{array} \quad \begin{array}{r} 0 \\ 8\overline{)4} \\ \underline{0} \\ 4 \end{array} \quad \begin{array}{r} 1 \\ 4\overline{)4} \\ \underline{4} \\ 0 \end{array} \quad \begin{array}{r} 0 \\ 2\overline{)0} \\ \underline{0} \\ 0 \end{array}$$

$36 = 100100_2$

15. $16^2 = 256$ and $16^3 = 4096$

$$\begin{array}{r} \text{(B)} \\ 11 \\ 256\overline{)2938} \\ \underline{2816} \\ 122 \end{array} \qquad \begin{array}{r} 7 \\ 16\overline{)122} \\ \underline{112} \\ 10 \\ \text{(A)} \end{array}$$

$2938 = B7A_{16}$

16.
$$\begin{array}{r} 1101_2 \\ \underline{+1011_2} \\ 11000_2 \end{array}$$

17.
$$\begin{array}{r} 45_6 \\ \underline{\times\, 23_6} \\ 223 \\ \underline{134} \\ 2003_6 \end{array}$$

18. $3_5 \times 1_5 = 3_5$

$3_5 \times 2_5 = 11_5$

$3_5 \times 3_5 = 14_5$

$3_5 \times 4_5 = 22_5$

$$\begin{array}{l} \;220_5 \\ 3_5\overline{)1210_5} \\ \underline{11} \\ \;\;11 \\ \;\;\underline{11} \\ \;\;\;00 \\ \;\;\;\underline{00} \\ \;\;\;\;0 \end{array}$$

19. $35 \times 28 = 28 + 56 + 896 = 980$

35 — 28
17 — 56
~~8 — 112~~
~~4 — 224~~
~~2 — 448~~
1 — 896

20. $43 \times 196 = 8428$

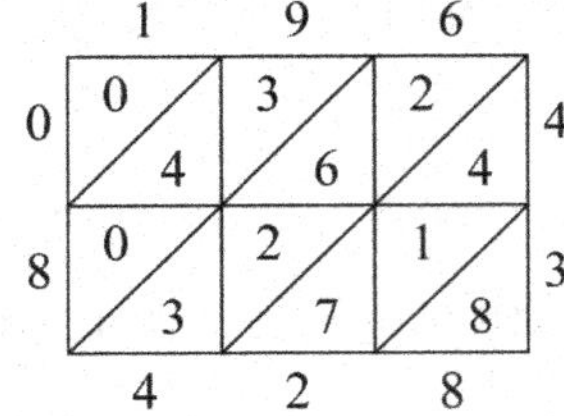

Chapter Five: Number Theory and the Real Number System

Section 5.1: Number Theory

1. Theory
3. Zero
5. Composite
7. Divisor
9. Conjecture
11. The prime numbers between 1 and 100 are: 2, 3, 5, 7, 11, 13, 17, 19, 23, 29, 31, 37, 41, 43, 47, 53, 59, 61, 67, 71, 73, 79, 83, 89, 97.

1	2	3	4	5	6	7	8	9	10
11	12	13	14	15	16	17	18	19	20
21	22	23	24	25	26	27	28	29	30
31	32	33	34	35	36	37	38	39	40
41	42	43	44	45	46	47	48	49	50
51	52	53	54	55	56	57	58	59	60
61	62	63	64	65	66	67	68	69	70
71	72	73	74	75	76	77	78	79	80
81	82	83	84	85	86	87	88	89	90
91	92	93	94	95	96	97	98	99	100

13. False. 8 is a factor of 64.
15. False; 35 is a multiple of 7.
17. True; since $32 \div 4 = 8$.
19. True; since $40 \div 5 = 8$.
21. False; 8 is a divisor of 72.
23. True; if a number is not divisible by 5, then it will not be divisible by $2 \cdot 5 = 10$.
25. True; since $2 \cdot 3 = 6$.
27. Divisible by 3 and 5.
29. Divisible by 2, 3, 4, 6, 8, and 9.
31. Divisible by 2, 3, 4, 5, 6, 8, and 10.
33. 60 (other answers are possible)
35. $12 = 2^2 \cdot 3$

$$\begin{array}{r|l} 2 & 12 \\ \hline 2 & 6 \\ \hline & 3 \end{array}$$

37. $180 = 2^2 \cdot 3^2 \cdot 5$

$$\begin{array}{r|l} 2 & 180 \\ \hline 2 & 90 \\ \hline 3 & 45 \\ \hline 3 & 15 \\ \hline & 5 \end{array}$$

39. $332 = 2^2 \cdot 83$

$$\begin{array}{r|l} 2 & 332 \\ \hline 2 & 166 \\ \hline & 83 \end{array}$$

41. $513 = 3^3 \cdot 19$

$$\begin{array}{r|l} 3 & 513 \\ \hline 3 & 171 \\ \hline 3 & 57 \\ \hline & 19 \end{array}$$

43. $1336 = 2^3 \cdot 167$

$$\begin{array}{r|l} 2 & 1336 \\ \hline 2 & 668 \\ \hline 2 & 334 \\ \hline & 167 \end{array}$$

45. Prime factorizations: $6 = 2 \cdot 3$, $21 = 3 \cdot 7$
    a) The common factor is 3, so the GCD is 3.
    b) The factors with the greatest exponent that appear in either number are 2, 3, and 7. The LCM is $2 \cdot 3 \cdot 7 = 42$.

47. Prime factorizations: $20 = 2^2 \cdot 5$, $35 = 5 \cdot 7$
    a) The common factor is 5, so the GCD is 5.
    b) The factors with the greatest exponent that appear in either number are 2^2, 5, and 7. The LCM is $2^2 \cdot 5 \cdot 7 = 140$.

49. Prime factorizations: $40 = 2^3 \cdot 5$, $900 = 2^2 \cdot 3^2 \cdot 5^2$
    a) The common factor is $2^2 \cdot 5$, so the GCD is 20.
    b) The factors with the greatest exponent that appear in either number are 2^3, 3^2, and 5. The LCM is $2^3 \cdot 3^2 \cdot 5^2 = 1800$.

51. Prime factorizations: $96 = 2^5 \cdot 3$, $212 = 2^2 \cdot 53$
    a) The common factor is 2^2, so the GCD is 4.
    b) The factors with the greatest exponent that appear in either number are 2^5, 3, and 5. The LCM is $2^5 \cdot 3 \cdot 53 = 5088$.

53. Prime factorizations: $24 = 2^3 \cdot 3$, $48 = 2^4 \cdot 3$, $128 = 2^7$
    a) The common factor is 2^3, so the GCD is 8.
    b) The factors with the greatest exponent that appear in any number are 2^7 and 3. The LCM is $2^7 \cdot 3 = 384$.

55. The LCM of $4 = 2^2$ and $14 = 2 \cdot 7$ is $2^2 \cdot 7 = 28$ days.

57. 2×60, 3×40, 4×30, 5×24, 6×20, 8×15, 10×12, 12×10, 15×8, 20×6, 24×5, 30×4, 40×3, or 60×2

59. The GCD of $70 = 2 \cdot 5 \cdot 7$ and $175 = 5^2 \cdot 7$ is $5 \cdot 7 = 35$ cars.

61. The GCD of $150 = 2 \cdot 3 \cdot 5^2$ and $180 = 2^2 \cdot 3^2 \cdot 5$ is $2 \cdot 3 \cdot 5 = 30$ trees.

63. The LCM of 5 and $6 = 2 \cdot 3$ is $2 \cdot 3 \cdot 5 = 30$ days.

65. a) The prime factorizations are $8 = 2^3$ and $9 = 3^2$. Yes, they are relatively prime.
    b) The prime factorizations are $15 = 3 \cdot 5$ and $24 = 2^3 \cdot 3$. No, they are not relatively prime.
    c) The prime factorizations are $39 = 3 \cdot 13$ and $52 = 2^2 \cdot 13$. No, they are not relatively prime.
    d) The prime factorizations are $117 = 3^2 \cdot 13$ and $178 = 2 \cdot 89$. Yes, they are relatively prime.

67. Use the list of primes generated in exercise 11. The next two sets of twin primes are 17, 19 and 29, 31.

69. Use the formula $2^{2^n} + 1$, where n is a natural number: $2^{2^1} + 1 = 5$, $2^{2^2} + 1 = 2^4 + 1 = 17$, and $2^{2^3} + 1 = 2^8 + 1 = 257$. These numbers are all prime.

71. A number is divisible by 14 if both 2 and 7 divide the number.

73. The GCD of 15 and 40 is 5.

$$15\overline{)40}^{\,2}\quad 30,\ 10 \qquad 10\overline{)15}^{\,1}\quad 10,\ 5 \qquad 5\overline{)10}^{\,2}\quad 10,\ 0$$

75. The GCD of 35 and 105 is 35.

$$35\overline{)105}^{\,3}\quad 105,\ 0$$

77. The GCD of 150 and 180 is 30.

$$150\overline{)180}^{\,1}\quad 150,\ 30 \qquad 30\overline{)150}^{\,5}\quad 150,\ 0$$

79. The proper factors of 28 are 1, 2, 4, 7, and 14.
Since $1+2+4+7+14=28$, 28 is a perfect number.

81. The proper factors of 72 are 1, 2, 3, 4, 6, 8, 9, 12, 18, 24, and 36.
Since $1+2+3+4+6+8+9+12+18+24+36=123$, 72 is not a perfect number.

83. a) $60=2^2\cdot 3^1\cdot 5^1$; Adding 1 to each exponent and then multiplying these numbers, we get $(2+1)(1+1)(1+1)$ $=3\cdot 2\cdot 2=12$ factors of 60.
    b) The factors of 60 are 1, 2, 3, 4, 5, 6, 10, 12, 15, 20, 30, and 60.

85. For any three consecutive natural numbers, one of the numbers is divisible by 2 and another number is divisible by 3. Therefore, the product of the three numbers would be divisible by 6.

87. $54,036=(54,000+36)$; $54,000\div 18=3000$; $36\div 18=2$; Thus, since $18\,|\,54,000$ and $18\,|\,36$, $18\,|\,54,036$.

89. $8=2+3+3$, $9=3+3+3$, $10=2+3+5$, $11=2+2+7$, $12=2+5+5$, $13=3+3+7$, $14=2+5+7$, $15=3+5+7$, $16=2+7+7$, $17=5+5+7$, $18=2+5+11$, $19=3+5+11$, $20=2+7+11$

91. a) $5=6-1$; $7=6+1$; $11=12-1$; $13=12+1$; $17=18-1$; $19=18+1$; $23=24-1$; $29=30-1$
    b) Conjecture: Every prime number greater than 3 differs by 1 from a multiple of 6.
    c) The conjecture should appear to be correct.

Section 5.2: The Integers

1. Whole

3. a) Positive
   b) Negative

5. a) $-3<2$
   b) $-3<-2$
   c) $-3<0$
   d) $-3>-4$

7. $-9,\ -6,\ -3,\ 0,\ 3,\ 6$

9. $-6,\ -5,\ -4,\ -3,\ -2,\ -1$

11. $-4+7=3$

13. $-9+6=-3$

15. $6+(-11)+0=-5+0=-5$

17. $(-3)+(-4)+9=-7+9=2$

19. $4-6=-2$

21. $-3-5=-3+(-5)=-8$

23. $-5-(-3)=-5+3=-2$

25. $14-20=14+(-20)=-6$

27. $-2\cdot 5=-10$

29. $(-4)(-4)=16$

31. $(-8)(-2)\cdot 6=16\cdot 6=96$

33. $(5\cdot 6)(-2)=(30)(-2)=-60$

35. $-12\div(-6)=2$

37. $\dfrac{20}{-20}=-1$

39. $56/-8=-7$

41. $-210/14 = -15$

43. a) $3^2 = 9$

b) $2^3 = 8$

45. a) $(-5)^2 = 25$

b) $-5^2 = -25$

47. a) $-2^4 = -161$

b) $(-2)^4 = 16$

49. a) $-4^3 = -64$

b) $(-4)^3 = -64$

51. $15 - 8 \cdot 2 = 15 - 16 = -1$

53. $(24 \div 4 \cdot 2) - (4 - 7 + 5) = (6 \cdot 2) - (-3 + 5) = 12 - 2 = 10$

55. $(-17 + 21)^2 \cdot 3 \div (13 - 18 \div 2) = (4)^2 \cdot 3 \div (13 - 9) = 16 \cdot 3 \div 4 = 48 \div 4 = 12$

57. $-3\left[(-5^2 + 5^3) \div (-2^2 + 2^3)\right] = -3[(-25 + 125) \div (-4 + 8)] = -3[100 \div 4] = -3[25] = -75$

59. $\dfrac{-\left[4 - (6 - 12)^2\right]}{[9 \div 3 + 4]^2 - 34/2} = \dfrac{-\left[4 - (-6)^2\right]}{[3 + 4]^2 - 17} = \dfrac{-[4 - 36]}{[7]^2 - 17} = \dfrac{-[-32]}{49 - 17} = \dfrac{32}{32} = 1$

61. $-600 + 200 - 400 - 300 = -1100$; 1100 feet under water

63. $29{,}035 - (-36{,}198) = 29{,}035 + 36{,}198 = 65{,}233$ feet

65. a) $+3 - (-8) = +3 + 8 = 11$; There is a 9-hour time difference.

b) $-1 - (-3) = -1 + 3 = 2$; There is a 2-hour time difference.

c) $-5 - (+9) = -5 + (-9) = -14$; There is a 14-hour time difference.

d) $-7 - (+8) = -7 + (-8) = -15$; There is a 15-hour time difference.

67. True

69. False; the difference of two negative integers may be positive, negative, or zero.

71. True

73. True; the quotient of two integers with unlike signs is a negative number.

75. False; the sum of a positive integer and a negative integer could be positive, negative, or zero.

77. Division by zero is undefined because $\dfrac{a}{0} = x$ always leads to a false statement, $a \neq 0$.

79. $\dfrac{-1 + 2 - 3 + 4 - 5 + \cdots - 99 + 100}{1 - 2 + 3 - 4 + 5 - \cdots + 99 - 100} = \dfrac{50}{-50} = -1$

81. $0 + 1 - 2 + 3 + 4 - 5 + 6 - 7 - 8 + 9 = 1$

Section 5.3: The Rational Numbers

1. Integers

3. Denominator

5. Improper

7. Repeating

9. Ten-thousandths

11. $\dfrac{6}{8} = \dfrac{6 \div 2}{8 \div 2} = \dfrac{3}{4}$

13. $\dfrac{24}{42} = \dfrac{24 \div 6}{42 \div 6} = \dfrac{4}{7}$

15. $\dfrac{95}{125} = \dfrac{95 \div 5}{125 \div 5} = \dfrac{19}{25}$

17. $2\dfrac{5}{6} = \dfrac{6 \cdot 2 + 5}{6} = \dfrac{17}{6}$

19. $-5\dfrac{3}{4} = -\dfrac{4 \cdot 5 + 3}{4} = -\dfrac{23}{4}$

21. $-4\frac{15}{16} = -\frac{16 \cdot 4 + 15}{16} = -\frac{79}{16}$

23. $\frac{3}{2}$

25. $\frac{15}{8}$

27. $\frac{13}{8} = 1\frac{5}{8}$

$$\begin{array}{r} 1 \\ 8\overline{)13} \\ \underline{8} \\ 5 \end{array}$$

29. $-\frac{46}{5} = -9\frac{1}{5}$

$$\begin{array}{r} 9 \\ 5\overline{)46} \\ \underline{45} \\ 1 \end{array}$$

31. $-\frac{878}{15} = -58\frac{8}{15}$

$$\begin{array}{r} 58 \\ 15\overline{)878} \\ \underline{870} \\ 8 \end{array}$$

33. $\frac{2}{5} = 0.4$

35. $\frac{1}{3} = 0.\overline{3}$

37. $\frac{3}{8} = 0.375$

39. $\frac{13}{6} = 2.1\overline{6}$

41. $0.6 = \frac{6}{10} = \frac{3}{5}$

43. $0.175 = \frac{175}{1000} = \frac{7}{40}$

45. $0.295 = \frac{295}{1000} = \frac{59}{200}$

47. $0.0131 = \frac{131}{10{,}000}$

49. Let $n = 0.\overline{1}$

$$\begin{aligned} 10n &= 1.\overline{1} \\ -\ n &= 0.\overline{1} \\ \hline 9n &= 1 \\ n &= \frac{1}{9} \end{aligned}$$

51. Let $n = 0.\overline{9}$

$$\begin{aligned} 10n &= 9.\overline{9} \\ -\ n &= 0.\overline{9} \\ \hline 9n &= 9 \\ n &= 1 \end{aligned}$$

53. Let $n = 1.\overline{36}$

$$\begin{aligned} 100n &= 136.\overline{36} \\ -\ n &= 1.\overline{36} \\ \hline 99n &= 135 \\ n &= \frac{135}{99} = \frac{15}{11} \end{aligned}$$

55. Let $n = 2.0\overline{5}$

$$\begin{aligned} 100n &= 205.\overline{5} \\ -\ 10n &= 20.\overline{5} \\ \hline 90n &= 185 \\ n &= \frac{185}{90} = \frac{37}{18} \end{aligned}$$

57. $\frac{2}{7} \cdot \frac{14}{15} = \frac{4}{15}$

59. $\left(\frac{-3}{8}\right)\left(\frac{-16}{15}\right) = \frac{2}{5}$

61. $\frac{7}{8} \div \frac{8}{7} = \frac{7}{8} \cdot \frac{7}{8} = \frac{49}{64}$

63. $\left(\frac{3}{5} \cdot \frac{4}{7}\right) \div \frac{1}{3} = \frac{12}{35} \cdot \frac{3}{1} = \frac{36}{35}$

65. $\left[\left(-\frac{2}{3}\right)\left(\frac{5}{8}\right)\right] \div \left(-\frac{7}{16}\right) = \left(-\frac{5}{12}\right)\left(-\frac{16}{7}\right) = \frac{20}{21}$

67. $\frac{1}{5} + \frac{2}{3} = \frac{1}{5} \cdot \frac{3}{3} + \frac{2}{3} \cdot \frac{5}{5} = \frac{3}{15} + \frac{10}{15} = \frac{13}{15}$

69. $\frac{2}{11} + \frac{5}{22} = \frac{2}{11} \cdot \frac{2}{2} + \frac{5}{22} = \frac{4}{22} + \frac{5}{22} = \frac{9}{22}$

71. $\frac{3}{14}+\frac{8}{21}=\frac{3}{14}\cdot\frac{3}{3}+\frac{8}{21}\cdot\frac{2}{2}=\frac{9}{42}+\frac{16}{42}=\frac{25}{42}$

73. $\frac{1}{12}+\frac{1}{48}+\frac{1}{72}=\frac{1}{12}\cdot\frac{12}{12}+\frac{1}{48}\cdot\frac{3}{3}+\frac{1}{72}\cdot\frac{2}{2}=\frac{12}{144}+\frac{3}{144}+\frac{2}{144}=\frac{17}{144}$

75. $\frac{1}{30}-\frac{3}{40}-\frac{7}{50}=\frac{1}{30}\cdot\frac{20}{20}-\frac{3}{40}\cdot\frac{15}{15}-\frac{7}{50}\cdot\frac{12}{12}=\frac{20}{600}-\frac{45}{600}-\frac{84}{600}=-\frac{109}{600}$

77. $\frac{2}{7}+\frac{1}{2}=\frac{2\cdot2+7\cdot1}{7\cdot2}=\frac{4+7}{14}=\frac{11}{14}$

79. $\frac{5}{6}-\frac{7}{8}=\frac{5\cdot8-6\cdot7}{6\cdot8}=\frac{40-42}{48}=-\frac{2}{48}=-\frac{1}{24}$

81. $\frac{1}{3}+\frac{1}{4}\cdot\frac{2}{5}=\frac{1}{3}+\frac{1}{10}=\frac{1}{3}\cdot\frac{10}{10}+\frac{1}{10}\cdot\frac{3}{3}=\frac{10}{30}+\frac{3}{30}=\frac{13}{30}$

83. $\frac{7}{8}-\frac{3}{4}\div\frac{5}{6}=\frac{7}{8}-\frac{3}{4}\cdot\frac{6}{5}=\frac{7}{8}-\frac{18}{20}=\frac{7}{8}\cdot\frac{5}{5}-\frac{18}{20}\cdot\frac{2}{2}=\frac{7\cdot5-18\cdot2}{40}=\frac{35-36}{40}=-\frac{1}{40}$

85. $\frac{7}{8}\div\frac{3}{4}\cdot\frac{3}{16}=\frac{7}{8}\cdot\frac{4}{3}\cdot\frac{3}{16}=\frac{7\cdot4\cdot3}{8\cdot3\cdot16}=\frac{7}{32}$

87. $\frac{1}{3}\cdot\frac{3}{7}+\frac{3}{5}\cdot\frac{10}{11}=\frac{3}{21}+\frac{30}{55}=\frac{1}{7}\cdot\frac{11}{11}+\frac{6}{11}\cdot\frac{7}{7}=\frac{11+42}{77}=\frac{53}{77}$

89. $\left(\frac{3}{4}+\frac{1}{6}\right)\div\left(2-\frac{7}{6}\right)=\left(\frac{3}{4}\cdot\frac{3}{3}+\frac{1}{6}\cdot\frac{2}{2}\right)\div\left(\frac{2}{1}\cdot\frac{6}{6}-\frac{7}{6}\right)=\left(\frac{9}{12}+\frac{2}{12}\right)\div\left(\frac{12}{6}-\frac{7}{6}\right)=\frac{11}{12}\div\frac{5}{6}=\frac{11}{12}\cdot\frac{6}{5}=\frac{11}{10}$

91. $10\frac{1}{5}-7\frac{3}{5}=\frac{5\cdot10+1}{5}-\frac{7\cdot5+3}{5}=\frac{51}{5}-\frac{38}{5}=\frac{51-38}{5}=\frac{13}{5}=2\frac{3}{5}$ minutes

93. $14\left(8\frac{5}{8}\right)=14\left(\frac{69}{8}\right)=\frac{966}{8}=\frac{966\div2}{8\div2}=\frac{483}{4}=120\frac{3}{4}$ in.

95. $2\frac{1}{4}+3\frac{7}{8}+4\frac{1}{4}=2\frac{4}{16}+3\frac{14}{16}+4\frac{4}{16}=9\frac{22}{16}=10\frac{6}{16}$ in.; $20\frac{5}{16}-10\frac{6}{16}=19\frac{21}{16}-10\frac{6}{16}=9\frac{15}{16}$ in.

Tony needs to buy $9\frac{15}{16}$ inches of PVC pipe.

97. $1-\left(\frac{1}{2}+\frac{2}{5}\right)=1-\left(\frac{5}{10}+\frac{4}{10}\right)=1-\frac{9}{10}=\frac{10}{10}-\frac{9}{10}$

$=\frac{1}{10}$

Student tutors represent $\frac{1}{10}$ of the budget.

99. $4\frac{1}{2}+30\frac{1}{4}+24\frac{1}{8}=4\frac{4}{8}+30\frac{2}{8}+24\frac{1}{8}=58\frac{7}{8}$ in.

101. a) $1\frac{49}{60}$, $2\frac{48}{60}$, $9\frac{6}{60}$, $6\frac{3}{60}$, $2\frac{9}{60}$, $\frac{22}{60}$

b) $1\frac{49}{60}+2\frac{48}{60}+9\frac{6}{60}+6\frac{3}{60}+2\frac{9}{60}+\frac{22}{60}=(1+2+9+6+2+0)+\frac{49+48+6+3+9+22}{60}=20+\frac{137}{60}$

$=22\frac{17}{60}$ or 22 hours, 17 minutes

103. $8\frac{3}{4}\text{ ft} = \frac{35}{4}\text{ ft} = \frac{35}{4}\cdot\frac{12}{1}\text{ in.} = 105\text{ in.}$

$\left[105-(3)\left(\frac{1}{8}\right)\right]\div 4 = \left[\frac{840}{8}-\frac{3}{8}\right]\div 4 = \frac{837}{8}\cdot\frac{1}{4} = \frac{837}{32} = 26\frac{5}{32}$; The length of each piece is $26\frac{5}{32}$ in.

105. a) $20+18\frac{3}{8}\div 2 = 20+9\frac{3}{16} = 29\frac{3}{16}$ in.

b) $26\frac{1}{4}+6\frac{3}{4} = 33$ in.

c) $26\frac{1}{4}+\left(6\frac{3}{4}-\frac{1}{4}\right) = 26\frac{1}{4}+6\frac{2}{4} = 32\frac{3}{4}$ in.

Oatmeal: $\left(\frac{1}{2}+1\right)\div 2 = \frac{3}{2}\cdot\frac{1}{2} = \frac{3}{4}$ cup

b) Water (or milk): $1+\frac{1}{2} = 1\frac{1}{2}$ cups

Oatmeal: $\frac{1}{2}+\frac{1}{2}\cdot\frac{1}{2} = \frac{3}{4}$ cup

107. $\frac{0.21+0.22}{2} = \frac{0.43}{2} = 0.215$

109. $\frac{-2.176+(-2.175)}{2} = \frac{-4.351}{2} = -2.1755$

111. $\left(\frac{1}{4}+\frac{3}{4}\right)\div 2 = \frac{4}{4}\cdot\frac{1}{2} = \frac{4}{8} = \frac{1}{2}$

113. $\left(\frac{1}{100}+\frac{1}{10}\right)\div 2 = \frac{11}{100}\cdot\frac{1}{2} = \frac{11}{200}$

115. a) Let $n = 0.\overline{9}$

$$\begin{aligned} 10n &= 9.\overline{9} \\ -\quad n &= 0.\overline{9} \\ \hline 9n &= 9 \\ n &= 1 \end{aligned}$$

b) $0.\overline{9}$

c) $\frac{1}{3}+\frac{2}{3} = \frac{3}{3} = 1,\ \frac{1}{3}+\frac{2}{3} = 0.\overline{3}+0.\overline{6} = 0.\overline{9}$

d) $0.\overline{9} = 1$

Section 5.4: The Irrational Numbers

1. Irrational
3. Radicand
5. Itself
7. Rationalized
9. $\sqrt{36} = 6$; Rational
11. Irrational; non-terminating, non-repeating decimal
13. Irrational; non-terminating, non-repeating decimal
15. Rational; quotient of two integers
17. $\frac{\sqrt{75}}{\sqrt{3}} = \sqrt{\frac{75}{3}} = \sqrt{25} = 5$; Rational
19. $\sqrt{0} = 0$
21. $\sqrt{25} = 5$
23. $-\sqrt{36} = -6$
25. $-\sqrt{100} = -10$
27. Rational, integer
29. Irrational; non-terminating, non-repeating decimal
31. Rational; terminating decimal
33. Rational; repeating decimal
35. Irrational; non-terminating, non-repeating decimal
37. $\sqrt{20} = \sqrt{4\cdot 5} = \sqrt{4}\cdot\sqrt{5} = 2\sqrt{5}$
39. $\sqrt{40} = \sqrt{4\cdot 10} = \sqrt{4}\cdot\sqrt{10} = 2\sqrt{10}$
41. $\sqrt{63} = \sqrt{9\cdot 7} = \sqrt{9}\cdot\sqrt{7} = 3\sqrt{7}$
43. $\sqrt{84} = \sqrt{4\cdot 21} = \sqrt{4}\cdot\sqrt{21} = 2\sqrt{21}$
45. $4\sqrt{6}+3\sqrt{6} = (4+3)\sqrt{6} = 7\sqrt{6}$
47. $$\begin{aligned} 5\sqrt{18}-7\sqrt{8} &= 5\sqrt{9\cdot 2}-7\sqrt{4\cdot 2} \\ &= 5\sqrt{9}\cdot\sqrt{2}-7\sqrt{4}\cdot\sqrt{2} \\ &= 5\cdot 3\cdot\sqrt{2}-7\cdot 2\cdot\sqrt{2} \\ &= 15\sqrt{2}-14\sqrt{2} \\ &= \sqrt{2} \end{aligned}$$

49. $4\sqrt{12}-7\sqrt{27}=4\sqrt{4\cdot 3}-7\sqrt{9\cdot 3}$
$=4\sqrt{4}\cdot\sqrt{3}-7\sqrt{9}\cdot\sqrt{3}$
$=4\cdot 2\cdot\sqrt{3}-7\cdot 3\cdot\sqrt{3}$
$=8\sqrt{3}-21\sqrt{3}$
$=-13\sqrt{3}$

51. $5\sqrt{7}+7\sqrt{28}-3\sqrt{75}$
$=5\sqrt{3}+7\sqrt{4\cdot 3}-3\sqrt{25\cdot 3}$
$=5\sqrt{3}+7\sqrt{4}\cdot\sqrt{3}-3\sqrt{25}\cdot\sqrt{3}$
$=5\sqrt{3}+7\cdot 2\cdot\sqrt{3}-3\cdot 5\cdot\sqrt{3}$
$=5\sqrt{3}+14\sqrt{3}-15\sqrt{3}$
$=4\sqrt{3}$

53. $\sqrt{2}\sqrt{18}=\sqrt{2\cdot 18}=\sqrt{36}=6$

55. $\sqrt{6}\sqrt{10}=\sqrt{6\cdot 10}=\sqrt{60}=\sqrt{4}\cdot\sqrt{15}=2\sqrt{15}$

57. $\frac{\sqrt{20}}{\sqrt{5}}=\sqrt{\frac{20}{45}}=\sqrt{4}=2$

59. $\frac{\sqrt{72}}{\sqrt{8}}=\sqrt{\frac{72}{8}}=\sqrt{9}=3$

61. $\frac{1}{\sqrt{5}}\cdot\frac{\sqrt{5}}{\sqrt{5}}=\frac{\sqrt{5}}{\sqrt{25}}=\frac{\sqrt{5}}{5}$

63. $\frac{\sqrt{3}}{\sqrt{7}}\cdot\frac{\sqrt{7}}{\sqrt{7}}=\frac{\sqrt{21}}{\sqrt{49}}=\frac{\sqrt{21}}{7}$

65. $\frac{\sqrt{20}}{\sqrt{3}}\cdot\frac{\sqrt{3}}{\sqrt{3}}=\frac{\sqrt{60}}{\sqrt{9}}=\frac{\sqrt{4}\sqrt{15}}{3}=\frac{2\sqrt{15}}{3}$

67. $\frac{\sqrt{10}}{\sqrt{6}}=\sqrt{\frac{10}{6}}=\sqrt{\frac{5}{3}}=\frac{\sqrt{5}}{\sqrt{3}}\cdot\frac{\sqrt{3}}{\sqrt{3}}=\frac{\sqrt{15}}{\sqrt{9}}=\frac{\sqrt{15}}{3}$

69. $\sqrt{37}$ is between 6 and 7 since $\sqrt{37}$ is between $\sqrt{36}=6$ and $\sqrt{49}=7$. $\sqrt{37}$ is between 6 and 6.5 since 37 is closer to 36 than to 49. Using a calculator $\sqrt{37}\approx 6.08$.

71. $\sqrt{97}$ is between 9 and 10 since $\sqrt{97}$ is between $\sqrt{81}=9$ and $\sqrt{100}=10$. $\sqrt{97}$ is between 9.5 and 10 since 97 is closer to 100 than to 81. Using a calculator $\sqrt{97}\approx 9.85$.

73. $\sqrt{170}$ is between 13 and 14 since $\sqrt{170}$ is between $\sqrt{169}=13$ and $\sqrt{196}=14$. $\sqrt{170}$ is between 13 and 13.5 since 170 is closer to 169 than to 196. Using a calculator $\sqrt{170}\approx 13.04$.

75. $H=2.9\sqrt{30}+20.1\approx 36.0$ inches

77. a) $t=\frac{\sqrt{100}}{4}=\frac{10}{4}=2.5$ sec

b) $t=\frac{\sqrt{400}}{4}=\frac{20}{4}=5$ sec

c) $t=\frac{\sqrt{900}}{4}=\frac{30}{4}=7.5$ sec

d) $t=\frac{\sqrt{1600}}{4}=\frac{40}{4}=10$ sec

79. False. $\sqrt{c}$ may be a rational number or an irrational number for a composite number c. For example, $\sqrt{25}$ is a rational number; $\sqrt{8}$ is an irrational number.

81. True

83. False. The product of a rational number and an irrational number may be a rational number or an irrational number.

85. $\pi+(-\pi)=0$

87. $\sqrt{2}\cdot\sqrt{3}=\sqrt{6}$

89. No. $\sqrt{2}\neq 1.414$ since $\sqrt{2}$ is an irrational number and 1.414 is a rational number.

91. No. 3.14 and $\frac{22}{7}$ are rational numbers, π is an irrational number.

93. $\sqrt{4\cdot 9}=\sqrt{36}=6$, $\sqrt{4}\cdot\sqrt{9}=2\cdot 3=6$, and $6=6$

95. a) $\sqrt{0.04} = 0.2$, which is a terminating decimal and thus it is rational.

b) $\sqrt{0.7} = \sqrt{\frac{7}{10}} = \sqrt{\frac{70}{100}} = \frac{\sqrt{70}}{10}$; $\sqrt{70}$ is irrational since the only integers with rational square roots are the perfect squares and 70 is not a perfect square. Thus $\sqrt{0.7} = \frac{\sqrt{70}}{10}$ is irrational.

97. a) $\left(44 \div \sqrt{4}\right) \div \sqrt{4} = 11$

b) $(44 \div 4) + \sqrt{4} = 13$

c) $4 + 4 + 4 + \sqrt{4} = 14$

d) $\sqrt{4}(4+4) + \sqrt{4} = 18$

Section 5.5: Real Numbers and Their Properties

1. Real
3. Closed
5. Commutative
7. Associative
9. Closed. The sum of two natural numbers is a natural number.
11. Not closed; $3 - 5 = -2$ is not a natural number.
13. Closed. The product of two integers is an integer.
15. Closed. The difference of two integers is an integer.
17. Closed. The sum of two rational numbers is a rational number.
19. Closed. The difference of two rational numbers is a rational number.
21. Not closed; $3 \div 0$ is not an irrational number.
23. Not closed; $\sqrt{3} \cdot \sqrt{3} = \sqrt{9} = 3$ is not an irrational number.
25. Closed. The sum of two real numbers is a real number.
27. Not closed; $3 \div 0$ is not a real number.
29. Commutative property of addition because $5 + x$ is changed to $x + 5$.
31. No. $3 \div 4 \neq 4 \div 3$
33. $(-3)(-4) = (-4)(-3) = 12$
35. $[(-2)+(-3)]+(-4) = (-5)+(-4) = -9$
    $(-2)+[(-3)+(-4)] = (-2)+(-2) = -9$
37. No. $(16 \div 8) \div 2 = 2 \div 2 = 1$
    $16 \div (8 \div 2) = 16 \div 4 = 4$
    $1 \neq 4$
39. No. $(81 \div 9) \div 3 = 9 \div 3 = 3$
    $81 \div (9 \div 3) = 81 \div 3 = 27$
    $3 \neq 27$
41. Distributive property
43. Associative property of multiplication
45. Associative property of addition
47. Commutative property of multiplication
49. Distributive property
51. Commutative property of addition
53. $3(b+7) = 3b + 21$
55. $6(c-2) = 6c - 12$
57. $-2(4x-1) = -8x + 2$
59. $32\left(\frac{1}{16}x - \frac{1}{32}\right) = 2x - 1$
61. $\sqrt{2}\left(\sqrt{8} - \sqrt{2}\right) = \sqrt{16} - \sqrt{4} = 4 - 2 = 2$
63. $5\left(\sqrt{2} + \sqrt{3}\right) = 5\sqrt{2} + 5\sqrt{3}$
65. Yes. You can either feed your cats first or give your cats water first.
67. No. The clothes must be washed first before being dried.

69. No. Pressing the keys will have no effect if there are no batteries in place.

71. Yes. The order does not matter.

73. Yes. The order does not matter

75. Answers will vary.

77. No. $0 \div a = 0 \quad (a \neq 0)$, but $a \div 0$ is undefined.

79. a) No. (*Man eating*) *tiger* is a tiger that eats men, and *man* (*eating tiger*) is a man that is eating a tiger.
    b) No. (*Horse riding*) *monkey* is a monkey that rides a horse, and *horse* (*riding monkey*) is a horse that rides a monkey.
    c) Answers will vary.

Section 5.6: Rules of Exponents and Scientific Notation

1. $x^{2+3} = x^5$

3. 1

5. $x^{2\cdot 3} = x^6$

7. a) $2^2 \cdot 2^3 = 2^{2+3} = 2^5 = 32$
   b) $(-2)^2 \cdot (-2)^3 = (-2)^{2+3} = (-2)^5 = -32$

9. a) $\dfrac{5^7}{5^5} = 5^{7-5} = 5^2 = 25$
   b) $\dfrac{(-5)^7}{(-5)^5} = (-5)^{7-5} = (-5)^2 = 25$

11. a) $6^0 = 1$
    b) $-6^0 = -1 \cdot 6^0 = -1 \cdot 1 = -1$

13. a) $(6x)^0 = 1$
    b) $6x^0 = 6 \cdot 1 = 6$

15. a) $3^{-3} = \dfrac{1}{3^3} = \dfrac{1}{27}$
    b) $7^{-2} = \dfrac{1}{7^2} = \dfrac{1}{49}$

17. a) $-9^{-2} = -\dfrac{1}{9^2} = -\dfrac{1}{81}$
    b) $(-9)^{-2} = \dfrac{1}{(-9)^2} = \dfrac{1}{81}$

19. a) $\left(2^3\right)^2 = 2^{3\cdot 2} = 2^6 = 64$
    b) $\left(3^2\right)^3 = 3^{2\cdot 3} = 3^6 = 729$

21. a) $4^3 \cdot 4^{-2} = 4^{3+(-2)} = 4^1 = 4$
    b) $2^{-2} \cdot 2^{-2} = 2^{-2+(-2)} = 2^{-4} = \dfrac{1}{2^4} = \dfrac{1}{16}$

23. $503{,}000 = 5.03 \times 10^5$

25. $0.00042 = 4.2 \times 10^{-4}$

27. $0.56 = 5.6 \times 10^{-1}$

29. $19{,}000 = 1.9 \times 10^4$

31. $0.000186 = 1.86 \times 10^{-4}$

33. $2.3 \times 10^3 = 2300$

35. $1.68 \times 10^{-3} = 0.00168$

37. $8.62 \times 10^{-5} = 0.0000862$

39. $2.01 \times 10^0 = 2.01$

41. $\left(4.1 \times 10^2\right)\left(2 \times 10^3\right) = (4.1 \times 2) \times \left(10^2 \times 10^3\right)$
    $= 8.2 \times 10^5 = 820{,}000$

43. $\left(5.1 \times 10^1\right)\left(3 \times 10^{-4}\right) = (5.1 \times 3) \times \left(10^1 \times 10^{-4}\right)$
    $= 15.3 \times 10^{-3}$
    $= 1.53 \times 10^{-2} = 0.0153$

45. $\left(\dfrac{7.5 \times 10^6}{3 \times 10^4}\right) = \left(\dfrac{7.5}{3}\right) \times \left(\dfrac{10^6}{10^4}\right) = 2.5 \times 10^2 = 250$

47. $\left(\dfrac{8.4 \times 10^{-6}}{4 \times 10^{-3}}\right) = \left(\dfrac{8.4}{4}\right) \times \left(\dfrac{10^{-6}}{10^{-3}}\right)$
    $= 2.1 \times 10^{-3} = 0.0021$

49. $(200{,}000)(3600) = \left(2.0 \times 10^5\right)\left(3.6 \times 10^3\right)$
    $= (2.0 \times 3.6) \times \left(10^5 \times 10^3\right)$
    $= 7.2 \times 10^8$

51. $(0.003)(0.00015) = (3.0\times10^{-3})(1.5\times10^{-4})$
$= (3.0\times1.5)\times(10^{-3}\times10^{-4})$
$= 4.5\times10^{-7}$

53. $\dfrac{5{,}600{,}000}{80{,}000} = \dfrac{5.6\times10^{6}}{8.0\times10^{4}} = \left(\dfrac{5.6}{8.0}\right)\times\left(\dfrac{10^{6}}{10^{4}}\right)$
$= 0.7\times10^{2} = 7.0\times10^{1}$

55. $\dfrac{0.00004}{200} = \dfrac{4.0\times10^{-5}}{2.0\times10^{2}} = \left(\dfrac{4.0}{2.0}\right)\times\left(\dfrac{10^{-5}}{10^{2}}\right)$
$= 2.0\times10^{-7}$

57. $3.6\times10^{-3};\ 1.7; 9.8\times10^{2};\ 1.03\times10^{4}$

59. $8.3\times10^{-5};\ 0.00079; 4.1\times10^{3};\ 40{,}000$

61. a) $\dfrac{17.82\times10^{12}}{318.6\times10^{6}} \approx 5.59322\times10^{4}$
$= \$55{,}932.20$

b) $\$68{,}018.58 - \$55{,}932.20 = \$12{,}086.38$

63. $\dfrac{328\times10^{6}}{7.69\times10^{12}} \approx 4.3\times10^{-5} = 0.043$

65. $\dfrac{20.66\times10^{12}}{328\times10^{6}} \approx 6.298780\times10^{4} = \$62{,}987.805$

67. $t = \dfrac{d}{r} = \dfrac{239{,}000}{200{,}00} = \dfrac{2.39\times10^{5}}{2.0\times10^{4}} = 1.195\times10^{1}$
$= 11.95$ hours

69. $\left(100{,}000\ \dfrac{\text{ft}^3}{\text{sec}}\right)\left(\dfrac{60}{1}\ \dfrac{\text{sec}}{\text{min}}\right)\left(\dfrac{60}{1}\ \dfrac{\text{min}}{\text{hr}}\right)(24\text{ hr})^{9}$
$= 8{,}640{,}000{,}000$
$= 8.64\times10\ \text{ft}^3$

71. $(5{,}800{,}000)(50) = (5.8\times10^{6})(5.0\times10^{1})$
$= 2.9\times10^{8}$ cells

73. $\dfrac{2\times10^{30}}{6\times10^{24}} \approx 3.33333\times10^{5} = 333{,}333$ times

75. 1000 times since $10^{3} = 1000$.

77. $(2.92\times10^{3})(0.97) \approx 2.832\times10^{3} = 2832$ people

79. a) $1{,}000{,}000 = 1\times10^{6},\ 1{,}000{,}000{,}000 = 1\times10^{9},\ 1{,}000{,}000{,}000{,}000 = 1\times10^{12}$

b) $\dfrac{1\times10^{6}}{1\times10^{3}} = 1\times10^{3} = 1000$ days or $\dfrac{1000}{365} \approx 2.74$ years

c) $\dfrac{1\times10^{9}}{1\times10^{3}} = 1\times10^{6} = 1{,}000{,}000$ days or $\dfrac{1{,}000{,}000}{365} \approx 2739.73$ years

d) $\dfrac{1\times10^{12}}{1\times10^{3}} = 1\times10^{9} = 1{,}000{,}000{,}000$ days or $\dfrac{1{,}000{,}000{,}000}{365} \approx 2{,}739{,}726.03$ years

e) $\dfrac{1\times10^{9}}{1\times10^{6}} = 1\times10^{3} = 1000$ times bigger

81. a) $E(0) = 2^{10}\cdot2^{0} = 2^{10}\cdot1 = 1024$ bacteria

b) $E\left(\frac{1}{2}\right) = 2^{10}\cdot2^{1/2} = 2^{10.5} \approx 1448$ bacteria

Section 5.7: Arithmetic and Geometric Sequences

1. Sequence
3. Arithmetic
5. Geometric
7. $a_1 = 3$
$a_2 = 3 + 4 = 7$
$a_3 = 7 + 4 = 11$
$a_4 = 11 + 4 = 15$
$a_5 = 15 + 4 = 19$

9. $a_1 = 25$
$a_2 = 25 + (-5) = 20$
$a_3 = 20 + (-5) = 15$
$a_4 = 15 + (-5) = 10$
$a_5 = 10 + (-5) = 5$

11. $a_1 = 5$
$a_2 = 5 + (-2) = 3$
$a_3 = 3 + (-2) = 1$
$a_4 = 1 + (-2) = -1$
$a_5 = -1 + (-2) = -3$

13. $a_n = a_1 + (n-1)d$
$a_{20} = (4) + (20-1)(3) = 61$

15. $a_n = a_1 + (n-1)d$
$a_{30} = (-20) + (30-1)(5) = 125$

17. $a_n = a_1 + (n-1)d$
$a_{20} = \left(\frac{4}{5}\right) + (20-1)(-1) = -\frac{91}{5}$

19. $d = a_2 - a_1 = 2 - 1 = 1$
$a_n = a_1 + (n-1)d$
$a_n = (1) + (n-1)(1)$
$a_n = 1 + n - 1$
$a_n = n$

21. $d = a_2 - a_1 = 3 - 1 = 2$
$a_n = a_1 + (n-1)d$
$a_n = (1) + (n-1)(2)$
$a_n = 1 + 2n - 2$
$a_n = 2n - 1$

23. $d = a_2 - a_1 = 8 - 5 = 3$
$a_n = a_1 + (n-1)d$
$a_n = (5) + (n-1)(3)$
$a_n = 5 + 3n - 3$
$a_n = 3n + 2$

25. $d = a_2 - a_1 = 2 - 1 = 1$
$a_n = a_1 + (n-1)d$
$a_{50} = (1) + (50-1)(1) = 50$
$s_n = \frac{n(a_1 + a_n)}{2}$
$s_{50} = \frac{50(1+50)}{2} = 1275$

27. $d = a_2 - a_1 = 3 - 1 = 2$
$a_n = a_1 + (n-1)d$
$a_{50} = (1) + (50-1)(2) = 99$
$s_n = \frac{n(a_1 + a_n)}{2}$
$s_{50} = \frac{50(1+99)}{2} = 2500$

29. $d = a_2 - a_1 = 6 - 11 = -5$
$a_n = a_1 + (n-1)d$
$a_8 = (11) + (8-1)(-5) = -24$
$s_n = \frac{n(a_1 + a_n)}{2}$
$s_8 = \frac{8(11 + (-24))}{2} = -52$

31. $a_1 = 2$
$a_2 = 2 \cdot 5 = 10$
$a_3 = 10 \cdot 5 = 50$
$a_4 = 50 \cdot 5 = 250$
$a_5 = 250 \cdot 5 = 1250$

33. $a_1 = 5$
$a_2 = 5 \cdot 2 = 10$
$a_3 = 10 \cdot 2 = 20$
$a_4 = 20 \cdot 2 = 40$
$a_5 = 40 \cdot 2 = 80$

35. $a_1 = -3$
$a_2 = -3 \cdot (-1) = 3$
$a_3 = 3 \cdot (-1) = -3$
$a_4 = -3 \cdot (-1) = 3$
$a_5 = 3 \cdot (-1) = -3$

37. $a_n = a_1 r^{n-1}$
$a_6 = 5(2)^{6-1} = 160$

39. $a_n = a_1 r^{n-1}$
$a_7 = 64\left(\frac{1}{2}\right)^{7-1} = 1$

41. $a_n = a_1 r^{n-1}$
$a_7 = -5(3)^{7-1} = -3645$

43. $r = \frac{a_2}{a_1} = \frac{9}{3} = 3$

$a_n = a_1 r^{n-1}$

$a_n = 3(3)^{n-1} = 3^n$

45. $r = \frac{a_2}{a_1} = \frac{1}{2}$

$a_n = a_1 r^{n-1}$

$a_n = 1\left(\frac{1}{2}\right)^{n-1} = \left(\frac{1}{2}\right)^{n-1}$

47. $r = \frac{a_2}{a_1} = \frac{-8}{-16} = \frac{1}{2}$

$a_n = a_1 r^{n-1}$

$a_n = -16\left(\frac{1}{2}\right)^{n-1}$

49. $s_n = \frac{a_1\left(1-r^n\right)}{1-r}$

$s_6 = \frac{3\left(1-2^6\right)}{1-2} = 189$

51. $s_n = \frac{a_1\left(1-r^n\right)}{1-r}$

$s_6 = \frac{-3\left(1-4^6\right)}{1-4} = -4095$

53. $s_n = \frac{a_1\left(1-r^n\right)}{1-r}$

$s_{15} = \frac{2\left(1-3^{15}\right)}{1-3} = 14,348,906$

55. $d = a_2 - a_1 = 2 - 1 = 1$

$a_n = a_1 + (n-1)d$

$a_{50} = (1) + (100-1)(1) = 100$

$s_n = \frac{n(a_1 + a_n)}{2}$

$s_{100} = \frac{100(1+100)}{2} = 5050$

57. $d = a_2 - a_1 = 4 - 2 = 2$

$a_n = a_1 + (n-1)d$

$a_{50} = (2) + (100-1)(2) = 200$

$s_n = \frac{n(a_1 + a_n)}{2}$

$s_{100} = \frac{100(2+200)}{2} = 10,100$

59. $d = a_2 - a_1 = 2 - 1 = 1$

$a_n = a_1 + (n-1)d$

$a_{12} = (1) + (12-1)(1) = 12$

$s_n = \frac{n(a_1 + a_n)}{2}$

$s_{12} = \frac{12(1+12)}{2} = 78$ times

61. a) 8 ft = 96 in.

$a_n = a_1 + (n-1)d$

$a_{12} = (96) + (12-1)(-3) = 63$ in.

b) $s_n = \frac{n(a_1 + a_n)}{2}$

$s_{12} = \frac{12(96+63)}{2} = 954$ in.

63. $a_n = a_1 r^{n-1}$

$a_{10} = 8000(1.08)^{10-1} \approx 15,992$ students

65. $a_n = a_1 r^{n-1}$

$a_{10} = 35,000(1.03)^{10-1} \approx \$45,667$

67. $a_n = a_1 r^{n-1}$

$a_{12} = 1.4(1.006)^{12-1} \approx 1.5$ billion people

69. $s_n = \frac{a_1\left(1-r^n\right)}{1-r}$; $s_6 = \frac{82\left(1-\left(\frac{1}{2}\right)^6\right)}{1-\left(\frac{1}{2}\right)} = 161.4375$

71. The smallest and largest possible numbers that are divisible by 6 are 12 and 1608, respectively.

$d = 6$

$a_n = a_1 + (n-1)d$

$1608 = 12 + (n-1)(6)$

$1608 = 12 + 6n - 6$

$1608 = 6n + 6$

$1602 = 6n$

$n = 267$

73. The total distance is 30 plus twice the sum of the terms of the geometric sequence having $a_1 = (30)(0.8) = 24$ and $r = 0.8$.

$$s_5 = \frac{a_1\left(1-r^n\right)}{1-r} = \frac{24\left(1-(0.8)^5\right)}{1-0.8} = \frac{24(1-0.32768)}{0.2} = 80.6784$$

The total distance is $30 + 2(80.6784) = 191.3568$ ft.

Section 5.8: Fibonacci Sequence

1. Fibonacci

3. Divine

5. Ratio

7. Fibonacci type sequence; $21+34=55,\ \ 34+55=89$

9. Not Fibonacci type; it is not true that each term is the sum of the two preceding terms.

11. Fibonacci type sequence; $1+2=3,\ \ 2+3=5$

13. Fibonacci type sequence; $40+65=105,\ \ 65+105=170$

15. 1, 1, 2, 3, 5, 8, 13, 21, 34, 55, 89, 144, 233, 377, 610

17. Answers will vary. Using the first ten Fibonacci numbers: $1+1+2+3+5+8+13+21+34+55=143$; $143 \div 11 = 13$

19. Answers will vary. Using 5, the result is $2(5)-8=10-8=2$, which is the second number preceding 5.

21. a) Answers will vary. Using 6 and 9, the sequence is 6, 9, 15, 24, 39, 63, 102, … .

b) $\frac{9}{6}=1.5,\ \frac{15}{9}\approx 1.667,\ \frac{24}{15}=1.6,\ \frac{39}{24}=1.625,\ \frac{63}{39}=1.615,\ \frac{102}{63}=1.619$; The ratios are approaching the golden number.

23. The sums of the numbers along the diagonals (parallel to the one shown) are Fibonacci numbers.

25. a) $\frac{\sqrt{5}+1}{2} \approx 1.618$

b) $\frac{\sqrt{5}-1}{2} \approx 0.618$

c) The results differ by 1.

27. Answers will vary.

29. $\frac{1}{89} \approx 0.0\underline{11235}9551$; The decimal expansion shows several terms of the Fibonacci sequence.

31. $\frac{55}{34} \approx 1.6176$ and $\frac{34}{21} \approx 1.619$

33. If $x = \frac{a}{b}$, then $\frac{1}{x} = \frac{b}{a}$ and $\frac{a+b}{a} = \frac{a}{a} + \frac{b}{a} = 1 + \frac{1}{x}$

$$\frac{a+b}{a} = \frac{a}{b} \Rightarrow 1 + \frac{1}{x} = x \Rightarrow x\left(1+\frac{1}{x}\right) = (x)x \Rightarrow x+1 = x^2 \Rightarrow x^2 - x - 1 = 0$$

$$x = \frac{-b \pm \sqrt{b^2 - 4(a)(c)}}{2(a)} = \frac{1 \pm \sqrt{1-4(1)(-1)}}{2(1)} = \frac{1 \pm \sqrt{5}}{2}$$

The positive solution, $\frac{1+\sqrt{5}}{2}$, is the golden ratio.

35. a) Answers will vary.
    b) Answers will vary.
    c) Answers will vary.

Review Exercises

1. Use the divisibility rules in section 5.1.
   56,340 is divisible by 2, 3, 4, 5, 6, 9, and 10.

2. Use the divisibility rules in section 5.1.
   400,644 is divisible by 2, 3, 4, 6, and 9

3. $840 = 2^3 \cdot 3 \cdot 5 \cdot 7$

$$\begin{array}{r|l} 2 & 840 \\ \hline 2 & 420 \\ \hline 2 & 210 \\ \hline 3 & 105 \\ \hline 5 & 35 \\ \hline & 7 \end{array}$$

4. $1452 = 2^2 \cdot 3 \cdot 11^2$

$$\begin{array}{r|l} 2 & 1452 \\ \hline 2 & 726 \\ \hline 3 & 363 \\ \hline 11 & 121 \\ \hline & 11 \end{array}$$

5. The GCD of $24 = 2^3 \cdot 3$ and $36 = 2^2 \cdot 3^2$ is $2^2 \cdot 3 = 12$. The LCM is $2^3 \cdot 3^2 = 72$.

6. The GCD of $63 = 3^2 \cdot 7$ and $108 = 2^2 \cdot 3^3$ is $3^2 = 9$. The LCM is $2^2 \cdot 3^3 \cdot 7 = 756$.

7. The GCD of $45 = 3^2 \cdot 5$ and $60 = 2^2 \cdot 3 \cdot 5$ is $2^2 \cdot 3^2 \cdot 5 = 180$ minutes.

8. $2 + (-6) = -4$

9. $-2 + 5 = 3$

10. $(-2) + (-4) = -6$

11. $4 - 8 = 4 + (-8) = -4$

12. $-5 - 4 = -5 + (-4) = -9$

13. $-3 - (-6) = -3 + 6 = 3$

14. $6 \cdot (-4) = -24$

15. $(-2)(-12) = 24$

16. $\dfrac{-35}{-7} = 5$

17. $\dfrac{12}{-6} = -2$

18. $8 + 5 \cdot 3 = 8 + 15 = 23$

19. $4 \cdot 3^2 - 2^3 \cdot 3 = 4 \cdot 9 - 8 \cdot 3 = 36 - 24 = 12$

20. $(64 \div 4 \cdot 2) - (25 - 7 \cdot 4) = (16 \cdot 2) - (25 - 28) = 32 - (-3) = 32 + 3 = 35$

21. $-2(5 \cdot 3^2 - 4^3 \div 8)^2 = -2(5 \cdot 9 - 64 \div 8)^2 = -2(45 - 8)^2 = -2(37)^2 = -2(1369) = -2738$

22. $\dfrac{11}{25} = 0.44$

23. $\dfrac{13}{4} = 3.25$

24. $\dfrac{6}{7} = 0.\overline{857142}$

25. $\dfrac{7}{12} = 0.58\overline{3}$

26. $4.2 = \dfrac{42}{10} = \dfrac{21}{5}$

27. Let $n = 0.\overline{6}$

$$\begin{aligned} 10n &= 6.\overline{6} \\ -n &= -0.\overline{6} \\ \hline 9n &= 6 \\ n &= \frac{6}{9} \\ n &= \frac{2}{3} \end{aligned}$$

28. Let $n = 0.\overline{51}$

$$100n = 51.\overline{51}$$
$$-n = -0.\overline{51}$$
$$99n = 51$$
$$n = \frac{51}{99}$$
$$n = \frac{17}{33}$$

29. $0.083 = \frac{83}{1000}$

30. $2\frac{4}{7} = \frac{2 \cdot 7 + 4}{7} = \frac{14+4}{7} = \frac{18}{7}$

31. $-3\frac{1}{4} = -\frac{3 \cdot 4 + 1}{4} = -\frac{12+1}{4} = -\frac{13}{4}$

32. $\frac{19}{5} = 3\frac{4}{5}$

$$5\overline{)19} \quad 3, \quad \underline{15}, \quad 4$$

33. $-\frac{136}{5} = -27\frac{1}{5}$

$$5\overline{)136} \quad 27, \quad \underline{135}, \quad 1$$

34. $\frac{4}{15} - \frac{2}{5} = \frac{4}{15} - \frac{2}{5} \cdot \frac{3}{3} = \frac{4}{15} - \frac{6}{15} = -\frac{2}{15}$

35. $\frac{1}{6} + \frac{5}{4} = \frac{1}{6} \cdot \frac{2}{2} + \frac{5}{4} \cdot \frac{3}{3} = \frac{2}{12} + \frac{15}{12} = \frac{17}{12}$

36. $\frac{7}{16} \cdot \frac{12}{21} = \frac{84}{336} = \frac{84}{4 \cdot 84} = \frac{1}{4}$

37. $\frac{5}{9} \div \frac{6}{7} = \frac{5}{9} \cdot \frac{7}{6} = \frac{35}{54}$

38. $\frac{1}{3} + \frac{2}{5} \cdot \frac{5}{11} = \frac{1}{3} + \frac{2}{11} = \frac{1}{3} \cdot \frac{11}{11} + \frac{2}{11} \cdot \frac{3}{3} = \frac{11+6}{33} = \frac{17}{33}$

39. $\frac{3}{4} - \frac{1}{8} \div \frac{5}{12} = \frac{3}{4} - \frac{1}{8} \cdot \frac{12}{5} = \frac{3}{4} - \frac{12}{40} = \frac{3}{4} \cdot \frac{10}{10} - \frac{12}{40}$
$= \frac{30-12}{40} = \frac{18}{40} = \frac{9}{20}$

40. $\frac{13}{16} \div \frac{7}{8} \cdot \frac{14}{15} = \frac{13}{16} \cdot \frac{8}{7} \cdot \frac{14}{15} = \frac{13}{14} \cdot \frac{14}{15} = \frac{13}{15}$

41. $\left(\frac{1}{4} + \frac{5}{6}\right) \div \left(3 - \frac{1}{6}\right) = \left(\frac{1}{4} \cdot \frac{3}{3} + \frac{5}{6} \cdot \frac{2}{2}\right) \div \left(\frac{3}{1} \cdot \frac{3}{3} - \frac{1}{6}\right)$
$= \left(\frac{3}{12} + \frac{10}{12}\right) \div \left(\frac{9}{3} - \frac{1}{6}\right)$
$= \frac{13}{12} \div \left(\frac{9}{3} \cdot \frac{2}{2} - \frac{1}{6}\right)$
$= \frac{13}{12} \div \left(\frac{18}{6} - \frac{1}{6}\right)$
$= \frac{13}{12} \div \frac{17}{6}$
$= \frac{13}{12} \cdot \frac{6}{17}$
$= \frac{13}{34}$

42. $\left(\frac{1}{8}\right)\left(17\tfrac{3}{4}\right) = \left(\frac{1}{8}\right)\left(\frac{71}{4}\right) = \frac{71}{32} = 2\frac{7}{32}$ teaspoons

43. $\sqrt{60} = \sqrt{4 \cdot 15} = \sqrt{4} \cdot \sqrt{15} = 2\sqrt{15}$

44. $\sqrt{2} - 4\sqrt{2} = (1-4)\sqrt{2} = -3\sqrt{2}$

45. $\sqrt{8} + 6\sqrt{2} = \sqrt{4 \cdot 2} + 6\sqrt{2} = 2\sqrt{2} + 6\sqrt{2} = 8\sqrt{2}$

46. $\sqrt{3} - 7\sqrt{27} = \sqrt{3} - 7\sqrt{9 \cdot 3} = \sqrt{3} - 7 \cdot 3 \cdot \sqrt{3}$
$= \sqrt{3} - 21\sqrt{3} = -20\sqrt{3}$

47. $\sqrt{28} + \sqrt{63} = 2\sqrt{7} + 3\sqrt{7} = 5\sqrt{7}$

48. $\sqrt{3} \cdot \sqrt{6} = \sqrt{18} = \sqrt{9 \cdot 2} = \sqrt{9} \cdot \sqrt{2} = 3\sqrt{2}$

49. $\sqrt{8} \cdot \sqrt{6} = \sqrt{48} = \sqrt{16 \cdot 3} = \sqrt{16} \cdot \sqrt{3} = 4\sqrt{3}$

50. $\frac{\sqrt{300}}{\sqrt{3}} = \sqrt{\frac{300}{3}} = \sqrt{100} = 10$

51. $\frac{4}{\sqrt{3}} \cdot \frac{\sqrt{3}}{\sqrt{3}} = \frac{4\sqrt{3}}{3}$

52. $\frac{\sqrt{7}}{\sqrt{5}} \cdot \frac{\sqrt{5}}{\sqrt{5}} = \frac{\sqrt{35}}{5}$

53. $3\left(2 + \sqrt{7}\right) = 6 + 3\sqrt{7}$

54. $\sqrt{3}\left(4 + \sqrt{6}\right) = 4\sqrt{3} + \sqrt{18} = 4\sqrt{3} + \sqrt{9 \cdot 2}$
$= 4\sqrt{3} + 3\sqrt{2}$

55. Commutative property of addition

56. Commutative property of multiplication

57. Associative property of addition

58. Distributive property

59. Associative property of multiplication

60. The natural numbers are not closed for subtraction; $2-3=-1$ and -1 is not a natural number.

61. The whole numbers are closed for multiplication.

62. The integers are not closed for division; $1 \div 2$ is not an integer.

63. The real numbers are closed for subtraction.

64. The irrational numbers are not closed for multiplication; $\sqrt{2}\cdot\sqrt{2}=2$ is not irrational.

65. $3^2 \cdot 3^3 = 3^{2+3} = 3^5 = 243$

66. $\dfrac{6^4}{6^2} = 6^{4-2} = 6^2 = 36$

67. $9^0 = 1$

68. $5^{-3} = \dfrac{1}{5^3} = \dfrac{1}{125}$

69. $(2^3)^4 = 2^{3\cdot 4} = 2^{12} = 4096$

70. $-8^0 = -1\cdot 8^0 = -1\cdot 1 = -1$

71. $-7^{-2} = -\dfrac{1}{7^2} = -\dfrac{1}{49}$

72. $3\cdot 2^{-3} = \dfrac{3}{2^3} = \dfrac{3}{8}$

73. 3.62×10^7

74. 1.58×10^{-5}

75. $280,000$

76. 0.000139

77. $\left(3\times 10^4\right)\left(2\times 10^{-9}\right) = (3)(2)\times(10^4)(10^{-9})$
$= 6.0\times 10^{-5}$

78. $\dfrac{1.5\times 10^{-3}}{5\times 10^{-4}} = \left(\dfrac{1.5}{5}\right)\left(\dfrac{10^{-3}}{10^{-4}}\right) = 0.3\times 10^1$
$= 3.0\times 10^0 = 3$

79. $(550,000)(2,000,000)$
$=\left(5.5\times 10^5\right)\left(2.0\times 10^6\right)$
$=(5.5)(2.0)\times\left(10^5\right)\left(10^6\right)$
$=11\times 10^{11}$
$=1.1\times 10^{12} = 1,100,000,000,000$

80. $\dfrac{8,400,000}{70,000} = \dfrac{8.4\times 10^6}{7.0\times 10^4} = \left(\dfrac{8.4}{7.0}\right)\left(\dfrac{10^6}{10^4}\right)$
$= 1.2\times 10^2 = 120$

81. $\dfrac{0.000002}{0.0000004} = \dfrac{2\times 10^{-6}}{4\times 10^{-7}} = \left(\dfrac{2}{4}\right)\left(\dfrac{10^{-6}}{10^{-7}}\right) = 0.5\times 10^1$
$= 5.0\times 10^0 = 5$

82. $\dfrac{1.49\times 10^{11}}{3.84\times 10^8} \approx 3.88\times 10^2 \approx 388$ times

83. $\dfrac{20,000,000}{3600} = \dfrac{2\times 10^7}{3.6\times 10^3} = \left(\dfrac{2}{3.6}\right)\left(\dfrac{10^7}{10^3}\right)$
$= 0.\bar{5}\times 10^4 \approx \$5555.56$/person

84. Arithmetic
$d = a_2 - a_1 = 9-3 = 6$
$a_5 = 21+6 = 27$
$a_6 = 27+6 = 33$

85. Geometric
$r = \dfrac{a_2}{a_1} = \dfrac{1}{\frac{1}{2}} = 2$
$a_5 = 4\cdot 2 = 8$
$a_6 = 8\cdot 2 = 16$

86. $a_n = a_1 + (n-1)d$
$a_9 = (-6)+(9-1)(2)$
$a_9 = 10$

87. $a_n = a_1 + (n-1)d$
$a_{10} = (-20)+(10-1)(5)$
$a_{10} = 25$

88. $a_n = a_1 r^{n-1}$
$a_5 = 3(2)^{5-1}$
$a_5 = 48$

89. $a_n = a_1 r^{n-1}$

$a_{10} = -1(3)^{10-1}$

$a_{10} = -19,683$

90. $d = a_2 - a_1 = 6 - 3 = 3$

$a_n = a_1 + (n-1)d$

$a_{50} = (3) + (50-1)(3) = 150$

$s_n = \dfrac{n(a_1 + a_n)}{2}$

$s_{50} = \dfrac{50(3+150)}{2} = 3825$

91. $d = a_2 - a_1 = 0.75 - 0.5 = 0.25$

$a_n = a_1 + (n-1)d$

$a_{20} = (0.5) + (20-1)(0.25) = 5.25$

$s_n = \dfrac{n(a_1 + a_n)}{2}$

$s_{20} = \dfrac{20(0.5+5.25)}{2} = 57.5$

92. $s_n = \dfrac{a_1\left(1-r^n\right)}{1-r}$

$s_4 = \dfrac{3\left(1-2^4\right)}{1-2} = 45$

93. $s_n = \dfrac{a_1\left(1-r^n\right)}{1-r}$

$s_6 = \dfrac{1\left(1-(-2)^6\right)}{1-(-2)} = -21$

94. Arithmetic

$d = a_2 - a_1 = 4 - 1 = 3$

$a_n = a_1 + (n-1)d$

$a_n = (1) + (n-1)(3)$

$a_n = 1 + 3n - 3$

$a_n = 3n - 2$

95. Geometric

$r = \dfrac{a_2}{a_1} = \dfrac{-2}{2} = -1$

$a_n = a_1 r^{n-1}$

$a_n = 2(-1)^{n-1}$

96. 1, 1, 2, 3, 5, 8, 13, 21, 34, 55, 89, 144, 233,377, 610

97. No, from the fifth term onward, each term is not the sum of the previous two terms.

98. Yes, each term is the sum of the previous two terms. The next two terms are $-3+(-5)=-8$ and $-5+(-8)=-13$.

Chapter Test

1. Use the divisibility rules in section 5.1.
20,270 is divisible by: 2, 5, and 10.

2. $825 = 3^2 \cdot 5 \cdot 7$

$$\begin{array}{r|l} 3 & 315 \\ 3 & 105 \\ 5 & 35 \\ & 7 \end{array}$$

3. $[(-3)+7]-(-4) = [4]+4 = 8$

4. $[(-70)(-5) \div (8-10)] = [350 \div (-2)] = -175$

5. $3\dfrac{2}{11} = \dfrac{3 \cdot 11 + 2}{11} = \dfrac{33+2}{11} = \dfrac{35}{11}$

6. $\dfrac{13}{25} = 0.52$

7. $6.45 = \dfrac{645}{100} = \dfrac{129}{20}$

8. $\dfrac{7}{20} - \dfrac{12}{25} \div \dfrac{9}{10} = \dfrac{7}{20} - \dfrac{12}{25} \cdot \dfrac{10}{9} = \dfrac{7}{20} - \dfrac{8}{15}$

$= \dfrac{21}{60} - \dfrac{32}{60} = -\dfrac{11}{60}$

9. $\sqrt{63} + \sqrt{28} = \sqrt{9}\sqrt{7} + \sqrt{4}\sqrt{7}$

$= 3\sqrt{7} + 2\sqrt{7} = 5\sqrt{7}$

10. $\dfrac{\sqrt{2}}{\sqrt{7}} = \dfrac{\sqrt{2}}{\sqrt{7}} \cdot \dfrac{\sqrt{7}}{\sqrt{7}} = \dfrac{\sqrt{14}}{\sqrt{49}} = \dfrac{\sqrt{14}}{7}$

11. Yes, the product of any two integers is an integer.

12. Distributive property of multiplication over addition

13. $\frac{4^5}{4^2} = 4^{5-2} = 4^3 = 64$

14. $4^3 \cdot 4^2 = 4^{3+2} = 4^5 = 1024$

15. $3^{-4} = \frac{1}{3^4} = \frac{1}{81}$

16. $\frac{7,200,000}{0.000009} = \frac{7.2\times10^6}{9.0\times10^{-6}} = \left(\frac{7.2}{9.0}\right)\left(\frac{10^6}{10^{-6}}\right)$

$= 0.8\times10^{12} = 8.0\times10^{11}$

17. $d = a_2 - a_1 = -6-(-2) = -4$

$a_n = a_1 + (n-1)d$

$a_n = (-2)+(n-1)(-4)$

$a_n = -2-4n+4$

$a_n = -4n+2$

18. $d = a_2 - a_1 = -5-(-2) = -3$

$a_n = a_1 + (n-1)d$

$a_{11} = (-2)+(11-1)(-3)$

$a_{11} = (-2)+(11-1)(-3) = -32$

$s_n = \frac{n(a_1 + a_n)}{2}$

$s_{11} = \frac{11(-2+(-32))}{2} = -187$

19. $r = \frac{a_2}{a_1} = \frac{6}{3} = 2$

$a_n = a_1 r^{n-1}$

$a_n = 3(2)^{n-1}$

20. 1, 1, 2, 3, 5, 8, 13, 21, 34, 55

Chapter Six: Algebra, Graphs, and Functions

Section 6.1: Order of Operations and Solving Linear Equations

1. Variable

3. Expression

5. Terms

7. Coefficient

9. Identity

11. $4x+9x=13x$

13. $-7x+3x-8=-4x-8$

15. $7x+3y-4x+8y=3x+11y$

17. $5x-7y+8-3z-11y-13=2x-18y-5$

19. $10x^2+x-21-11x^2+7x+16=-x^2+8x-5$

21. $3(t+3)+5(t-2)+1$
$=3t+9+5t-10+1$
$=8t$

23. $6.2x-8.3+7.1x=13.3x-8.3$

25. $\frac{3}{5}x+\frac{3}{10}x-8$
$=\frac{6}{10}x+\frac{3}{10}x-8$
$=\frac{9}{10}x-8$

27. $0.2(x+4)+1.2(x-3)$
$=0.2x+0.8+1.2x-3.6$
$=1.4x-2.8$

29. $\frac{2}{3}(x+6)+\frac{1}{6}(x+6)$
$=\frac{2}{3}x+4+\frac{1}{6}x+1$
$=\frac{4}{6}x+4+\frac{1}{6}x+1$
$=\frac{5}{6}x+5$

31. $x-7=(4)-7=-3$

33. $-3x+7=-3(-2)+7=6+7=13$

35. $x^2=8^2=64$

37. $-x^2=-(-5)^2=-25$

39. $x^2-5x+12=(3)^2-5(3)+12$
$=9-15+12=6$

41. $-2x^2+5x-9$
$=-2(3)^2+5(3)-9$
$=-2(9)+15-9$
$=-18+6=-12$

43. x^3-3x^2+7x-5
$=(2)^3-3(2)^2+7(2)-5$
$=8-3(4)+14-5$
$=8-12+9=5$

45. $4x+5=17$
$4(3)+5=17$
$12+5=17$
$17=17$
Yes, 3 is a solution.

47. $8x-7=5x+1$
$8(2)-7=5(2)+1$
$16-7=10+1$
$9\neq 11$
No, –2 is not a solution.

49. $3(2x-7)+1=-4(3x+1)+20$
$3(2(2)-7)+1=-4(3(2)+1)+20$
$3(4-7)+1=-4(6+1)+20$
$3(-3)+1=-4(7)+20$
$-9+1=-28+20$
$-8=-8$
Yes, 2 is a solution.

51. $y+7=9$
$y+7-7=9-7$
$y=2$

53. $\frac{x}{9}=-5$
$9\left(\frac{x}{9}\right)=9\left(\frac{-5}{1}\right)$
$x=-45$

55. $$\begin{aligned} 7x &= -63 \\ \frac{7x}{7} &= \frac{-63}{7} \\ x &= -9 \end{aligned}$$

57. $$\begin{aligned} 16 &= -3t - 2 \\ 16 + 2 &= -3t - 2 + 2 \\ 18 &= -3t \\ \frac{18}{-3} &= \frac{-3t}{-3} \\ -6 &= t \end{aligned}$$

59. $$\begin{aligned} 6t - 8 &= 4t - 2 \\ 6t - 4t - 8 &= 4t - 4t - 2 \\ 2t - 8 &= -2 \\ 2t - 8 + 8 &= -2 + 8. \\ 2t &= 6 \\ \frac{2t}{2} &= \frac{6}{2} \\ t &= 3 \end{aligned}$$

61. $$\begin{aligned} 3x + 2 - 6x &= -x - 15 + 8 - 5x \\ -3x + 2 &= -6x - 7 \\ -3x + 6x + 2 &= -6x + 6x - 7 \\ 3x + 2 &= -7 \\ 3x + 2 - 2 &= -7 - 2 \\ 3x &= -9 \\ \frac{3x}{3} &= \frac{-9}{3} \\ x &= -3 \end{aligned}$$

63. $$\begin{aligned} 4(3n+1) &= 5(4n-6) + 9n \\ 12n + 4 &= 20n - 30 + 9n \\ 12n + 4 &= 29n - 30 \\ 12n - 12n + 4 &= 29n - 12n - 30 \\ 4 &= 17n - 30 \\ 4 + 30 &= 17n - 30 + 30 \\ 34 &= 17n \\ \frac{17n}{17} &= \frac{34}{17} \\ 2 &= n \end{aligned}$$

73. $$\begin{aligned} 2(x+3) - 4 &= 2(x-4) \\ 2x + 6 - 4 &= 2x - 8 \\ 2x + 2 &= 2x - 8 \\ 2x - 2x + 2 &= 2x - 2x - 8. \\ 2 &\neq -8 \end{aligned}$$

No solution

65. $$\begin{aligned} \frac{1}{2}x + \frac{1}{3} &= \frac{2}{3} \\ 6\left(\frac{1}{2}x + \frac{1}{3}\right) &= 6\left(\frac{2}{3}\right) \\ 3x + 2 &= 4 \\ 3x + 2 - 2 &= 4 - 2 \\ 3x &= 2 \\ \frac{3x}{3} &= \frac{2}{3} \\ x &= \frac{2}{3} \end{aligned}$$

67. $$\begin{aligned} \frac{x}{4} + 2x &= \frac{1}{3} \\ 12\left(\frac{x}{4} + 2x\right) &= 12\left(\frac{1}{3}\right) \\ 3x + 24x &= 4 \\ 27x &= 4 \\ \frac{27x}{27} &= \frac{4}{27} \\ x &= \frac{4}{27} \end{aligned}$$

69. $$\begin{aligned} 0.9x - 1.2 &= 2.4 \\ 0.9x + 1.2 - 1.2 &= 2.4 + 1.2 \\ 0.9x &= 3.6 \\ \frac{0.9x}{0.9} &= \frac{3.6}{0.9} \\ x &= 4 \end{aligned}$$

71. $$\begin{aligned} 0.2x + 1.3 &= 0.4x - 4.5 \\ 10(0.2x + 1.3) &= 10(0.4x - 4.5) \\ 2x + 13 &= 4x - 45 \\ 2x - 2x + 13 &= 4x - 2x - 45 \\ 13 &= 2x - 45 \\ 13 + 45 &= 2x - 45 + 45 \\ 58 &= 2x \\ \frac{58}{2} &= \frac{2x}{2} \\ x &= 29 \end{aligned}$$

75. $$\begin{aligned} 4(x-4) + 12 &= 4(x-1) \\ 4x - 16 + 12 &= 4x - 4 \\ 4x - 4 &= 4x - 4 \end{aligned}$$

All real numbers

77. $s = 0.08(79) = \$6.32$ in sales tax

79. a) $F=\frac{9}{5}(0)+32=0+32=32°\text{F}$

b) $F=\frac{9}{5}(100)+32=180+32=212°\text{F}$

c) $F=\frac{9}{5}(35)+32=63+32=95°\text{F}$

d) $F=\frac{9}{5}(-5)+32=-9+32=23°\text{F}$

81. $(-1)^n=1$ for any even number, n, since there will be an even number of factors of (-1), and when these are multiplied, the product will always be 1.

83. $1^n=1$ for all natural numbers, since 1 multiplied by itself any number of times will always be 1.

85. a)

$$\begin{aligned}P&=14.70+0.43x\\148&=14.70+0.43x\\148-14.70&=14.70-14.70+0.43x\\133.3&=0.43x\\\frac{133.3}{0.43}&=\frac{0.43x}{0.43}\\x&=310\text{ ft below sea level}\end{aligned}$$

b)

$$\begin{aligned}P&=14.70+0.43x\\128.65&=14.70+0.43x\\128.65-14.70&=14.70-14.70+0.43x\\113.95&=0.43x\\\frac{113.95}{0.43}&=\frac{0.43x}{0.43}\\x&=265\text{ ft below sea level}\end{aligned}$$

Section 6.2: Formulas

1. Subscripts

3. $P=4s=4(10)=40$

5. $P=2l+2w=2(15)+2(8)$
$=30+16=46$

7.
$$\begin{aligned}F&=ma\\40&=m(5)\\\frac{40}{5}&=\frac{5m}{5}\\8&=m\end{aligned}$$

9.
$$\begin{aligned}A&=2\pi rh+2\pi r^2=2\pi(2)(3)+2\pi(2)^2\\&=12\pi+8\pi\\&=20\pi\approx 62.83\end{aligned}$$

11.
$$\begin{aligned}K&=\frac{1}{2}mv^2\\4500&=\frac{1}{2}m(30)^2\\4500&=450m\\\frac{4500}{450}&=\frac{450m}{450}\\10&=m\end{aligned}$$

13.
$$\begin{aligned}T&=\frac{PV}{k}\\\frac{80}{1}&=\frac{P(20)}{0.5}\\80(0.5)&=20P\\40&=20P\\\frac{40}{20}&=\frac{20P}{20}\\2&=P\end{aligned}$$

15.
$$\begin{aligned}V&=-\frac{1}{2}at^2\\2304&=-\frac{1}{2}a(12)^2\\\frac{2304}{1}&=-\frac{144a}{2}\\2304&=-72a\\\frac{2304}{-72}&=\frac{-72a}{-72}\\-32&=a\end{aligned}$$

17.
$$\begin{aligned}V&=\pi r^2h\\942&=\pi(5)^2h\\942&=25\pi h\\\frac{942}{25\pi}&=\frac{25\pi h}{25\pi}\\11.99&\approx h\end{aligned}$$

19.
$$\begin{aligned}F&=\frac{9}{5}C+32\\F&=\frac{9}{5}(7)+32\\F&=\frac{63}{5}+32=12.6+32=44.6\end{aligned}$$

21. $m = \frac{y_2 - y_1}{x_2 - x_1}$

$m = \frac{8-(-4)}{-3-(-5)}$

$m = \frac{8+4}{-3+5} = \frac{12}{2} = 6$

23. $x = \frac{-b+\sqrt{b^2-4ac}}{2a}$

$x = \frac{-(-5)+\sqrt{(-5)^2-4(2)(-12)}}{2(2)}$

$x = \frac{5+\sqrt{25+96}}{4}$

$x = \frac{5+\sqrt{121}}{4} = \frac{5+11}{4} = \frac{16}{4} = 4$

25. $s = -16t^2 + v_0 t + s_0$

$s = -16(4)^2 + 30(4) + 150$

$s = -16(16) + 120 + 150$

$s = -256 + 120 + 150 = 14$

27. $P = \frac{nRT}{V}$

$63 = \frac{(27)R(2)}{6}$

$63 = \frac{54R}{6}$

$63 = 9R$

$\frac{63}{9} = \frac{9R}{9}$

$7 = R$

29. $A = P\left(1+\frac{r}{n}\right)^{nt}$

$A = 1200\left(1+\frac{0.04}{2}\right)^{(2)(5)}$

$A = 1200(1+0.02)^{10}$

$A = 1200(1.02)^{10} \approx 1462.79$

31. $6x + 3y = 9$

$6x - 6x + 3y = -6x + 9$

$3y = -6x + 9$

$\frac{3y}{3} = \frac{-6x+9}{3}$

$y = \frac{-6x+9}{3} = -2x + 3$

33. $x - 6y = 12$

$x - x - 6y = -x + 12$

$-6y = -x + 12$

$\frac{-6y}{-6} = \frac{-x+12}{-6}$

$y = \frac{-x+12}{-6}$

$y = \frac{-x}{-6} + \frac{12}{-6} = \frac{1}{6}x - 2$

35. $2x - 3y + 6 = 0$

$2x - 3y + 6 - 6 = 0 - 6$

$2x - 3y = -6$

$2x - 2x - 3y = -2x - 6$

$-3y = -2x - 6$

$\frac{-3y}{-3} = \frac{-2x-6}{-3}$

$y = \frac{-2x-6}{-3}$

$y = \frac{-2x}{-3} + \frac{-6}{-3} = \frac{2}{3}x + 2$

37. $d = rt$

$\frac{d}{t} = \frac{rt}{t}$

$r = \frac{d}{t}$

39. $p = a + b + c$

$p - b = a + b - b + c$

$p - b = a + c$

$p - b - c = a + c - c$

$a = p - b - c$

41. $C = 2\pi r$

$$\frac{C}{2\pi} = \frac{2\pi r}{2\pi}$$

$$r = \frac{C}{2\pi}$$

43. $A = \frac{1}{2}bh$

$$2(A) = 2\left(\frac{1}{2}bh\right)$$

$$2A = bh$$

$$\frac{2A}{h} = \frac{bh}{h}$$

$$b = \frac{2A}{h}$$

45. $y = a + bx$

$$y - a = a - a + bx$$

$$y - a = bx$$

$$\frac{y-a}{b} = \frac{bx}{b}$$

$$\frac{y-a}{b} = x$$

47. $P = 2l + 2w$

$$P - 2l = 2l - 2l + 2w$$

$$P - 2l = 2w$$

$$\frac{P-2l}{2} = \frac{2w}{2}$$

$$w = \frac{P-2l}{2}$$

49. $F = \frac{9}{5}C + 32$

$$F - 32 = \frac{9}{5}C + 32 - 32$$

$$F - 32 = \frac{9}{5}C$$

$$\frac{5}{9}(F-32) = \frac{5}{9}\left(\frac{9}{5}C\right)$$

$$C = \frac{5}{9}(F-32)$$

51. $A = P(1+rt)$

$$A = P + Prt$$

$$A - P = P - P + Prt$$

$$A - P = Prt$$

$$\frac{A-P}{Pt} = \frac{Prt}{Pt}$$

$$r = \frac{A-P}{Pt}$$

53. $d = rt$

$$403 = 62t$$

$$\frac{403}{62} = t$$

$$t = 6.5 \text{ hours}$$

55. $i = prt$

$$128 = 800(r)(2)$$

$$128 = 1600r$$

$$\frac{128}{1600} = \frac{1600r}{1600}$$

$$r = 0.08 = 8\%$$

57. a) $6 \text{ ft} = 6(12) = 72 \text{ in.}$

$$B = \frac{703w}{h^2}$$

$$B = \frac{703(200)}{(72)^2}$$

$$B = \frac{140{,}600}{5184} \approx 27.12$$

b) $B = \frac{703w}{h^2}$

$$26 = \frac{703w}{(72)^2}$$

$$26 = \frac{703w}{5184}$$

$$134{,}784 = 703w$$

$$\frac{134{,}784}{703} = \frac{703w}{703}$$

$$w \approx 191.73 \text{ lb}$$

He would have to lose $200 - 191.73 = 8.27$ lb.

Section 6.3: Applications of Algebra

1. Ratio

3. $x + 3$

5. $11x$

7. $6 - 4y$

9. $8w+9$

11. $4x+6$

13. $\frac{z}{13}-4$

15. $\frac{12-s}{4}$

17. $3(x+7)$

19. Let $x=$ the number
$x+8=$ the sum of a number and 8
$$x+8=23$$
$$x+8-9=23-9$$
$$x=15$$

21. Let $x=$ the number
$9x=$ the product of 9 and a number
$$9x=54$$
$$\frac{9x}{9}=\frac{54}{9}$$
$$x=6$$

23. Let $x=$ the number
$4x-10=$ ten less than four times a number
$$4x-10=42$$
$$4x-10+10=42+10$$
$$4x=52$$
$$\frac{4x}{4}=\frac{52}{4}$$
$$x=13$$

25. Let $x=$ the number
$4x+12=$ twelve more than 4 times a number
$$4x+12=32$$
$$4x+12-12=32-12$$
$$4x=20$$
$$\frac{4x}{4}=\frac{20}{4}$$
$$x=5$$

27. Let $x=$ the number
$x+6=$ the number increased by 6
$2x-3=$ three less than twice the number
$$x+6=2x-3$$
$$x-x+6=2x-x-3$$
$$6=x-3$$
$$6+3=x-3+3$$
$$9=x$$

29. Let $x=$ the number
$x+10=$ the number increased by 10
$2(x+3)=$ two times the sum of the number and 3
$$x+10=2(x+3)$$
$$x+10=2x+6$$
$$x-x+10=2x-x+6$$
$$10=x+6$$
$$10-6=x+6-6$$
$$4=x$$

31. Let $x=$ the number of miles driven
$0.50x=$ the reimbursement for mileage
$$150+0.50x=255$$
$$0.50x=105$$
$$\frac{0.50x}{0.50}=\frac{105}{0.50}$$
$$x=210 \text{ miles}$$

33. Let $x=$ the pretax rental cost.
$x+0.05x=$ total rental cost
$$x+0.05x=42$$
$$1.05x=42$$
$$\frac{1.05x}{1.05}=\frac{42}{1.05}$$
$$x=40$$
The pretax rental cost is \$40 per half hour.

35. Let $x=$ Tito's dollar sales
$0.06x=$ the amount Tito made on commission
$$400+0.06x=790$$
$$400-400+0.06x=790-400$$
$$0.06x=390$$
$$\frac{0.06x}{0.06}=\frac{390}{0.06}$$
$$x=\$6500$$

37. Let $x=$ amount invested in mutual funds
$3x=$ amount invested in stocks
$$x+3x=20{,}000$$
$$4x=20{,}000$$
$$\frac{4x}{4}=\frac{20{,}000}{4}$$
$$x=5000$$
\$5000 was invested in mutual funds and $3(5000)=\$15{,}000$ was invested in stocks.

39. Let $w =$ width
$w+2 =$ length

$$2w+2l=P$$
$$2w+2(w+2)=52$$
$$2w+2w+4=52$$
$$4w+4=52$$
$$4w+4-4=52-4$$
$$4w=48$$
$$\frac{4w}{4}=\frac{48}{4}$$
$$w=12$$

The width is 12 ft and the length is $12+2=14$ ft.

41. Let $w =$ width
$2w =$ length of entire enclosed region
$3w+2(2w) =$ total amount of fencing

$$3w+2(2w)=140$$
$$3w+4w=140$$
$$7w=140$$
$$\frac{7w}{7}=\frac{140}{7}$$
$$w=20$$

The width is 20 ft and the length is $2(20)=40$ ft.

43. a) Let $c =$ cost of the water bill

$$\frac{\$5.87}{1000 \text{ gallons}}=\frac{c}{4327}$$
$$1000c=5.87(4327)$$
$$1000c=25{,}399.49$$
$$\frac{1000c}{1000}=\frac{25{,}399.49}{1000}$$
$$c\approx \$25.40$$

b) $$\frac{\$5.87}{1000 \text{ gallons}}=\frac{28.71}{g}$$
$$5.87g=28.71(1000)$$
$$5.87g=28{,}710$$
$$\frac{5.87g}{5.87}=\frac{28{,}710}{5.87}$$
$$g=4890.97 \text{ gallons}$$

45. $$\frac{1}{1{,}800{,}000}=\frac{16.4}{x}$$
$$x=16.4(1{,}800{,}000)$$
$$x=29{,}520{,}000 \text{ households}$$

47. a) $$\frac{40}{12}=\frac{x}{480}$$
$$40(480)=12x$$
$$\frac{40(480)}{12}=\frac{12x}{12}$$
$$x=1600 \text{ lb}$$

b) $$\frac{1}{12}=\frac{x}{480}$$
$$12x=480$$
$$x=\frac{480}{12}=40 \text{ bags}$$

49. $$\frac{40}{0.6}=\frac{250}{x}$$
$$40x=0.6(250)$$
$$40x=150$$
$$\frac{40x}{40}=\frac{150}{40}$$
$$x=3.75 \text{ m}\ell$$

51. $$\frac{40}{1}=\frac{35}{x}$$
$$40x=35$$
$$\frac{40x}{40}=\frac{35}{40}$$
$$x=0.875 \text{ cc}$$

53. a) Let $x =$ the number of months
$0.10(100)=\$10$ saved per year
$$10x=45$$
$$x=4.5=4\tfrac{1}{2} \text{ months}$$

b) $25-18=7$ years
$7(12)(10)-45=840-45=\$795$ total savings

Section 6.4: Variation

1. Direct
3. Joint
5. Direct
7. Inverse

9. Direct

11. Inverse

13. Direct

15. Inverse

17. Direct

19. Direct

21. Answers will vary.

23. a) $y = kx$

b) $y = 9(12) = 108$

25. a) $m = \dfrac{k}{n^2}$

b) $m = \dfrac{20}{(10)^2} = \dfrac{20}{100} = 0.20$

27. a) $A = \dfrac{kB}{C}$

b) $A = \dfrac{(5)5}{10} = 2.5$

29. a) $F = kDE$

b) $F = 7(3)(10) = 210$

31. a) $H = kL$

b) $12 = k(40)$

$\dfrac{12}{40} = \dfrac{40k}{40}$

$k = 0.3$

$H = 0.3L$

$H = 0.3(10) = 3$

33. a) $q = \dfrac{k}{w^2}$

b) $9 = \dfrac{k}{5^2}$

$9 = \dfrac{k}{25}$

$225 = k$

$q = \dfrac{225}{w^2}$

$q = \dfrac{225}{(3)^2} = \dfrac{225}{9} = 25$

35. a) $J = k\sqrt{b}$

b) $2 = k\sqrt{36}$

$2 = 6k$

$\dfrac{2}{6} = \dfrac{6k}{6}$

$k = \dfrac{1}{3}$

$J = \dfrac{1}{3}\sqrt{b}$

$J = \dfrac{1}{3}\sqrt{81} = \dfrac{1}{3}(9) = 3$

37. a) $Z = kWY$

b) $12 = k(9)(4)$

$12 = 36k$

$\dfrac{12}{36} = \dfrac{36k}{36}$

$k = \dfrac{1}{3}$

$Z = \dfrac{1}{3}WY$

$Z = \dfrac{1}{3}(50)(6) = \dfrac{300}{3} = 100$

39. a) $F = \dfrac{kM_1M_2}{d^2}$

b) $20 = \dfrac{k(5)(10)}{(0.2)^2}$

$20 = \dfrac{50k}{0.04}$

$50k = 0.8$

$k = \dfrac{0.8}{50} = 0.016$

$F = \dfrac{0.016M_1M_2}{d^2}$

$F = \dfrac{0.016(10)(20)}{(0.4)^2} = \dfrac{3.2}{0.16} = 20$

41. a) $p = kn$

b) $450 = k(8)$

$\dfrac{450}{8} = \dfrac{8k}{8}$

$k = 56.25$

$p = 56.25n$

$p = 56.25(18) = \$1012.50$

43. a) $t = \frac{k}{T}$

b) $2 = \frac{k}{75}$

$k = 75(2) = 150$

$t = \frac{150}{T}$

$t = \frac{150}{80} = 1.875$ minutes

45. a) $d = kt^2$

b) $64 = k(2)^2$

$64 = 4k$

$k = \frac{64}{4} = 16$

$d = 16t^2$

$d = 16(3)^2 = 16(9) = 144$ feet

47. a) $R = \frac{kL}{A}$

b) $0.2 = \frac{k(200)}{0.05}$

$200k = 0.01$

$k = \frac{0.01}{200} = 0.00005$

$R = \frac{0.00005L}{A}$

$R = \frac{0.00005(5000)}{0.01} = \frac{0.25}{0.01} = 25$ ohms

49. a) $W = kI^2R$

b) $6 = k(0.2)^2(150)$

$6k = 6$

$k = 1$

$W = 1(0.3)^2(100)$

$W = (0.09)(100) = 9$ watts

51. a) $y = kx$

$y = 2x$

$\frac{y}{2} = \frac{2x}{2}$

$x = \frac{y}{2} = 0.5y$; Varies directly

b) $k = 0.5$

53. a) y will double.

$y = kx$; $k(2x) = 2(kx) = 2y$

b) y will be half as large.

$y = kx$; $k\left(\frac{1}{2}x\right) = \frac{1}{2}(kx) = \frac{1}{2}y$

55. $I = \frac{k}{d^2}$

$\frac{1}{16} = \frac{k}{(4)^2}$

$\frac{1}{16} = \frac{k}{16}$

$k = 1$

$I = \frac{1}{d^2}$

$I = \frac{1}{(3)^2} = \frac{1}{9}$

Section 6.5: Solving Linear Inequalities

1. Inequality

3. Compound

5. All

7. Closed

9. $x \geq 4$; 

11. $x + 9 \geq 4$

$x + 9 - 9 \geq 4 - 9$

$x \geq -5$

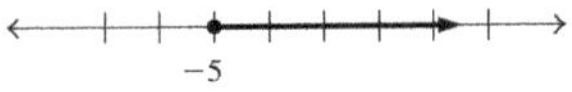

13. $-6x \leq 36$

$\frac{-6x}{-6} \geq \frac{36}{-6}$

$x \geq -6$

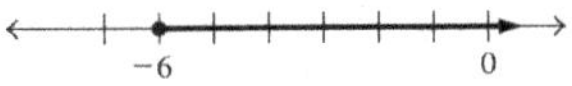

15. $\frac{-x}{3} \geq 3$

$-3\left(\frac{-x}{3}\right) \leq -3(3)$

$x \leq -9$

17.

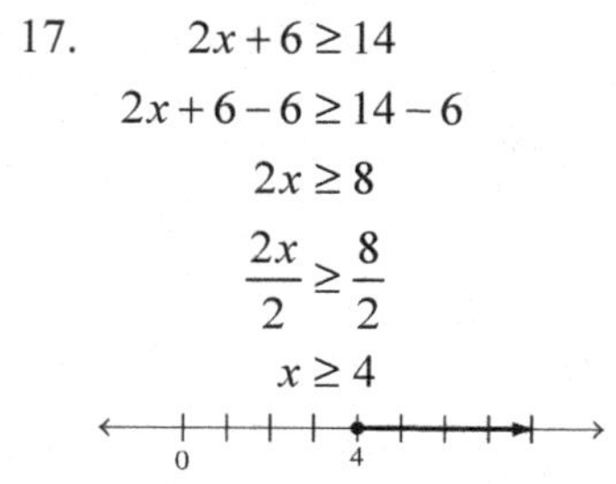

$$2x+6 \ge 14$$
$$2x+6-6 \ge 14-6$$
$$2x \ge 8$$
$$\frac{2x}{2} \ge \frac{8}{2}$$
$$x \ge 4$$

19.

$$-2x+5 \le 4x-25$$
$$-2x-4x+5 \le 4x-4x-25$$
$$-6x+5 \le -25$$
$$-6x+5-5 \le -25-5$$
$$-6x \le -30$$
$$\frac{-6x}{-6} \ge \frac{-30}{-6}$$
$$x \ge 5$$

21.

$$4(2x-1) < 2(4x-3)$$
$$8x-4 < 8x-6$$
$$8x-8x-4 < 8x-8x-6$$
$$-4 < -6$$

False, no solution

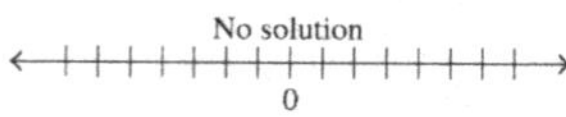

23.

$$5(2x-7) \le 2(5x+4)$$
$$10x-35 \le 10x+8$$
$$10x-10x-35 \le 10x-10x+8$$
$$-35 \le 8$$

True, all real numbers.

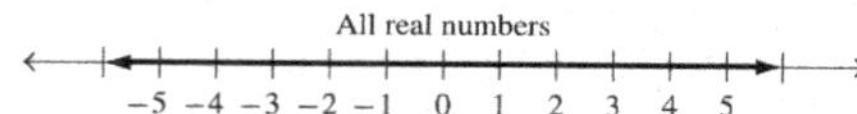

37.

$$-15 < -4x-3$$
$$-12 < -4x$$
$$\frac{-12}{-4} > \frac{-4x}{-4}$$
$$3 > x$$

25. $-1 \le x \le 3$;

27.

$$2 < x-4 \le 6$$
$$2+4 < x-4+4 \le 6+4$$
$$6 < x \le 10$$

29.

$$\frac{1}{2} < \frac{x+4}{2} \le 4$$
$$2\left(\frac{1}{2}\right) < 2\left(\frac{x+4}{2}\right) \le 2(4)$$
$$1 < x+4 \le 8$$
$$1-4 < x+4-4 \le 8-4$$
$$1 < x+4 \le 8$$
$$-3 < x \le 4$$

31. $x > 1$;

33.

$$-4x \le 36$$
$$\frac{-4x}{-4} \ge \frac{36}{-4}$$
$$x \ge -9$$

35.

$$-\frac{x}{4} \ge 2$$
$$(-4)\left(-\frac{x}{4}\right) \le (-4)(2)$$
$$x \le -8$$

39.

$$3(x+4) \ge 4x+13$$
$$3x+12 \ge 4x+13$$
$$3x-4x+12 \ge 4x-4x+13$$
$$-x+12 \ge 13$$
$$-x+12-12 \ge 13-12$$
$$-x \ge 1$$
$$\frac{-x}{-1} \le \frac{1}{-1}$$
$$x \le -1$$

41. $$\begin{aligned} 5(x+4)-6 &\le 2x+8 \\ 5x+20-6 &\le 2x+8 \\ 5x+14 &\le 2x+8 \\ 5x-2x+14 &\le 2x-2x+8 \\ 3x+14 &\le 8 \\ 3x+14-14 &\le 8-14 \\ 3x &\le -6 \\ \frac{3x}{3} &\le \frac{-6}{3} \\ x &\le -2 \end{aligned}$$

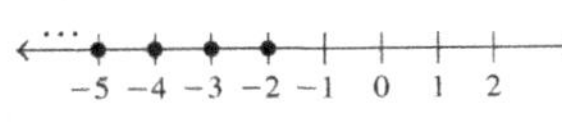

43. $$\begin{aligned} 1 > -x > -6 \\ \frac{1}{-1} < \frac{-x}{-1} < \frac{-6}{-1} \\ -1 < x < 6 \end{aligned}$$

45. $$\begin{aligned} 0.3 \le \frac{x+2}{10} \le 0.5 \\ 10(0.3) \le 10\left(\frac{x+2}{10}\right) \le 10(0.5) \\ 3 \le x+2 \le 5 \\ 3-2 \le x+2-2 \le 5-2 \\ 1 \le x \le 3 \end{aligned}$$

47. a) 2015, 2016
b) 2010, 2011, 2012
c) 2010, 2011, 2012
d) 2014, 2015, 2016

49. Let $x =$ the number of months
Option A: $150+20x$;
Option B: $35x$

$$\begin{aligned} 35x &> 150+20x \\ 35x-20x &> 150+20x-20x \\ 15x &> 150 \\ x &> 10 \end{aligned}$$

It would take 11 months for Option A to cost less that Option B.

51. Let $x =$ number hours worked as cashier

$$\begin{aligned} 7.25x+662.5 &\le 3200 \\ 7.25x &\le 2537.5 \\ \frac{7.25x}{7.25} &\le \frac{2537.5}{7.25} \\ x &\le 350 \end{aligned}$$

Julie can work a maximum of 350 hours as a cashier.

53. Let $x =$ the length of time Tom can park.

$$\begin{aligned} 1.75+0.50(x-1) &\le 4.25 \\ 1.75+0.50x-0.50 &\le 4.25 \\ 0.50x &\le 3 \\ \frac{0.50x}{0.50} &\le \frac{3}{0.50} \\ x &\le 6 \end{aligned}$$

Tom can park for at most 6 hours.

55. $$\begin{aligned} 36 < 84-32t < 68 \\ 36-84 < 84-84-32t < 68-84 \\ -48 < -32t < -16 \\ \frac{-48}{-32} > \frac{-32t}{-32} > \frac{-16}{-32} \\ 1.5 > t > 0.5 \\ 0.5 < t < 1.5 \end{aligned}$$

The time will be between 0.5 seconds and 1.5 seconds.

57. Let $x =$ Hyun's grade on fourth test

$$\begin{aligned} 93 &\le \frac{87+91+98+x}{4} \\ 93 &\le \frac{276+x}{4} \\ 4(93) &\le 4\left(\frac{276+x}{4}\right) \\ 372 &\le 276+x \\ 372-276 &\le 276-276+x \\ 96 &\le x \end{aligned}$$

Hyun must earn a grade of 96% or higher on the fourth test.

59. Let $x =$ the number of gallons

Minimum coverage per can: $250x = 2750$

$$x = \frac{2750}{250} = 11$$

Maximum coverage per can:
$400x = 2750$

$$x = \frac{2750}{400} = 6.875 \text{ cans}$$

The number of gallons, x, can be represented by $6.875 \le x \le 11$.

61. Student's answer:

$$-\frac{1}{3}x \le 4$$
$$-3\left(-\frac{1}{3}x\right) \le -3(4)$$
$$x \le -12$$

Correct answer:

$$-\frac{1}{3}x \le 4$$
$$-3\left(-\frac{1}{3}x\right) \ge -3(4)$$
$$x \ge -12$$

Yes, -12 is in both solution sets.

Section 6.6: Graphing Linear Equations

1. Graph

3. x

5. Plotting points, using intercepts, and using the slope and y-intercept

7. m

9. – 15.

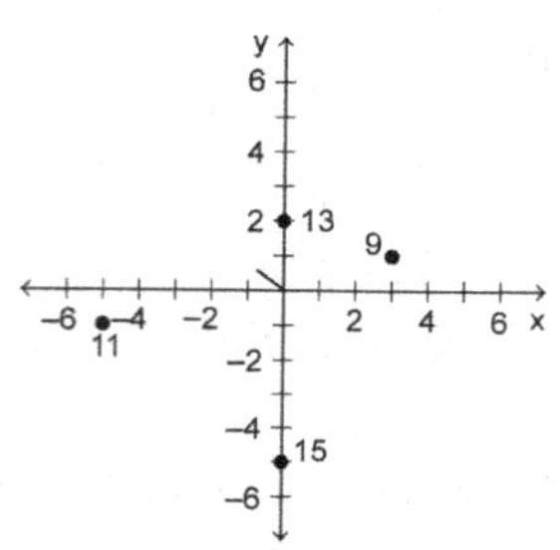

17. – 23.

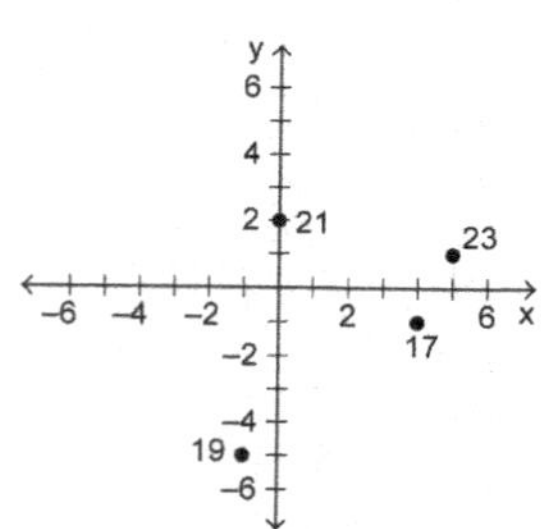

25. (0, 2)

27. (–2, 0)

29. (–3, –4)

31. (2, –2)

33. (2, 2)

35. Substitute $(0,2)$ into $2x+3y=6$.

$$2x+3y=6$$
$$2(0)+3(2)=6$$
$$6=6$$

Therefore, $(0,2)$ satisfies $2x+3y=6$.

Substitute $(3,0)$ into $2x+3y=6$.

$$2x+3y=6$$
$$2(3)+3(0)=6$$
$$6=6$$

Therefore, $(3,0)$ satisfies $2x+3y=6$.

Substitute $2,3)$ into $2x+3y=6$.

$$2x+3y=6$$
$$2(2)+3(3)=6$$
$$13 \ne 6$$

Therefore, $(2,3)$ does not satisfy $2x+3y=6$.

37. Substitute $(8,7)$ into $3x-2y=10$.

$$3x-2y=10$$
$$3(8)-2(7)=10$$
$$10=10$$

Therefore, $(8,7)$ satisfies $3x-2y=10$.

Substitute $(-1,4)$ into $3x-2y=10$.

$$3x-2y=10$$
$$3(-1)-2(4)=10$$
$$-11\neq 10$$

Therefore, $(-1,4)$ does not satisfy $3x-2y=10$.

Substitute $\left(\frac{10}{3},0\right)$ into $3x-2y=10$.

$$3x-2y=10$$
$$3\left(\tfrac{10}{3}\right)-2(0)=10$$
$$10=10$$

Therefore, $\left(\frac{10}{3},0\right)$ satisfies $3x-2y=10$.

39. Substitute $(1,-1)$ into $6y=3x+6$.

$$6y=3x+6$$
$$6(-1)=3(1)+6$$
$$-6\neq 9$$

Therefore, $(1,-1)$ does not satisfy $6y=3x+6$.

Substitute $(4,3)$ into $6y=3x+6$.

$$6y=3x+6$$
$$6(3)=3(4)+6$$
$$18=18$$

Therefore, $(4,3)$ satisfies $6y=3x+6$.

Substitute $(2,5)$ into $6y=3x+6$.

$$6y=3x+6$$
$$6(5)=3(2)+6$$
$$30\neq 12$$

Therefore, $(2,5)$ does not satisfy $6y=3x+6$.

41. Substitute $(8,0)$ into $y=-\frac{3}{8}x+3$.

$$y=-\frac{3}{8}x+3$$
$$0=-\frac{3}{8}(8)+3$$
$$0=0$$

Therefore, $(8,0)$ satisfies $y=-\frac{3}{8}x+3$.

Substitute $(-8,9)$ into $y=-\frac{3}{8}x+3$.

$$y=-\frac{3}{8}x+3$$
$$9=-\frac{3}{8}(-8)+3$$
$$9\neq 6$$

Therefore, $(-8,9)$ does not satisfy $y=-\frac{3}{8}x+3$.

Substitute $(0,3)$ into $y=-\frac{3}{8}x+3$.

$$y=-\frac{3}{8}x+3$$
$$3=-\frac{3}{8}(3)+3$$
$$3=3$$

Therefore, $(0,3)$ satisfies $y=-\frac{3}{8}x+3$.

43.

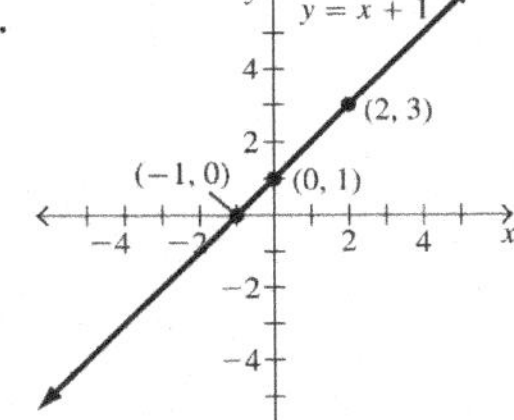

45.

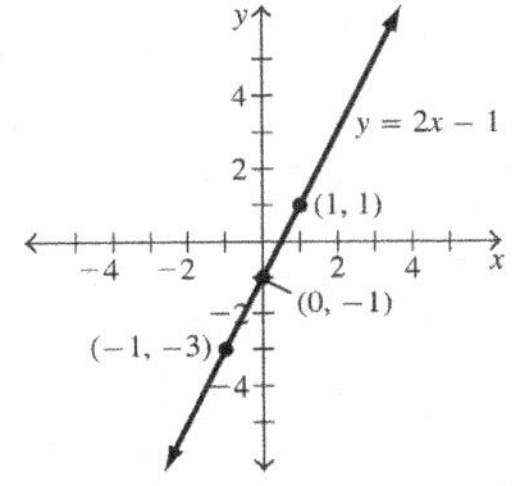

47.

(0, 6)
y + 3x = 6
(1, 3)
(2, 0)

49.

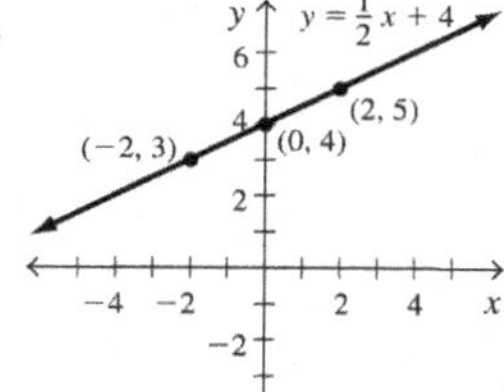

51.

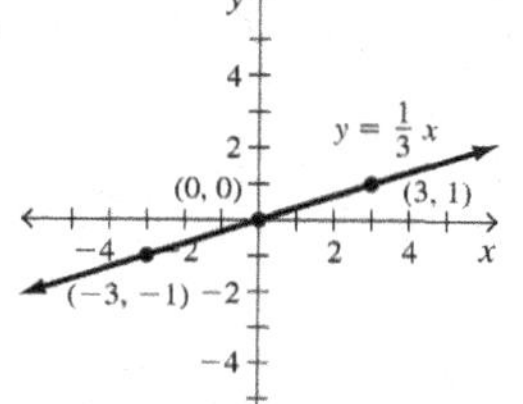

53.

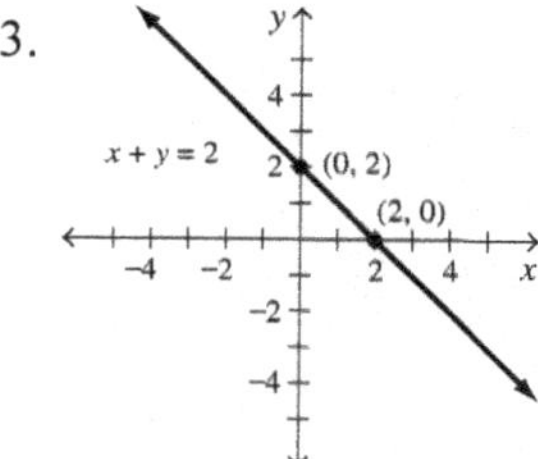

55.

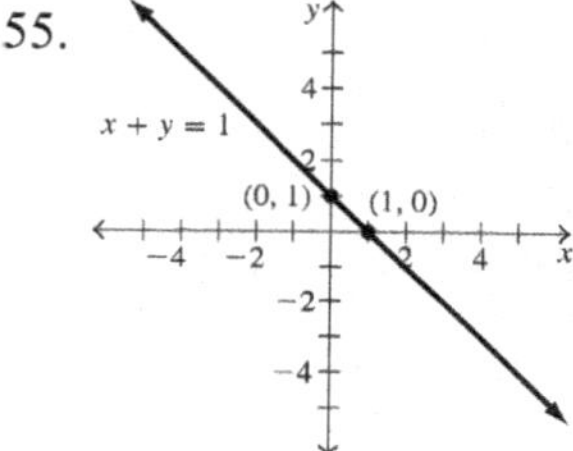

57.

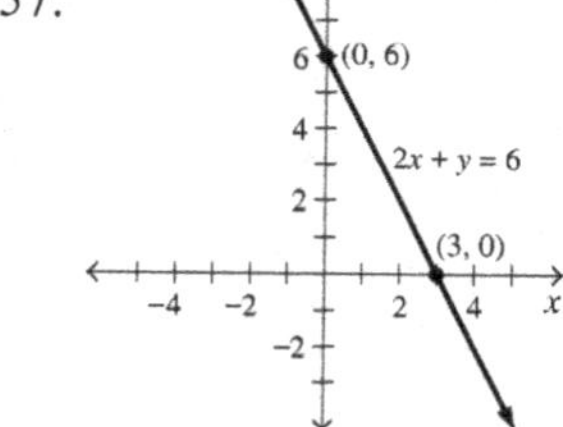

59.

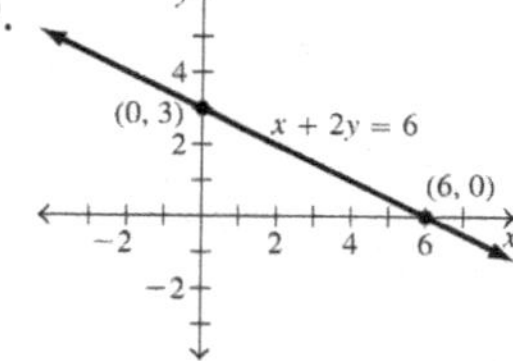

61.

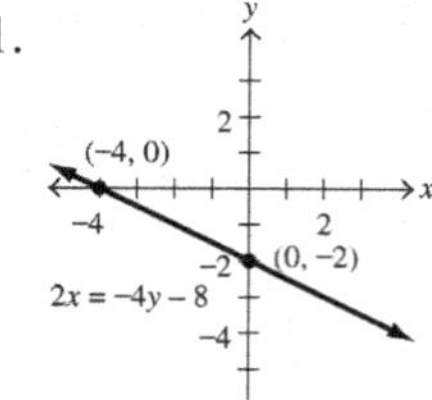

63. $m = \dfrac{9-5}{5-3} = \dfrac{4}{2} = 2$

65. $m = \dfrac{8-1}{8-5} = \dfrac{7}{3}$

67. $m = \dfrac{5-5}{4-1} = \dfrac{0}{3} = 0$

69. $m = \dfrac{3-(-3)}{8-8} = \dfrac{6}{0}$; Undefined

71.

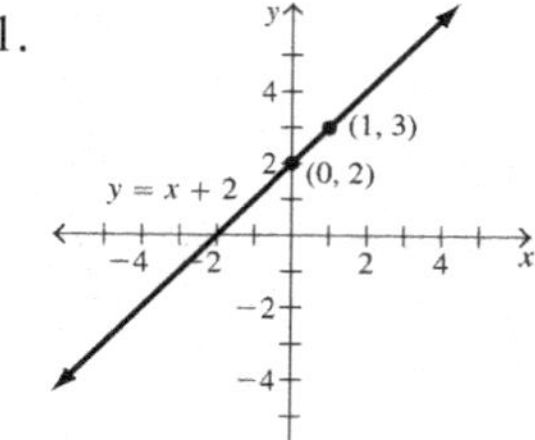

73.

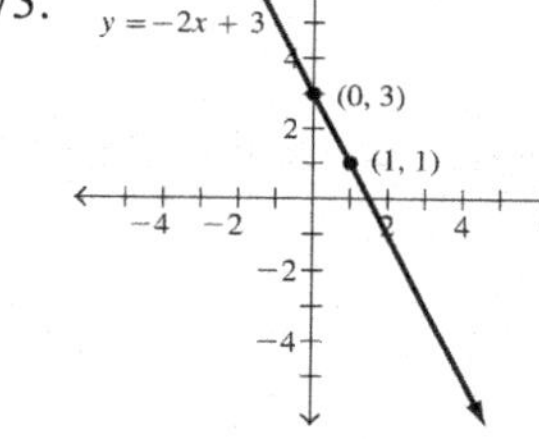

75.

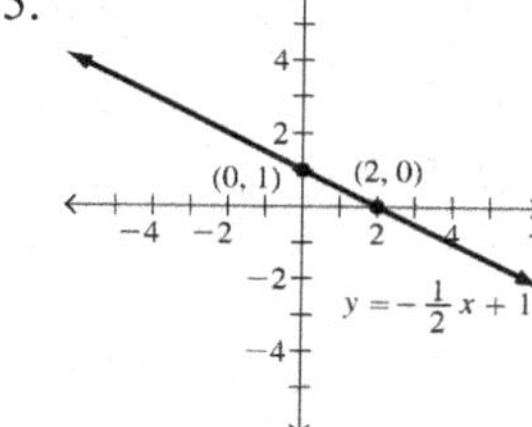

77.

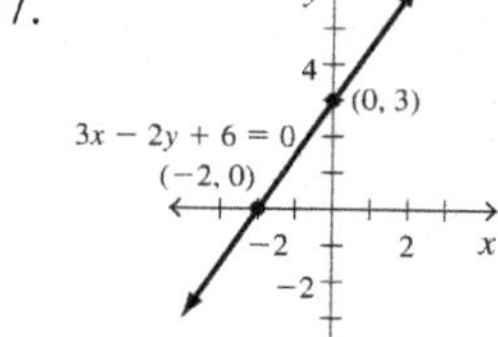

79. The y-intercept is 3; thus $b = 3$. The slope is negative since the graph falls from left to right. The change in y is 3, while the change in x is 4. Thus m, the slope, is $-\frac{3}{4}$. The equation is $y = -\frac{3}{4}x + 3$.

81. The y-intercept is 2; thus $b = 2$. The slope is positive since the graph rises from left to right. The change in y is 3, while the change in x is 1. Thus m, the slope, is $\frac{3}{1} = 3$. The equation is $y = 3x + 2$.

83. Since the line is vertical, its slope is undefined.

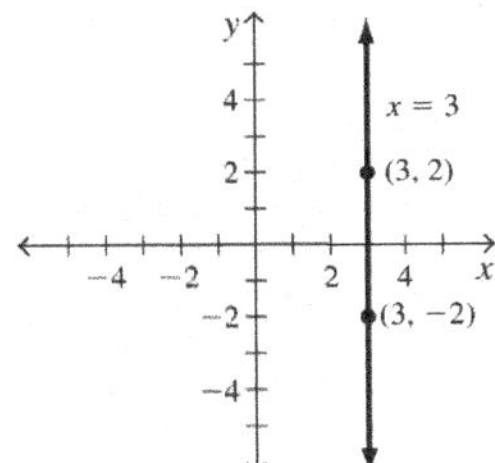

85. Since the line is horizontal, its slope is 0.

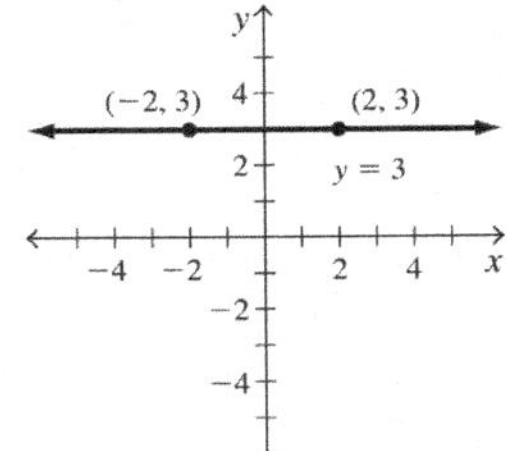

87. The line is horizontal, with a y-intercept of $(0, -2)$. The equation of the line is $y = -2$.

89. The line is vertical, with an x-intercept of $(-3, 0)$. The equation of the line is $x = -3$.

91. a) $D(4, 2)$

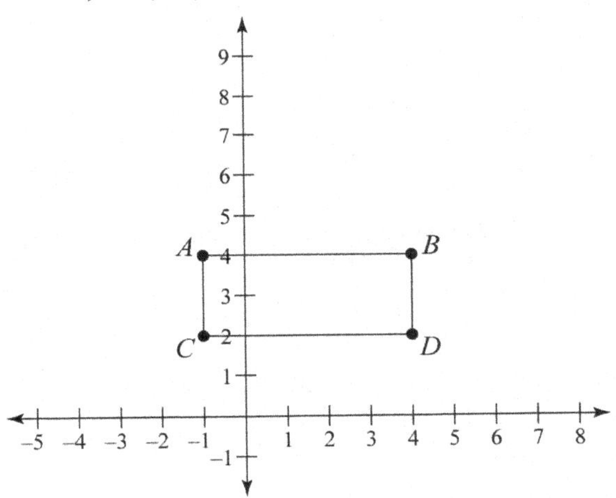

b) $A = lw = 5(2) = 10$ square units

93. The two possible parallelograms are shown below.

$D(7, 2)$

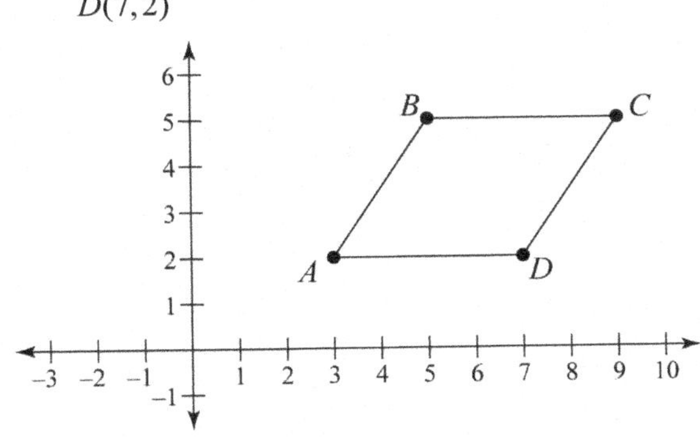

$D(-1, 2)$

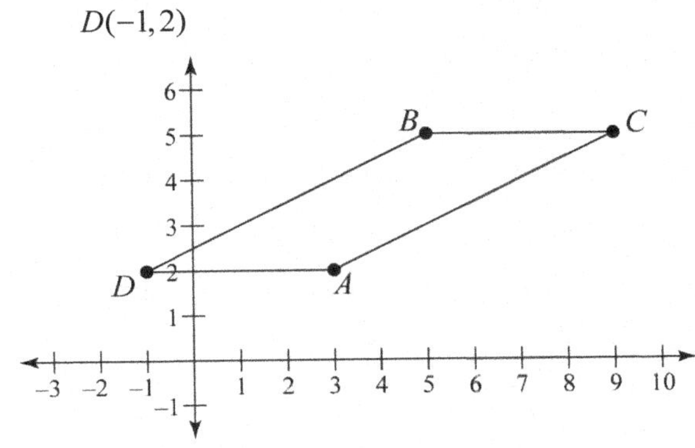

95. For the line joining points P and Q to be parallel to the x-axis, both ordered pairs must have the same y-value, so $b=-3$.

97. For the line joining points P and Q to be parallel to the y-axis, both ordered pairs must have the same x-value.

$$3b-1=8$$
$$3b=9$$
$$b=3$$

99. a)

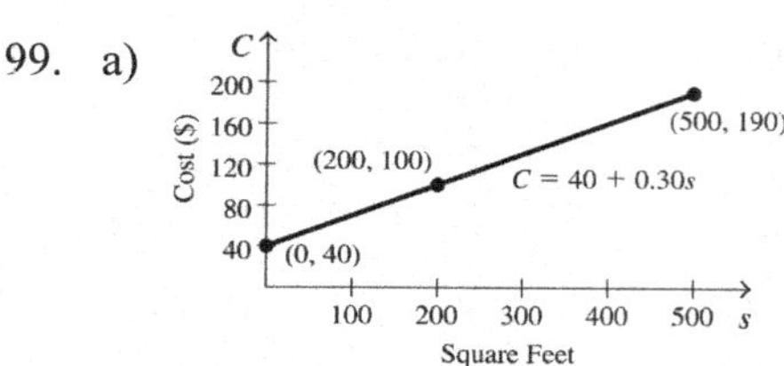

b) \$130

c) $70=40+0.3s$

$30=0.3s$

$s=100$

She hung 100 square feet of wallpaper.

101. a)

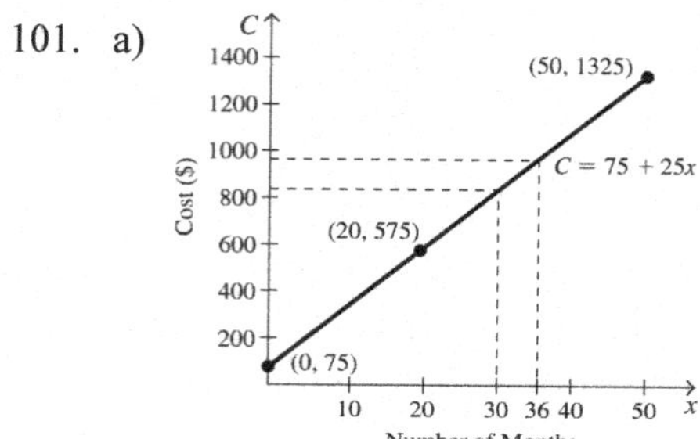

b) \$825

c) 36 months

103. a) $m=\dfrac{19-9}{5-0}=\dfrac{10}{5}=2$

b) $y=2x+9$

c) $y=2(3)+9=6+9=15$ defects

d) $17=2x+9$

$17-9=2x+9-9$

$8=2x$

$x=4$ workers

105. a) $m=\dfrac{34.7-20}{7-0}=2.1$

b) $y=2.1x+20$

c) $y=2.1(6)+20=32.6$

d) $24=2.1x+20$

$4=2.1x$

$x\approx 1.9$

The percentage was 24 in the year $2010+2=2012$.

107. a) Solve the equations for y to put them in slope-intercept form. Then compare the slopes and y-intercepts. If the slopes are equal but the y-intercepts are different, then the lines are parallel.

b)

$$2x-3y=6$$
$$2x-2x-3y=6-2x$$
$$-3y=6-2x$$
$$\frac{-3y}{-3}=\frac{6-2x}{-3}$$
$$y=\frac{2}{3}x-2$$

$$4x=6y+6$$
$$4x-6=6y+6-6$$
$$4x-6=6y$$
$$\frac{6y}{6}=\frac{4x-6}{6}$$
$$y=\frac{2}{3}x-1$$

Since the two equations have the same slope, $m=\frac{2}{3}$, and different y-intercepts, the graphs of the equations are parallel lines.

Section 6.7: Solving Systems of Linear Equations

1. System
3. Inconsistent
5. Dependent
7. One
9. Infinite

11. $y = 4x - 2$ $\quad y = -x + 8$

$6 = 4(2) - 2$ $\quad 6 = -(2) + 8$

$6 = 8 - 2$ $\quad 6 = 6$

$6 = 6$

Therefore, (2, 6) is a solution.

$y = 4x - 2$ $\quad y = -x + 8$

$-10 = 4(3) - 2$ $\quad -10 = -(3) + 8$

$-10 = 12 - 2$ $\quad -10 \neq 5$

$-10 \neq 10$

Therefore, (3, –10) is not a solution.

$y = 4x - 2$ $\quad y = -x + 8$

$7 = 4(1) - 2$ $\quad 7 = -(1) + 8$

$7 = 4 - 2$ $\quad 7 = 7$

$7 \neq 2$

Therefore (1, 7) is not a solution.

13. (0,3)

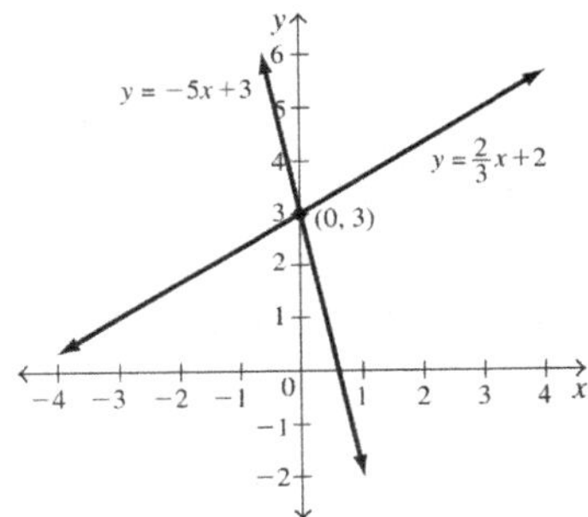

15. (2,1)

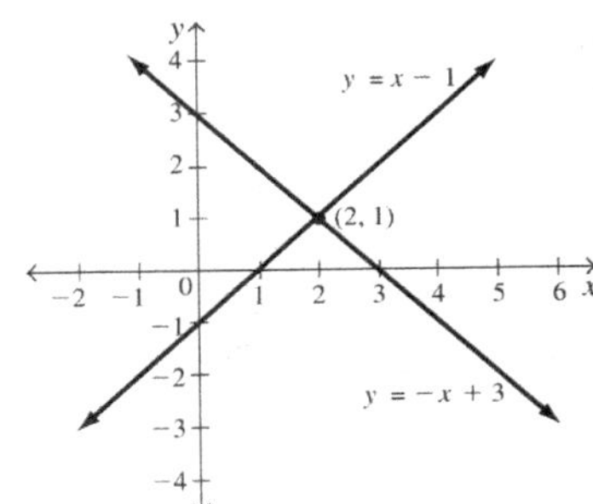

17. (−2,3)

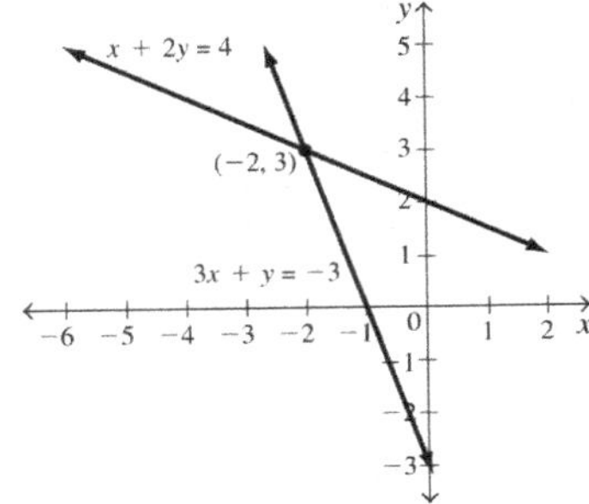

19. No solution; inconsistent system

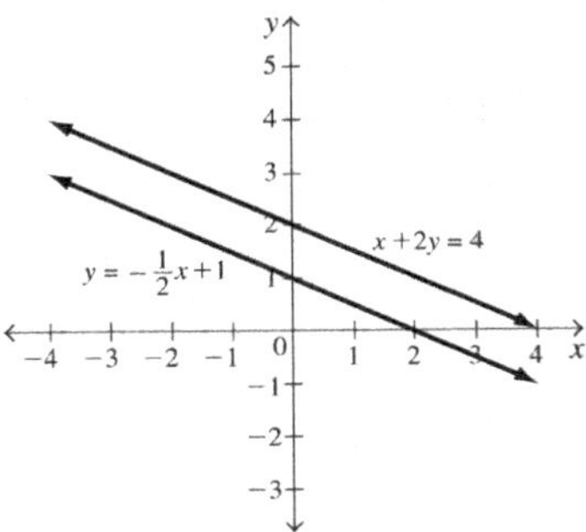

21. Infinite number of solutions; dependent system

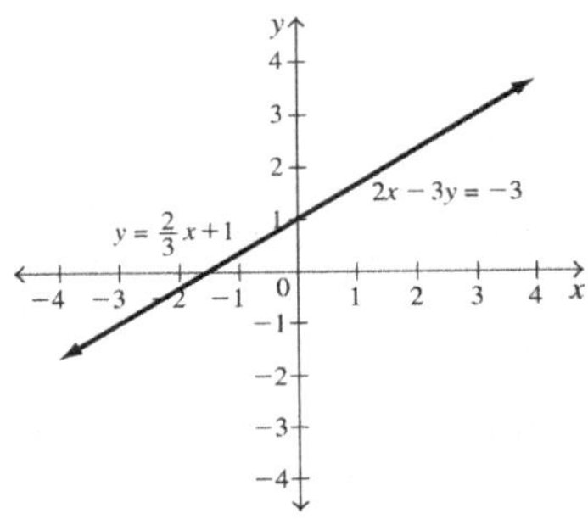

23. $y = x + 9$

$y = -x + 11$

Substitute $x + 9$ for y in the second equation and solve for x.

$(x + 9) = -x + 11$

$2x + 9 = 11$

$2x = 2$

$x = 1$

Substitute 1 for x in the first equation.

$y = (1) + 9 = 10$

The solution is (1, 10). The system is consistent.

25. $6x + 3y = 3$

$x - 3y = 4$

Solve the second equation for x.

$x - 3y = 4$

$x = 3y + 4$

Substitute $3y + 4$ for x in the original first equation and solve for y.

$6(3y + 4) + 3y = 3$

$18y + 24 + 3y = 3$

$21y + 24 = 3$

$21y = -21$

$y = -1$

25. (continued)

Substitute –1 for y in the original second equation and solve for x.

$$x-3(-1)=4$$
$$x+3=4$$
$$x=1$$

The solution is (1, –1). The system is consistent.

27. $y-x=4$

$x-y=3$

Solve the first equation for y.

$$y-x=4$$
$$y=x+4$$

Substitute $x+4$ for y in the original second equation and solve for x.

$$x-(x+4)=3$$
$$4=3 \text{ False}$$

The system has no solution and is inconsistent.

29. $y=-\frac{2}{3}x+3$

$x+3y=6$

Substitute $-\frac{2}{3}x+3$ for y in the second equation and solve for x.

$$x+3\left(-\tfrac{2}{3}x+3\right)=6$$
$$x-2x+9=6$$
$$-x+9=6$$
$$-x=-3$$
$$x=3$$

Substitute 3 for x in the first equation.

$$y=-\tfrac{2}{3}(3)+3=-2+3=1$$

The solution is (3, 1). The system is consistent.

31. $x-y=-2$

$y=3x-4$

Substitute $3x-4$ for y in the first equation and solve for x.

$$x-(3x-4)=-2$$
$$x-3x+4=-2$$
$$-2x+4=-2$$
$$-2x=-6$$
$$x=3$$

Substitute 3 for x in the second equation

$$y=3(3)-4=9-4=5$$

The solution is (3, 5). The system is consistent.

33. $3x+y=15$

$x=-\frac{1}{3}y+5$

Substitute $-\frac{1}{3}y+5$ for x in the first equation and solve for y.

$$3\left(-\tfrac{1}{3}y+5\right)+y=15$$
$$-y+15+y=15$$
$$15=15 \text{ True}$$

The system has an infinite number of solutions and is consistent.

35. $x+y=4$

$$\begin{array}{r} 3x-y=8 \\ \hline 4x=12 \\ x=3 \end{array}$$

Substitute 3 for x in the first equation.

$$(3)+y=4$$
$$y=1$$

The solution is (3, 1). The system is consistent.

37. $4x+3y=-1$

$2x-\ \ y=-13$

Multiply the second equation by –2.

$$\begin{array}{r} 4x+3y=-1 \\ -4x+2y=26 \\ \hline 5y=25 \\ y=5 \end{array}$$

Substitute 5 for y in the original second equation.

$$2x-(5)=-13$$
$$2x=-8$$
$$x=-4$$

The solution is (–4, 5). The system is consistent.

39. $2x+5y=6$

$4x+3y=-2$

Multiply the first equation by –2.

$$\begin{array}{r} -4x-10y=-12 \\ 4x+\ \ 3y=-2 \\ \hline -7y=-14 \\ y=2 \end{array}$$

39. (continued)

Substitute 2 for y in the second original equation.

$$4x+3(2)=-2$$
$$4x+6=-2$$
$$4x=-8$$
$$x=-2$$

The solution is (–2, 2). The system is consistent.

41. $5x-2y=16$

$4x+3y=-1$

Multiply the first equation by 3 and the second equation by 2.

$$\begin{array}{r} 15x-6y=48 \\ \underline{8x+6y=-2} \\ 23x=46 \\ x=2 \end{array}$$

Substitute 2 for x in the second original equation.

$$4(2)+3y=-1$$
$$8+3y=-1$$
$$3y=-9$$
$$y=-3$$

The solution is (2, –3). The system is consistent.

43. $3x-2y=7$

$4y-6x=-14$

Rearrange the equations.

$3x-2y=7$

$-6x+4y=-14$

Multiply the first equation by 2.

$$\begin{array}{r} 6x-4y=14 \\ \underline{-6x+4y=-14} \\ 0=0 \text{ True} \end{array}$$

The system has an infinite number of solutions and is consistent.

45. $2x-7y=1$

$4x-14y=3$

Multiply the first equation by –2.

$$\begin{array}{r} -4x+14y=-2 \\ \underline{4x-14y=3} \\ 0=1 \text{ False} \end{array}$$

The system has no solution and is inconsistent.

47. $3x+2y=8$

$5x-3y=2$

Multiply the first equation by 3 and the second equation by 2.

$$\begin{array}{r} 9x+6y=24 \\ \underline{10x-6y=4} \\ 19x=28 \\ x=\frac{28}{19} \end{array}$$

Use elimination again to solve for y.

Multiply the first equation by 5 and the second equation by –3.

$$\begin{array}{r} 15x+10y=40 \\ \underline{-15x+9y=-6} \\ 19y=34 \\ y=\frac{34}{19} \end{array}$$

The solution is $\left(\frac{28}{19}, \frac{34}{19}\right)$.

The system is consistent.

49. a) Let $x=$ number of miles

$C=$ cost

U-Haul: $C=0.79x+30$

Discount: $C=0.85x+24$

b)

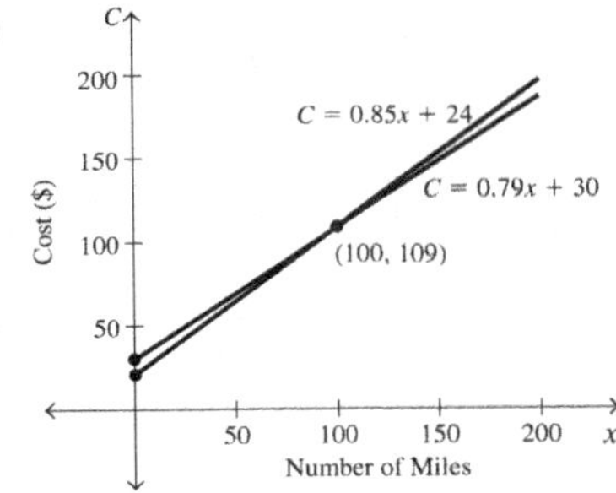

c) The cost graphs intersect when $x=100$, so 100 miles is when the rental costs will be equal.

51. a) $C = 15x + 400$
$R = 25x$

b) 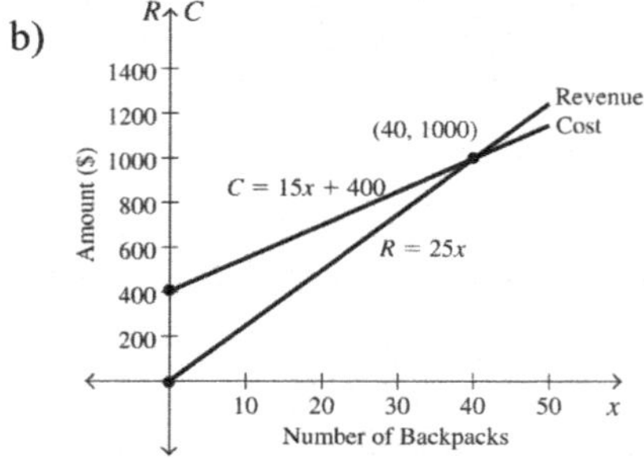

c) The cost and revenue graphs intersect when $x = 40$, so 40 is the number of backpacks Benjamin's must sell to break even.

d) $P = R - C = 25x - (15x + 400) = 10x - 400$

e) $P = 10(30) - 400 = 300 - 400 = -\$100$ (loss)

f) $1000 = 10x - 400$
$10x = 1400$
$x = 140$ backpacks

53. a) $C = 15x + 4050$
$R = 40x$

b) 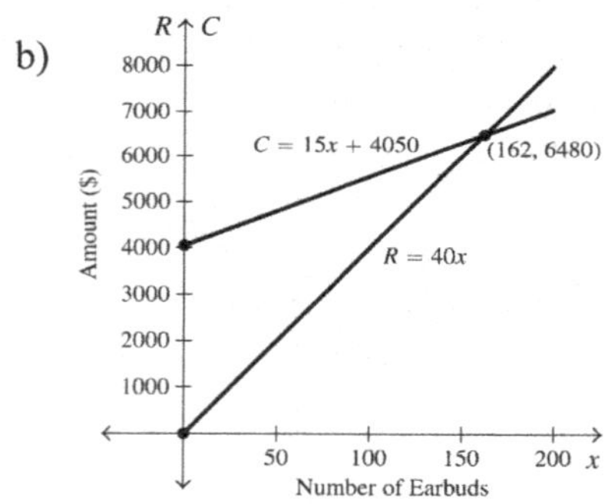

c) The cost and revenue graphs intersect when $x = 162$, so 162 is the number of units the manufacturer must sell to break even.

d) $P = R - C = 40x - (15x + 4050)$
$= 25x - 4050$

e) $P = 25(155) - 4050 = 3875 - 4050 = -\$175$ (loss)

f) $575 = 25x - 4050$
$25x = 4625$
$x = 185$ units

55. Let $x =$ shares of Under Armour stock
$y =$ shares of Hershey Company stock.

$$x + y = 50$$
$$73x + 85y = 4070$$

Multiply the first equation by (–73).

$$-73x - 73y = -3650$$
$$73x + 85y = 4070$$
$$12y = 420$$
$$y = 35$$

Substitute 35 for y in $x + y = 50$.

$$x + 35 = 50$$
$$x = 15$$

He owns 15 shares of Under Armour and 35 shares of Hershey Company.

57. Let $x =$ liters of 10% solution
$15 - x =$ liters of 40% solution

$$x + y = 15 \Rightarrow y = 15 - x$$
$$0.40x + 0.10y = 0.25(15)$$

Substitute $15 - x$ for y in the second equation and solve for x.

$$0.10x + 0.40(15 - x) = 0.25(15)$$
$$0.10x + 6 - 0.40x = 3.75$$
$$-0.30x = -2.25$$
$$x = 7.5$$

Substitute 7.5 for x in $y = 15 - x$.

$$y = 15 - (7.5) = 7.5$$

He should mix 7.5 liters of 10% solution with 7.5 liters of 40% solution.

59. a) Let $y =$ weekly salary
$x =$ weekly sales in dollars
Harbor Sales: $y = 249 + 0.15x$
Pampered Traveler: $y = 300 + 0.12x$

$$249 + 0.15x = 300 + 0.12x$$
$$0.15x = 51 + 0.12x$$
$$0.03x = 51$$
$$x = 1700$$

Nathan's weekly sales volume would need to be \$1700 for the total income from both companies to be equal.

b) Harbor Sales: $s = 249 + 0.15(2500) = \$624$
Pampered Traveler: $s = 300 + 0.12(2500)$
$= \$600$
Harbor Sales would give the greater salary.

61. a) Let $x =$ square feet installed
$y =$ total cost

Home Depot: $y = 2.65x + 468.75$
Hardwood Guys: $y = 3.10x + 412.50$

$$2.65x + 468.75 = 3.10x + 412.50$$
$$2.65x + 56.25 = 3.10x$$
$$56.25 = 0.45x$$
$$x = 125$$

The costs are equal when Roberto has 125 square feet of flooring installed.

b) Home Depot: $y = 2.65(196) + 468.75$
$= \$988.15$

Hardwood Guys: $y = 3.10(196) + 412.50$
$= \$1020.10$

Home Depot will be less expensive.

63. Steve's Garage: $C = 200 + 50x$
Greg's Garage: $C = 375 + 25x$

$$200 + 50x = 375 + 25x$$
$$50x = 175 + 25x$$
$$25x = 175$$
$$x = 7$$

The total cost would be the same for both garages for 7 hours of labor.

65. a) Two lines with different slopes are not parallel, nor are they on the same line, and therefore have exactly one point of intersection giving one solution.

b) Two lines with the same slope and different y-intercepts are distinct parallel lines so they system will have no solution.

c) Two lines with the same slopes and y-intercepts have infinitely many solutions, each point on the line.

67. Using $x = \frac{1}{u}$ and $y = \frac{1}{v}$, the system of equations becomes the following.

$$x + 2y = 8$$
$$3x - y = 3$$

Multiply the second equation by 2.

$$x + 2y = 8$$
$$\underline{6x - 2y = 6}$$
$$7x = 14$$
$$x = 2$$

Substitute 2 for x in the first equation.

$$2 + 2y = 8$$
$$2y = 6$$
$$y = 3$$

$$x = \frac{1}{u} \Rightarrow u = \frac{1}{x}, \text{ so } u = \frac{1}{2}.$$
$$y = \frac{1}{v} \Rightarrow v = \frac{1}{y}, \text{ so } v = \frac{1}{3}.$$

The solution is $\left(\frac{1}{2}, \frac{1}{3}\right)$.

69. a) First equation: $(2) + (1) + (4) = 7$
$7 = 7$

Second equation: $(2) - (1) + 2(4) = 9$
$9 = 9$

Third equation: $-(2) + 2(1) + (4) = 4$
$4 = 4$

$(2, 1, 4)$ is a solution.

b) Add the original first and third equations to eliminate x.

$$x + y + z = 7$$
$$\underline{-x + 2y + z = 4}$$
$$3y + 2z = 11$$

Add the original second and third equations to eliminate x.

$$x - y + 2z = 9$$
$$\underline{-x + 2y + z = 4}$$
$$y + 3z = 13$$

69. (continued)

Use the new equations to create a system of equations.

$$3y+2z=11$$
$$y+3z=13$$

Multiply the new second equation by (–3) to eliminate y.

$$3y+2z=11$$
$$\underline{-3y-9z=-39}$$
$$-7z=-28$$
$$z=4$$

Substitute 4 for z in the new second equation.

$$y+3(4)=13$$
$$y+12=13$$
$$y=1$$

Substitute 1 for y and 4 for z in the original first equation.

$$x+(1)+(4)=7$$
$$x+5=7$$
$$x=2$$

The solution is $(2,1,4)$.

71. Answers will vary. One possible solution is given below.

$$y=2x+2$$
$$\tfrac{1}{2}y=x+1$$

If we multiply the second equation by 2, we get the equation $y=2x+2,$ which is the same as the first equation. Two lines that lie on top of one another have an infinite number of solutions.

73. a) Answers will vary.
b) Answers will vary.
c) Answers will vary.
d) Answers will vary.

Section 6.8: Linear Inequalities in Two Variables and Systems of Linear Inequalities

1. Half-plane
3. Solid
5. Ordered
7. Feasible
9. Objective

11. a) $3x-4y>12$

$$3(0)-4(0)>12$$
$$0>12 \text{ False}$$

$(0,0)$ is not a solution.

b) $2x+3y\le 6$

$$2(0)+3(0)\le 6$$
$$0\le 6 \text{ True}$$

$(0,0)$ is a solution.

c) $y\ge \frac{2}{3}x-5$

$$(0)\ge \frac{2}{3}(0)-5$$
$$0\ge -5 \text{ True}$$

$(0,0)$ is a solution.

d) $y<-\frac{5}{4}x-3$

$$0<-\frac{5}{4}(0)-3$$
$$0<-3 \text{ False}$$

$(0,0)$ is not a solution.

13. Graph $y = x+1$. Since the original statement is greater than, a dashed line is drawn. Since the point (0, 0) does not satisfy the inequality $y > x+1$, all points in the half-plane above the line $y = x+1$ are in the solution set.

Graph for Exercise 13

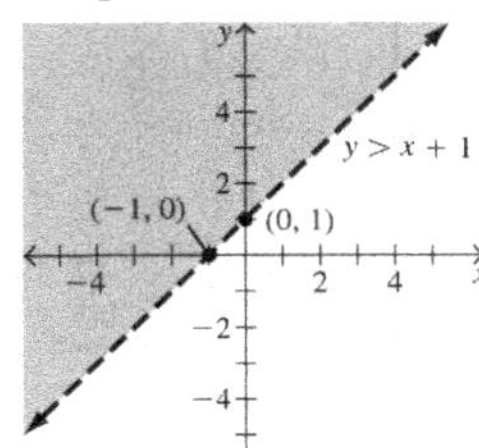

Graph for Exercise 15

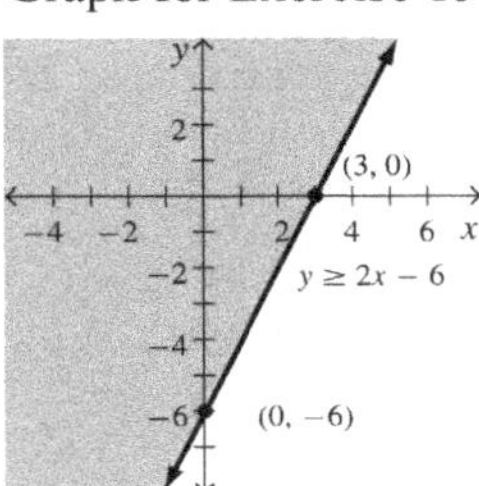

15. Graph $y = 2x-6$. Since the original statement is greater than or equal to, a solid line is drawn. Since the point (0, 0) satisfies the inequality $y \geq 2x-6$, all points on the line and in the half-plane above the line $y = 2x-6$ are in the solution set.

17. Graph $2x-3y=6$. Since the original statement is strictly greater than, a dashed line is drawn. Since the point (0, 0) does not satisfy the inequality $2x-3y>6$, all points in the half-plane below the line $2x-3y=6$ are in the solution set.

Graph for Exercise 17

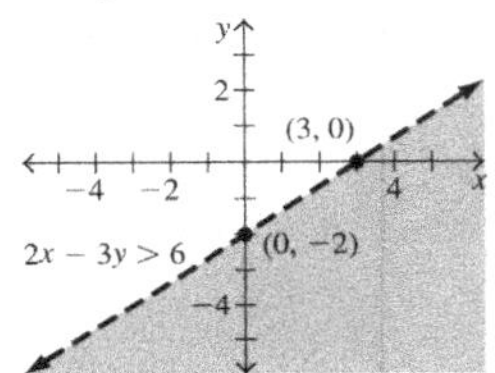

Graph for Exercise 19

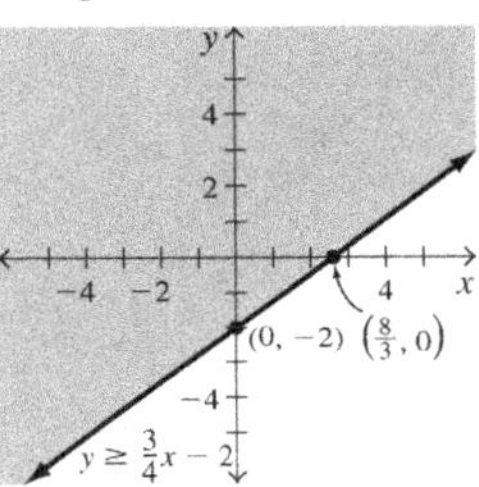

19. Graph $y = \frac{3}{4}x-2$. Since the original statement is less than or equal to, a solid line is drawn. Since the point (0, 0) satisfies the inequality $y \geq \frac{3}{4}x-2$, all points in the half-plane above the line $y = \frac{3}{4}x-2$ are in the solution set.

21. Graph $3x+2y=6$. Since the original statement is strictly less than, a dashed line is drawn. Since the point (0, 0) satisfies the inequality $3x+2y<6$, all points in the half-plane to the left of the line $3x+2y=6$ are in the solution set.

Graph for Exercise 21

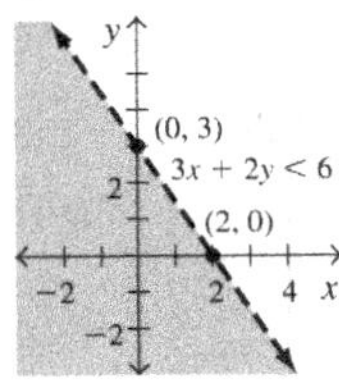

Graph for Exercise 23

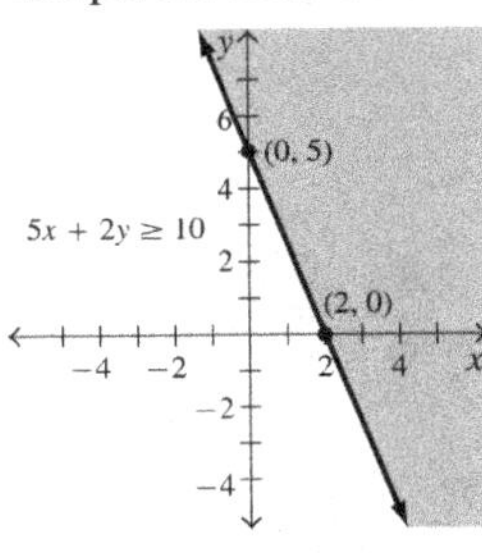

23. Graph $5x+2y=10$. Since the original statement is greater than or equal to, a solid line is drawn. Since the point (0, 0) does not satisfy the inequality $5x+2y \geq 10$, all points on the line and in the half-plane above the line $5x+2y=10$ are in the solution set.

25. Graph $y = -2x + 1$. Since the original statement is greater than or equal to, a solid line is drawn. Since the point (0, 0) does not satisfy the inequality $y \geq -2x + 1$, all points on the line and in the half-plane above the line $y = -2x + 1$ are in the solution set.

Graph for Exercise 25

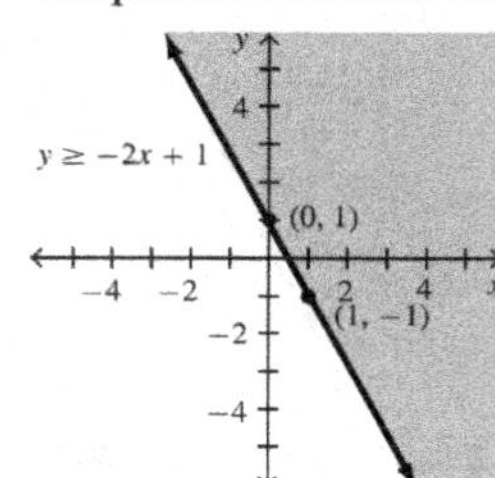

Graph for Exercise 27

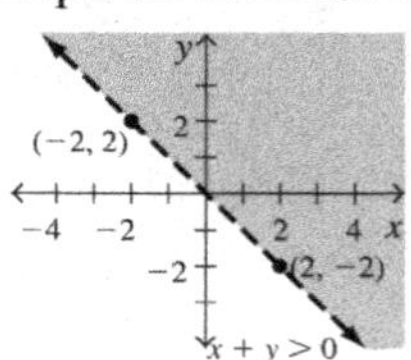

27. Graph $x + y = 0$. Since the original statement is strictly greater than, a dashed line is drawn. Since the point (1, 1) satisfies the inequality $x + y > 0$, all points in the half-plane above the line $x + y = 0$ are in the solution set.

29.

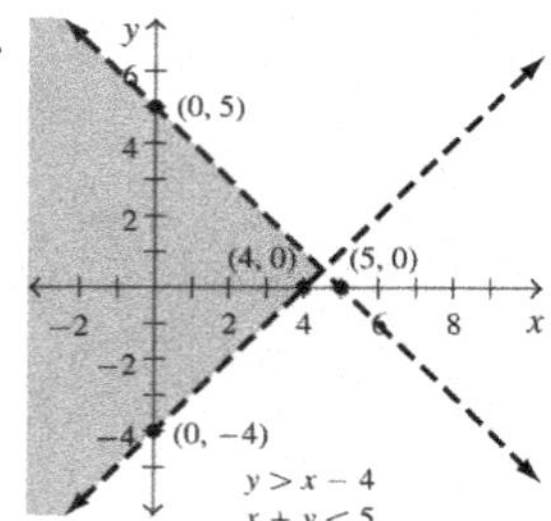

31.

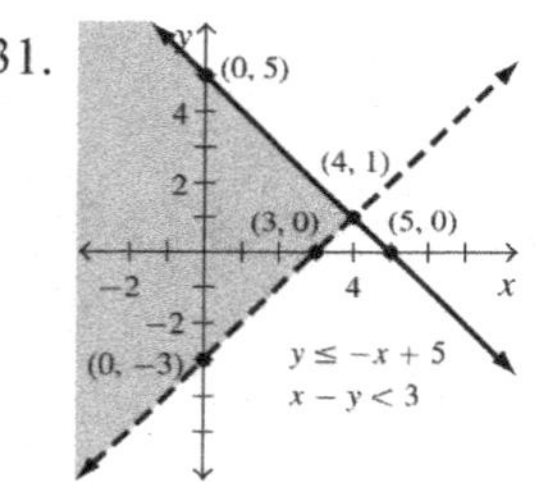

33.

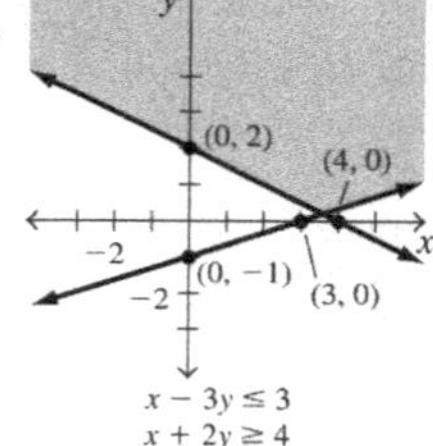

35.

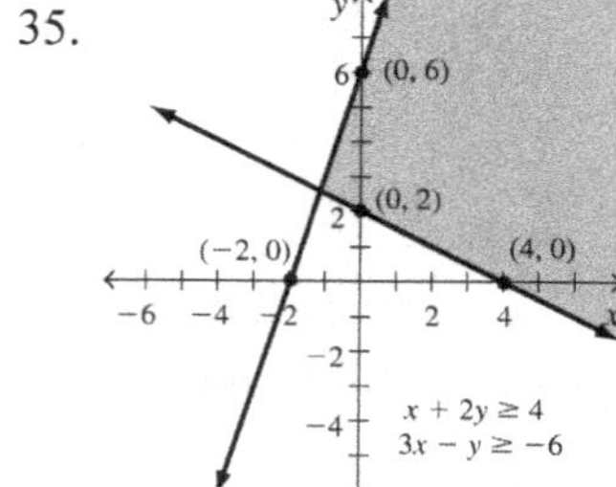

37.

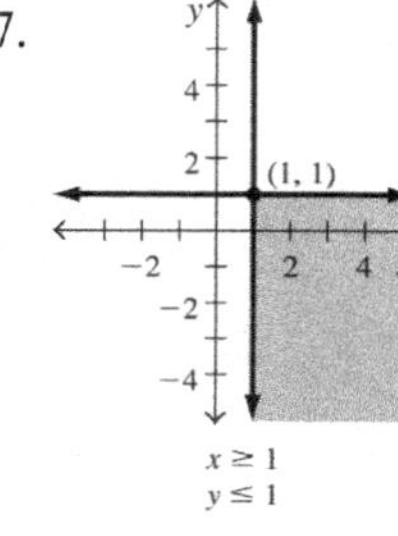

39. At $(0,0)$, $K = 6(0) + 4(0) = 0$
At $(0,4)$, $K = 6(0) + 4(4) = 16$
At $(2,3)$, $K = 6(2) + 4(3) = 24$
At $(5,0)$, $K = 6(5) + 4(0) = 30$

The maximum value is 30 at $(5,0)$; the minimum value is 0 at $(0,0)$.

41. At $(10,20)$, $K = 2(10) + 3(20) = 80$
At $(10,40)$, $K = 2(10) + 3(40) = 140$
A $t(50,30)$, $K = 2(50) + 3(30) = 190$
At $(50,10)$, $K = 2(50) + 3(10) = 130$
At $(20,10)$, $K = 2(20) + 3(10) = 70$

The maximum value is 190 at $(50,30)$; the minimum value is 70 at $(20,10)$.

43. a)

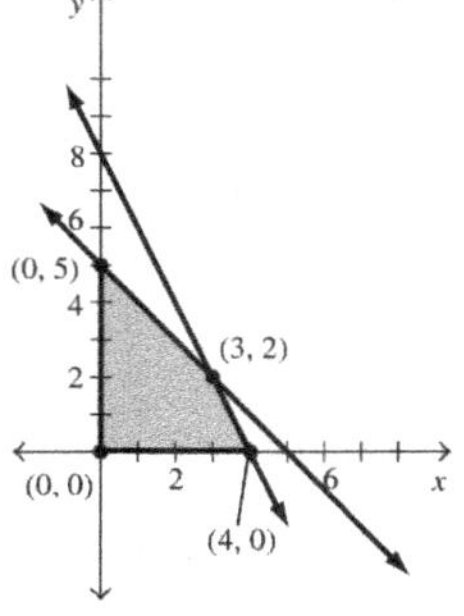

b) $P = 4x + 3y$

At $(0,0)$, $P = 4(0) + 3(0) = 0$

At $(0,5)$, $P = 4(0) + 3(5) = 15$

At $(3,2)$, $P = 4(3) + 3(2) = 18$

At $(4,0)$, $P = 4(4) + 3(0) = 16$

The minimum value is 0 at $(0,0)$; the maximum value is 18 at $(3,2)$.

45. a)

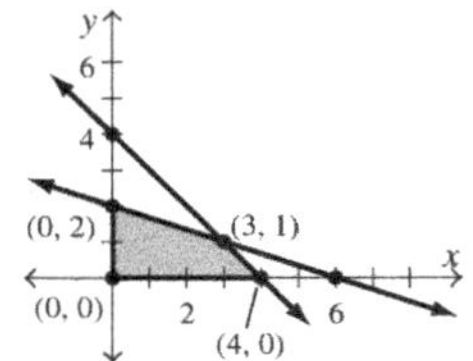

b) $P = 7x + 6y$

At $(0,0)$, $P = 7(0) + 6(0) = 0$

At $(0,2)$, $P = 7(0) + 6(2) = 12$

At $(3,1)$, $P = 7(3) + 6(1) = 27$

At $(4,0)$, $P = 7(4) + 6(0) = 28$

The minimum value is 0 at $(0,0)$; the maximum value is 28 at $(4,0)$.

47. a)

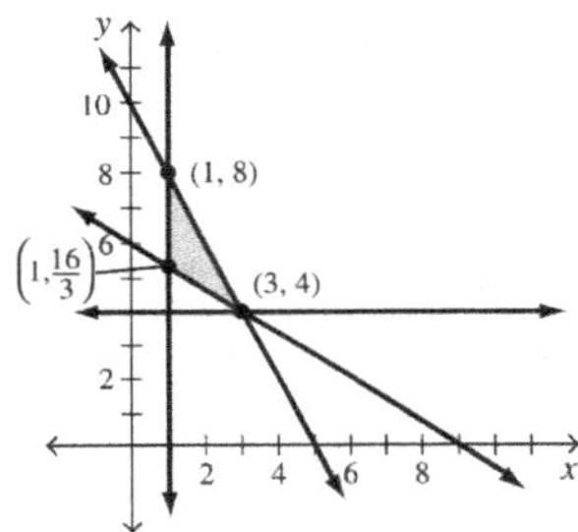
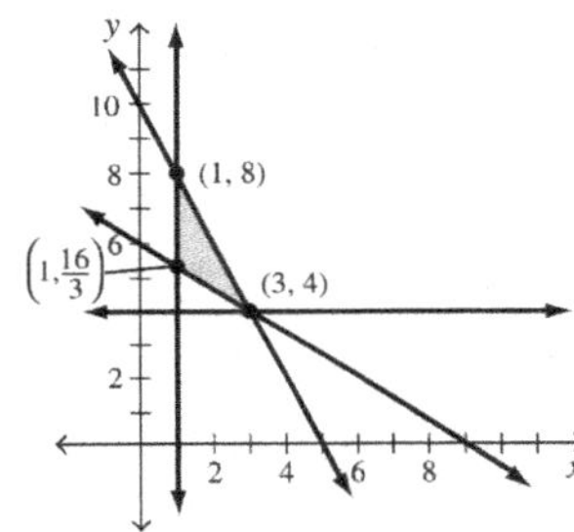

b) $P = 2.20x + 1.65y$

At $(3,4)$, $P = 2.20(3) + 1.65(4) = 13.2$

At $(1,8)$, $P = 2.20(1) + 1.65(8) = 15.4$

At $\left(1, \frac{16}{3}\right)$, $P = 2.20(1) + 1.65\left(\frac{16}{3}\right) = 11$

The maximum value is 15.4 at $(1,8)$; the minimum value is 11 at $\left(1, \frac{16}{3}\right)$.

49. a) $x + y < 500$

$x \geq 150$

$y \geq 150$

b)

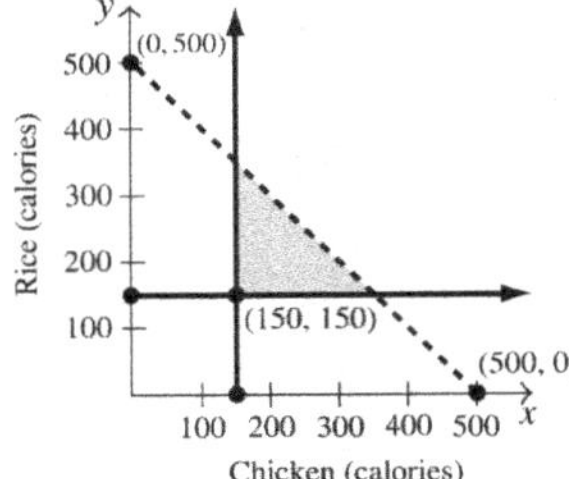

c) One example is (220, 220): $\frac{220}{180}(3) \approx 3.7$ oz of chicken; $\frac{220}{200}(8) = 8.8$ oz of rice

51. a) Let $x =$ number of skateboards
$y =$ number of in-line skates

$$x + y \le 20$$
$$x \ge 3$$
$$x \le 6$$
$$y \ge 2$$

b) $P = 25x + 20y$

c)

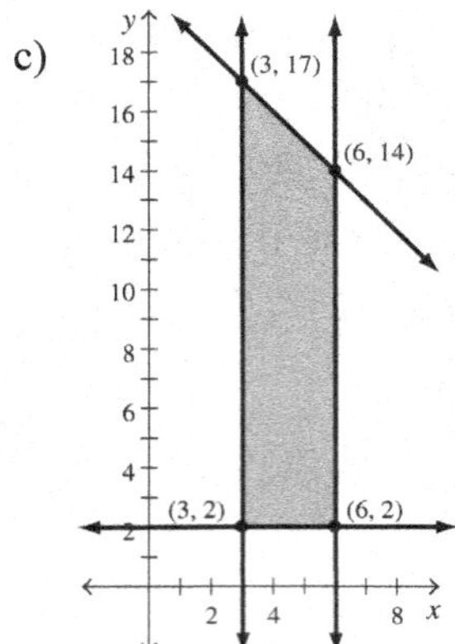

d) $(3,2),\ (3,17),\ (6,14),\ (6,2)$

e) At $(3,2)$, $P = 25(3) + 20(2) = 115$
At $(3,17)$, $P = 25(3) + 20(17) = 415$
At $(6,14)$, $P = 25(6) + 20(14) = 430$
At $(6,2)$, $P = 25(6) + 20(2) = 190$

Six skateboards and 14 pairs of in-line skates

f) The maximum profit is \$430.

53. a) Let $x =$ hours machine I operates
$y =$ hours machine II operates

$$3x + 4y \ge 60$$
$$10x + 5y \ge 100$$
$$x \ge 0$$
$$y \ge 0$$

b) $C = 28x + 33y$

c)

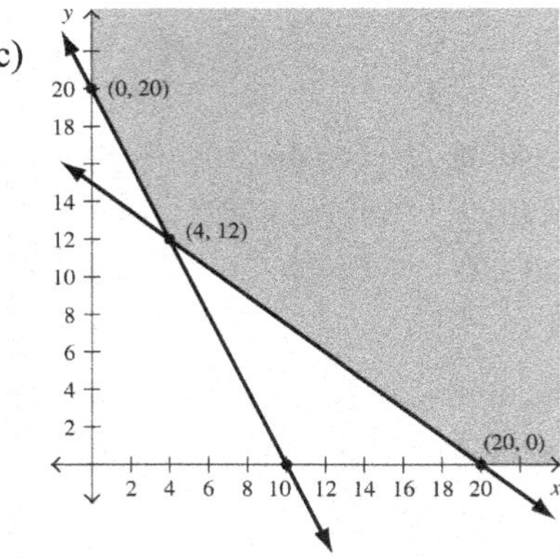

d) $(0,20),\ (20,0),\ (4,12)$

e) At $(0,20)$, $C = 28(0) + 33(20) = 660$
At $(20,0)$, $C = 28(20) + 33(0) = 560$
At $(4,12)$, $C = 28(4) + 33(12) = 508$

4 hours for Machine I and 12 hours for Machine II

f) The minimum cost is \$508.

55. Let $x =$ pounds of all-beef hot dogs
$y =$ pounds of regular hot dogs

$$x + \frac{1}{2}y \le 200$$
$$\frac{1}{2}y \le 150$$
$$x \ge 0$$
$$y \ge 0$$

$P = 0.30y + 0.40x$

At $(0,0)$, $P = 0.30(0) + 0.40(0) = 0$
At $(0,300)$, $P = 0.30(0) + 0.40(300) = 120$
At $(50,300)$, $P = 0.30(50) + 0.40(300) = 135$
At $(200,0)$, $P = 0.30(200) + 0.40(0) = 60$

The manufacturer should make 50 lb of the all-beef hot dogs and 300 lb of the regular hot dogs for a profit of \$110.

57. a) No, if the lines are parallel there may not be a solution to the system.

b) Answers may vary.

$$x + y > 4$$
$$x + y < 1$$

This system has no solution.

59. Inequalities (a), (b), and (d) have the same graph, since they are equivalent.

a)
$$\begin{aligned} 2x - y &< 8 \\ 2x - 2x - y &< -2x + 8 \\ -y &< -2x + 8 \\ \frac{-y}{-1} &> \frac{-2x}{-1} + \frac{2}{-1} \\ y &> 2x - 8 \end{aligned}$$

b)
$$\begin{aligned} -2x + y &> -8 \\ -2x + 2x + y &> 2x - 8 \\ y &> 2x - 8 \end{aligned}$$

c)
$$\begin{aligned} 2x - 4y &< 16 \\ 2x - 2x - 4y &< -2x + 16 \\ -4y &< -2x + 16 \\ \frac{-4y}{-4} &> \frac{-2x}{-4} + \frac{16}{-4} \\ y &> \frac{1}{2}x - 4 \end{aligned}$$

d) $y > 2x - 8$

Section 6.9: Solving Quadratic Equations by Using Factoring and by Using the Quadratic Formula

1. Binomial

3. FOIL

5. Quadratic

7. $(x+3)(x-11) = x^2 - 11x + 3x - 33$
$\quad = x^2 - 8x - 33$

9. $(2x+3)(3x-1) = 6x^2 - 2x + 9x - 3$
$\quad = 6x^2 + 7x - 3$

11. $(6x-7)(8x+9) = 48x^2 + 54x - 56x - 63$
$\quad = 48x^2 - 2x - 63$

13. $x^2 + 5x + 6 = (x+2)(x+3)$

15. $x^2 - x - 6 = (x-3)(x+2)$

17. $x^2 + 3x - 10 = (x+5)(x-2)$

19. $x^2 - 2x - 3 = (x+1)(x-3)$

21. $x^2 - 9x + 18 = (x-6)(x-3)$

23. $x^2 + 3x - 28 = (x+7)(x-4)$

25. $2x^2 + 7x + 3 = (2x+1)(x+3)$

27. $3x^2 + x - 2 = (3x-2)(x+1)$

29. $3x^2 - 17x + 10 = (3x-2)(x-5)$

31. $4x^2 + 11x + 6 = (4x+3)(x+2)$

33. $4x^2 - 11x + 6 = (4x-3)(x-2)$

35. $8x^2 - 2x - 3 = (4x-3)(2x+1)$

37. $(x-6)(5x-4) = 0$
$x - 6 = 0$ or $5x - 4 = 0$
$x = 6$ $\quad 5x = 4$
$\quad x = \frac{4}{5}$

39. $(3x+4)(2x-1) = 0$
$3x + 4 = 0$ or $2x - 1 = 0$
$3x = -4$ $\quad 2x = 1$
$x = -\frac{4}{3}$ $\quad x = \frac{1}{2}$

41. $x^2 + 8x + 15 = 0$
$(x+5)(x+3) = 0$
$x + 5 = 0$ or $x + 3 = 0$
$x = -5$ $\quad x = -3$

43. $x^2 - 8x + 7 = 0$
$(x-1)(x-7) = 0$
$x - 1 = 0$ or $x - 7 = 0$
$x = 1$ $\quad x = 7$

45. $x^2 - 15 = 2x$
$x^2 - 2x - 15 = 0$
$(x+3)(x-5) = 0$
$x + 3 = 0$ or $x - 5 = 0$
$x = -3$ $\quad x = 5$

47. $x^2 = 4x - 3$
$x^2 - 4x + 3 = 0$
$(x-1)(x-3) = 0$
$x - 1 = 0$ or $x - 3 = 0$
$x = 1$ $\quad x = 3$

49.
$$6x^2+19x+15=0$$
$$(2x+3)(3x+5)=0$$
$$2x+3=0 \quad \text{or} \quad 3x+5=0$$
$$2x=-3 \qquad 3x=-5$$
$$x=-\frac{3}{2} \qquad x=-\frac{5}{3}$$

51.
$$3x^2+10x=8$$
$$3x^2+10x-8=0$$
$$(x+4)(3x-2)=0$$
$$x+4=0 \quad \text{or} \quad 3x-2=0$$
$$x=-4 \qquad 3x=2$$
$$x=\frac{2}{3}$$

53.
$$3x^2=-5x+2$$
$$3x^2+5x-2=0$$
$$(x+2)(3x-1)=0$$
$$x+2=0 \quad \text{or} \quad 3x-1=0$$
$$x=-2 \qquad 3x=1$$
$$x=\frac{1}{3}$$

55.
$$3x^2=-5x-2$$
$$3x^2+5x+2=0$$
$$(x+1)(3x+2)=0$$
$$x+1=0 \quad \text{or} \quad 3x+2=0$$
$$x=-1 \qquad 3x=-2$$
$$x=-\frac{2}{3}$$

57.
$$20x^2=15-3x$$
$$20x^2+13x-15=0$$
$$(4x+5)(5x-3)=0$$
$$4x+5=0 \quad \text{or} \quad 5x-3=0$$
$$4x=-5 \qquad 5x=3$$
$$x=-\frac{5}{4} \qquad x=\frac{3}{5}$$

59. $x^2+2x-3=0;\ \ a=1,\ \ b=2,\ \ c=-3$
$$x=\frac{-b\pm\sqrt{b^2-4ac}}{2a}=\frac{-(2)\pm\sqrt{(2)^2-4(1)(-3)}}{2(1)}=\frac{-2\pm\sqrt{4+12}}{2}=\frac{-2\pm\sqrt{16}}{2}=\frac{-2\pm4}{2}$$
$$x=\frac{-2-4}{2}=-3 \text{ or } x=\frac{-2+4}{2}=1$$

61. $x^2-3x-18=0;\ \ a=1,\ \ b=-3,\ \ c=-18$
$$x=\frac{-b\pm\sqrt{b^2-4ac}}{2a}=\frac{-(-3)\pm\sqrt{(-3)^2-4(1)(-18)}}{2(1)}=\frac{3\pm\sqrt{9+72}}{2}=\frac{3\pm\sqrt{81}}{2}=\frac{3\pm9}{2}$$
$$x=\frac{3-9}{2}=-3 \text{ or } x=\frac{3+9}{2}=6$$

63. $x^2-4x+2=0;\ \ a=1,\ \ b=-4,\ \ c=2$
$$x=\frac{-b\pm\sqrt{b^2-4ac}}{2a}=\frac{-(-4)\pm\sqrt{(-4)^2-4(1)(2)}}{2(1)}=\frac{4\pm\sqrt{16-8}}{2}=\frac{4\pm\sqrt{8}}{2}=\frac{4\pm2\sqrt{2}}{2}=\frac{2\left(2\pm\sqrt{2}\right)}{2}$$
$$x=2\pm\sqrt{2}$$

65. $x^2-2x+3=0;\ \ a=1,\ \ b=-2,\ \ c=3$
$$x=\frac{-b\pm\sqrt{b^2-4ac}}{2a}=\frac{-(-2)\pm\sqrt{(-2)^2-4(1)(3)}}{2(1)}=\frac{2\pm\sqrt{4-12}}{2}=\frac{2\pm\sqrt{-8}}{2}$$
No real solution

67. $3x^2 + 9x + 5 = 0;\ a = 3,\ b = 9,\ c = 5$

$$x = \frac{-b \pm \sqrt{b^2 - 4ac}}{2a} = \frac{-(9) \pm \sqrt{(9)^2 - 4(3)(5)}}{2(3)} = \frac{-9 \pm \sqrt{81 - 60}}{6} = \frac{-9 \pm \sqrt{21}}{6}$$

69. $3x^2 - 8x + 1 = 0;\ a = 3,\ b = -8,\ c = 1$

$$x = \frac{-b \pm \sqrt{b^2 - 4ac}}{2a} = \frac{-(-8) \pm \sqrt{(-8)^2 - 4(3)(1)}}{2(3)} = \frac{8 \pm \sqrt{64 - 12}}{6} = \frac{8 \pm \sqrt{52}}{6} = \frac{8 \pm 2\sqrt{13}}{6} = \frac{2\left(4 \pm \sqrt{13}\right)}{6} = \frac{4 \pm \sqrt{13}}{3}$$

71. $4x^2 - 5x - 3 = 0;\ a = 4,\ b = -5,\ c = -3$

$$x = \frac{-b \pm \sqrt{b^2 - 4ac}}{2a} = \frac{-(-5) \pm \sqrt{(-5)^2 - 4(4)(-3)}}{2(4)} = \frac{5 \pm \sqrt{25 + 48}}{8} = \frac{5 \pm \sqrt{73}}{8}$$

73. $2x^2 + 7x + 5 = 0;\ a = 2,\ b = 7,\ c = 5$

$$x = \frac{-b \pm \sqrt{b^2 - 4ac}}{2a} = \frac{-(7) \pm \sqrt{(7)^2 - 4(2)(5)}}{2(2)} = \frac{-7 \pm \sqrt{49 - 40}}{4} = \frac{-7 \pm \sqrt{9}}{4} = \frac{-7 \pm 3}{4}$$

$$x = \frac{-7 - 3}{4} = -\frac{5}{2} \text{ or } x = \frac{-7 + 3}{4} = -1$$

75. $3x^2 = 9x - 5$

$3x^2 - 9x + 5 = 0; a = 3, b = -9, c = 5$

$$x = \frac{-b \pm \sqrt{b^2 - 4ac}}{2a} = \frac{-(-9) \pm \sqrt{(-9)^2 - 4(3)(5)}}{2(3)} = \frac{9 \pm \sqrt{81 - 60}}{6} = \frac{9 \pm \sqrt{21}}{6}$$

77 $4x^2 + 6x = -5$

$4x^2 + 6x + 5 = 0; a = 4, b = 6, c = 5$

$$x = \frac{-b \pm \sqrt{b^2 - 4ac}}{2a} = \frac{-(6) \pm \sqrt{(6)^2 - 4(4)(5)}}{2(4)} = \frac{-6 \pm \sqrt{36 - 80}}{8} = \frac{-6 \pm \sqrt{-44}}{8}\text{; No real solution}$$

79. $h = 2(0.60)^2 + 80(0.60) + 40 = 0.72 + 48 + 40 = 88.72,$ or 88.72 minutes

81. Let $x =$ width of strip of grass
$20 - 2x =$ width of flower garden
$30 - 2x =$ length of flower garden

Area of backyard: $lw = 30(20) = 600\text{ m}^2$

Area of flower garden: $lw = (30 - 2x)(20 - 2x)$

Area of grass: $600 - (30 - 2x)(20 - 2x)$

$600 - (30 - 2x)(20 - 2x) = 336$

$600 - \left(600 - 100x + 4x^2\right) = 336$

$600 - 600 + 100x - 4x^2 = 336$

$-4x^2 + 100x = 336$

$4x^2 - 100x + 336 = 0$

$x^2 - 25x + 84 = 0$

$(x - 21)(x - 4) = 0$

$x - 21 = 0$ or $x - 4 = 0$

$x = 21$ $\quad x = 4$

$x \neq 21$ since the width of the backyard is 20 m.

Width of flower garden $= 20 - 2x = 20 - 2(4) = 20 - 8 = 12$ m

Length of flower garden $= 30 - 2x = 30 - 2(4) = 30 - 8 = 22$ m

83. $$-16t^2+128t=256$$
$$-16t^2+128t-256=0$$
$$16(t-4)(t-4)=0$$
$$t-4=0$$
$$t=4$$
At 4 seconds, the ball will be 256 ft above ground.

85. a) Since the equation is equal to 6 and not 0, the zero-factor property cannot be used.

b) $(x-4)(x-7)=6$

$x^2-11x+28=6$

$x^2-11x+22=0$

$a=1,\ b=-11,\ c=22;\ x=\dfrac{-(-11)\pm\sqrt{(-11)^2-4(1)(22)}}{2(1)}=\dfrac{11\pm\sqrt{121-88}}{2}=\dfrac{11\pm\sqrt{33}}{2}$

87. Other answers are possible.

$(x-(-1))(x-3)=0$

$(x+1)(x-3)=0$

$x^2-2x-3=0$

Section 6.10: Functions and Their Graphs

1. Relation
3. Domain
5. Upward
7. $x=-\dfrac{b}{2a}$
9. This is a function since each value of x is paired with a unique value of y.
11. This is not a function since $x=2$ is paired with four different values of y.
13. This is a function since each value of x is paired with a unique value of y.
15. This is not a function since $x=1$ is paired with four different values of y.
17. This is a function since each value of x is paired with a unique value of y.
Domain: $x=-2,-1,1,2,3$; Range: $y=-1,1,2,3$
19. This is a function since each vertical line intersects the graph at only one point.
Domain: all real numbers; Range: all real numbers
21. This is not a function since $x=-1$ is not paired with a unique value of y.
23. This is a function since each vertical line intersects the graph at only one point.
Domain: $\mathbb{R}$; Range: $y\geq -4$
25. This is not a function since it is possible to draw a vertical line that intersects the graph at more than one point.
27. This is a function since each vertical line intersects the graph at only one point.
Domain: $\mathbb{R}$; Range: $\mathbb{R}$

31. $f(x)=x+3$

$f(5)=5+3=8$

33. $f(x)=-3x+3$

$f(-1)=-3(-1)+3=3+3=6$

35. $f(x)=x^2-3x+1$

$f(4)=(4)^2-3(4)+1=16-12+1=5$

37. $f(x)=-x^2-2x+1$

$f(4)=-(4)^2-2(4)+1=-16-8+1=-23$

39. $f(x) = 4x^2 - 6x - 9$

$f(-3) = 4(-3)^2 - 6(-3) - 9 = 36 + 18 - 9 = 45$

41.

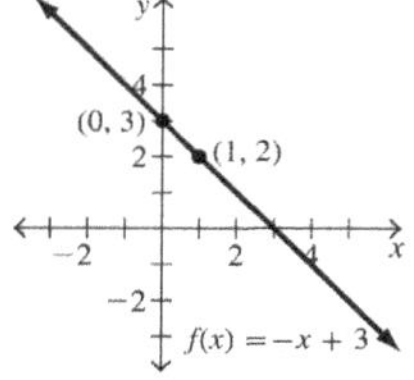

43.

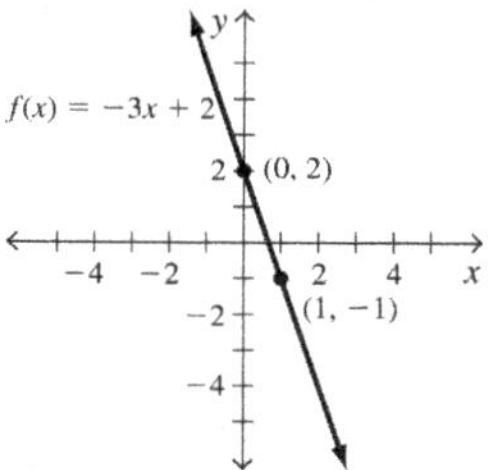

45.

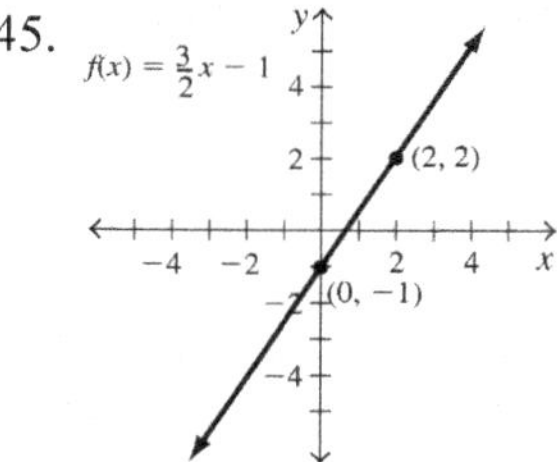

47. $y = x^2 - 1$

a) $a = 1 > 0$, opens upward

b) $x = 0$

c) $(0,-1)$

d) $(0,-1)$

e) $(1,0),(-1,0)$

f)

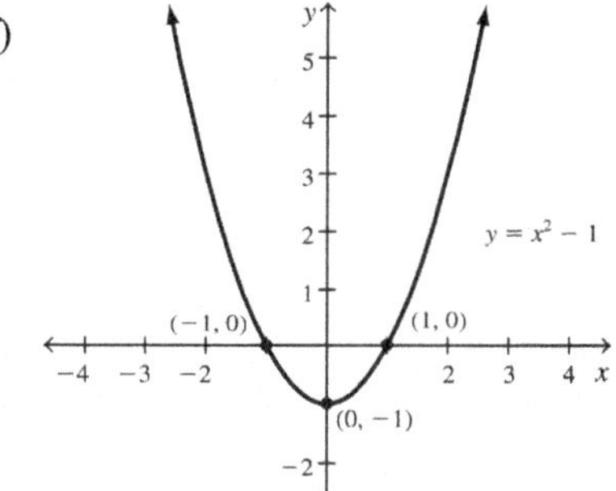

g) Domain: $\mathbb{R}$; Range: $y \geq -1$

49. $y = -x^2 + 4$

a) $a = -1 < 0$, opens downward

b) $x = 0$

c) $(0,4)$

d) $(0,4)$

e) $(-2,0),(2,0)$

f)

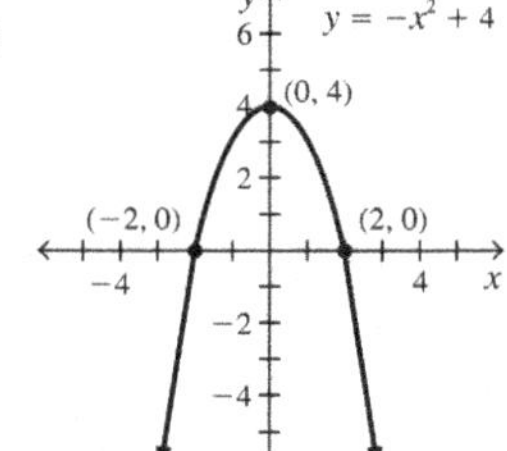

g) Domain: $\mathbb{R}$; Range: $y \leq 4$

51. $y = -2x^2 - 8$

a) $a = -2 < 0$, opens downward

b) $x = 0$

c) $(0,-8)$

d) $(0,-8)$

e) no x-intercepts

f)

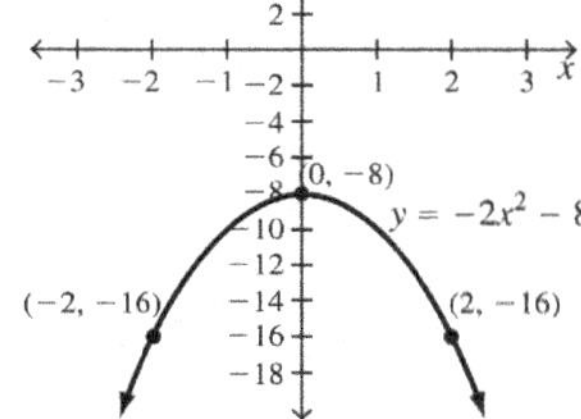

g) Domain: $\mathbb{R}$; Range: $y \leq -8$

53. $f(x) = x^2 + 4x - 4$

a) $a = 1 > 0$, opens upward

b) $x = -2$

c) $(-2, -8)$

d) $(0, -4)$

e) $(0.83, 0), (-4.83, 0)$

f) 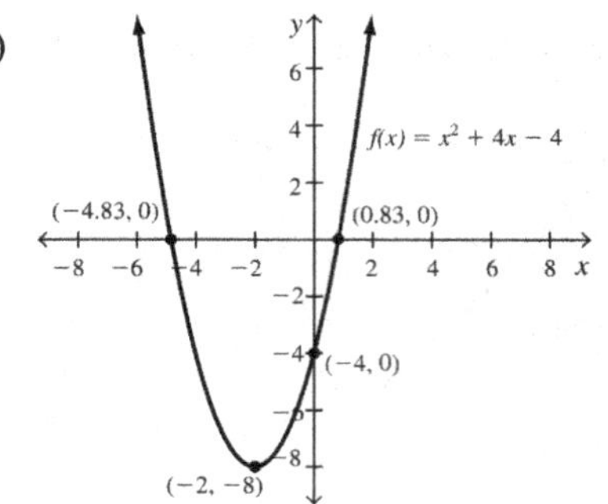

g) Domain: $\mathbb{R}$; Range: $y \geq -8$

55. $y = x^2 + 5x + 6$

a) $a = 1 > 0$, opens upward

b) $x = -\dfrac{5}{2}$

c) $(-2.5, -0.25)$

d) $(0, 6)$

e) $(-3, 0), (-2, 0)$

f) 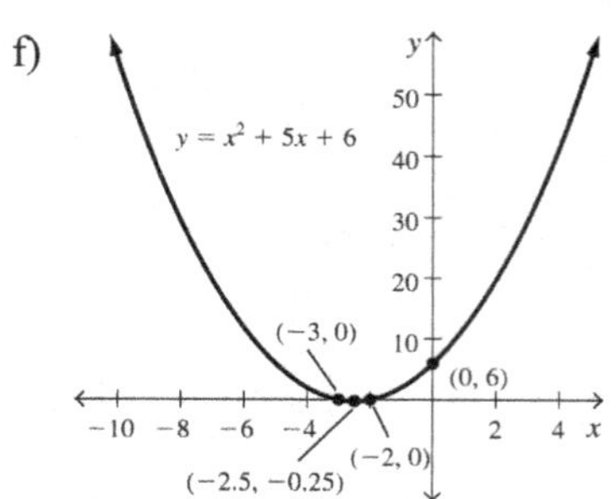

g) Domain: $\mathbb{R}$; Range: $y \geq -0.25$

57. $f(x) = -x^2 + 4x - 6$

a) $a = -1 < 0$, opens downward

b) $x = 2$

c) $(2, -2)$

d) $(0, -6)$

e) no x-intercepts

f) 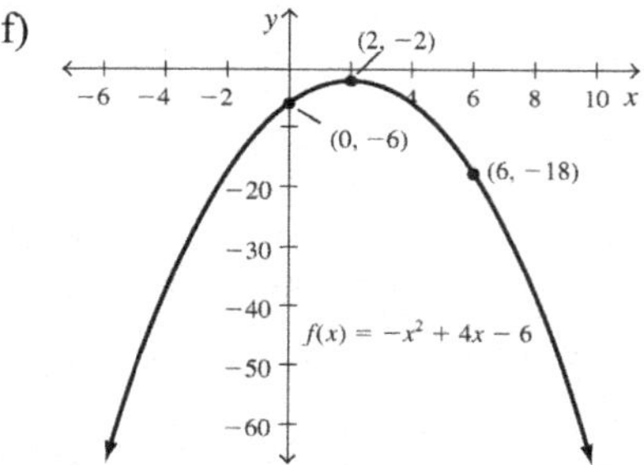

g) Domain: $\mathbb{R}$; Range: $y \leq -2$

59. $f(x) = -2x^2 + 3x - 2$

a) $a = -2 < 0$, opens downward

b) $x = \dfrac{3}{4}$

b) $\left(\dfrac{3}{4}, -\dfrac{7}{8}\right)$

b) $(0, -2)$

e) None

f) 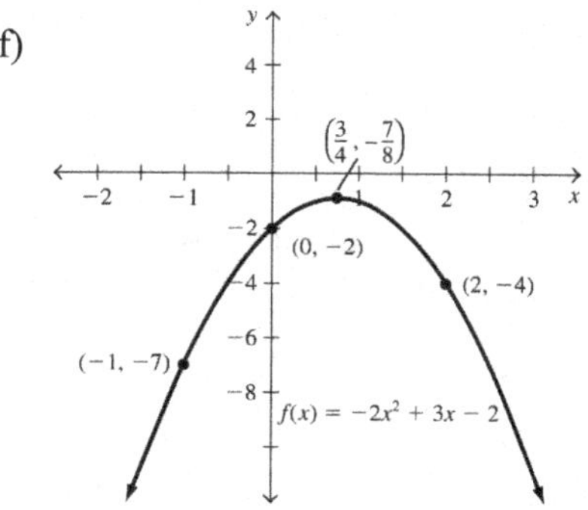

g) Domain: $\mathbb{R}$; Range: $y \leq -\dfrac{7}{8}$

61. Domain: $\mathbb{R}$; Range: $y > 0$

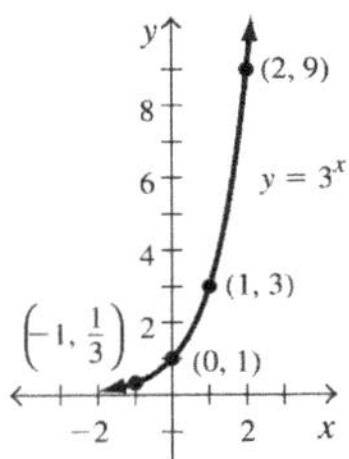

63. Domain: $\mathbb{R}$; Range: $y > 0$

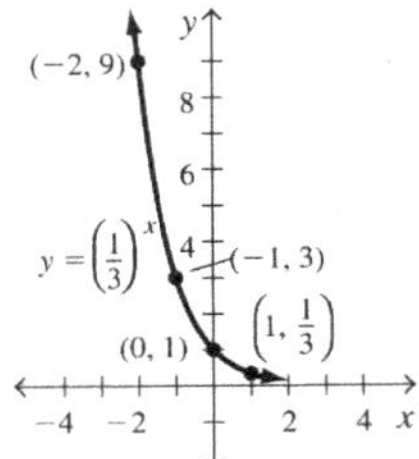

65. Domain: $\mathbb{R}$; Range: $y > 1$

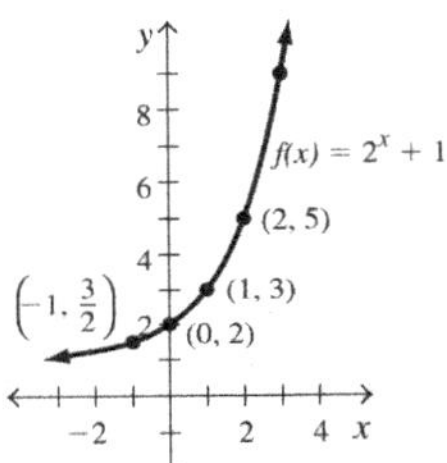

67. Domain: $\mathbb{R}$; Range: $y > 1$

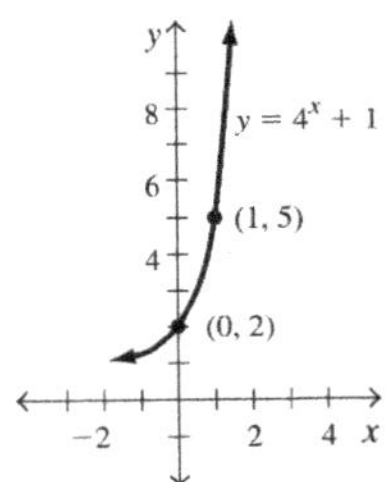

69. Domain: $\mathbb{R}$; Range: $y > 0$

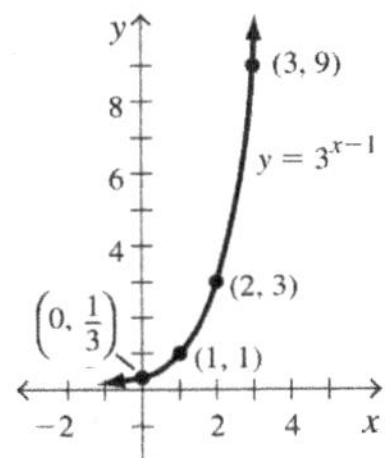

71. $p(20,000) = 3.5(20,000) - 15,000$
$= 70,000 - 15,000$
$= \$55,000$

73. a) $t(3) = -0.65(3)^2 + 5.16(3) + 64.42 \approx 74°\text{ F}$

b) 2 P.M.

c) $x = -\frac{b}{2a} = -\frac{5.16}{2(-0.65)} \approx 3.97 \approx 4$ hours, or 2 P.M.

$t(4) = -0.65(4)^2 + 5.16(4) + 64.42$
$= 74.66 \approx 75°\text{F}$

75. $y = 2000(3)^x = 2000(3)^5 = 486,000$ bacteria

77. a) $P(10) = 4000(1.3)^{0.1(10)} = 5200$ people

b) $P(50) = 4000(1.3)^{0.1(50)} = 14,851.72$
$\approx 14,852$
people

79. $P(8) = 1.39e^{0.006(8)} \approx 1.46$ billion people

81. $V(390) = 24e^{0.08(394)} \approx \$1.17 \times 10^{15}$

83. $R(1000) = 10e^{-0.000428(1000)} \approx 6.52$ grams

85. a) $d_{19} = 21.9(2)^{(20-19)/12} \approx 23.2$ cm

b) $d_4 = 21.9(2)^{(20-4)/12} \approx 55.2$ cm

c) $d_0 = 21.9(2)^{(20-0)/12} \approx 69.5$ cm

87. a) $f(20) = -0.85(20) + 187 = 170$ beats per minute

b) $f(30) = -0.85(30) + 187 = 161.5 \approx 162$ beats per minute

c) $f(50) = -0.85(50) + 187 = 144.5 \approx 145$ beats per minute

d) $f(60) = -0.85(60) + 187 = 136$ beats per minute

e) $-0.85x + 187 = 85$
$-0.85x = -102$
$x = 120$ years of age

Review Exercises

1. $x^2+10x=(3)^2+10(3)=9+30=39$

2. $-x^2-5=-(-2)^2-5=-4-5=-9$

3. $3x^2-2x+7=3(2)^2-2(2)+7=12-4+7=15$

4. $4x^3+7x^2-3x+1=4(-1)^3+7(-1)^2-3(-1)+1$
$=-4+7+3+1=7$

5. $3x-9+x+6=4x-3$

6. $2x+3(x-2)+7x=2x+3x-6+7x$
$=12x-6$

7. $$\begin{aligned}4t+8&=-12\\4t+8-8&=-12-8\\4t&=-20\\\frac{4t}{4}&=\frac{-20}{4}\\t&=-5\end{aligned}$$

8. $$\begin{aligned}4(x-2)&=3+5(x+4)\\4x-8&=3+5x+20\\4x-8&=5x+23\\4x-4x-8&=5x-4x+23\\-8&=x+23\\-8-23&=x+23-23\\-31&=x\end{aligned}$$

9. $$\begin{aligned}\frac{x+5}{2}&=\frac{x-3}{4}\\4(x+5)&=2(x-3)\\4x+20&=2x-6\\4x-2x+20&=2x-2x-6\\2x+20&=-6\\2x+20-20&=-6-20\\2x&=-26\\\frac{2x}{2}&=\frac{-26}{2}\\x&=-13\end{aligned}$$

10. $$\begin{aligned}\frac{x+2}{5}&=\frac{x+1}{4}\\4(x+2)&=5(x+1)\\4x+8&=5x+5\\4x-4x+8&=5x-4x+5\\8&=x+5\\8-5&=x+5-5\\x&=3\end{aligned}$$

11. $F=ma=(40)(5)=200$

12. $V=lwh=(10)(7)(3)=210$

13. $$\begin{aligned}z&=\frac{\overline{x}-\mu}{\sigma/\sqrt{n}}\\2&=\frac{\overline{x}-100}{3/\sqrt{16}}\\\frac{2}{1}&=\frac{\overline{x}-100}{3/4}\\2\left(\frac{3}{4}\right)&=1(\overline{x}-100)\\\frac{3}{2}&=\overline{x}-100\\\frac{3}{2}+100&=\overline{x}-100+100\\\overline{x}&=\frac{3}{2}+\frac{200}{2}=\frac{203}{2}=101.5\end{aligned}$$

14. $$\begin{aligned}E&=mc^2\\400&=m(4)^2\\400&=16m\\\frac{400}{16}&=\frac{16m}{16}\\25&=m\end{aligned}$$

15. $$\begin{aligned}8x-4y&=16\\8x-8x-4y&=-8x+16\\-4y&=-8x+16\\\frac{-4y}{-4}&=\frac{-8x+16}{-4}\\y&=\frac{-8x}{-4}+\frac{16}{-4}=2x-4\end{aligned}$$

16. $$\begin{aligned} 2x+9y &= 17 \\ 2x-2x+9y &= -2x+17 \\ 9y &= -2x+17 \\ \frac{9y}{9} &= \frac{-2x+17}{9} \\ y &= \frac{-2x+17}{9} \text{ or } y = \frac{-2x}{9}+\frac{17}{9} \end{aligned}$$

17. $$\begin{aligned} A &= lw \\ \frac{A}{l} &= \frac{lw}{l} \\ \frac{A}{l} &= w \end{aligned}$$

18. $$\begin{aligned} L &= 2(wh+lh) \\ L &= 2wh+2lh \\ L-2wh &= 2wh-2wh+2lh \\ L-2wh &= 2lh \\ \frac{L-2wh}{2h} &= \frac{2lh}{2h} \\ l &= \frac{L-2wh}{2h} \text{ or } l = \frac{L}{2h}-\frac{2wh}{2h} = \frac{L}{2h}-w \end{aligned}$$

19. $5+3x$

20. $\frac{9}{q}-15$

21. Let $x =$ the number
$3+7x =$ three increased by 7 times a number
$$\begin{aligned} 3+7x &= 17 \\ 3-3+7x &= 17-3 \\ 7x &= 14 \\ \frac{7x}{7} &= \frac{14}{7} \\ x &= 2 \end{aligned}$$

22. Let $x =$ the number
$3x =$ the product of 3 and the number
$3x+8 =$ the product of 3 and the number increased by 8
$x-6 = 6$ less than the number
$$\begin{aligned} 3x+8 &= x-6 \\ 3x-x+8 &= x-x-6 \\ 2x+8 &= -6 \\ 2x+8-8 &= -6-8 \\ 2x &= -14 \\ \frac{2x}{2} &= \frac{-14}{2} \\ x &= -7 \end{aligned}$$

23. Let $x =$ the number
$x-4 =$ the difference of the number and 4
$5(x-4) = 5$ times the difference of the number and 4
$$\begin{aligned} 5(x-4) &= 45 \\ 5x-20 &= 45 \\ 5x-20+20 &= 45+20 \\ 5x &= 65 \\ \frac{5x}{5} &= \frac{65}{5} \\ x &= 13 \end{aligned}$$

24. Let $x =$ the number
$10x+14$ 14 more than 10 times the number
$x+12 =$ the sum of the number and 12
$8(x+12) = 8$ times the sum of the number and 12
$$\begin{aligned} 10x+14 &= 8(x+12) \\ 10x+14 &= 8x+96 \\ 10x-8x+14 &= 8x-8x+96 \\ 2x+14 &= 96 \\ 2x+14-96 &= 96-96 \\ 2x &= 82 \\ \frac{2x}{2} &= \frac{82}{2} \\ x &= 41 \end{aligned}$$

25. Let $x =$ the amount invested in bonds
$2x =$ the amount invested in mutual funds
$$\begin{aligned} x+2x &= 15,000 \\ 3x &= 15,000 \\ \frac{3x}{3} &= \frac{15,000}{3} \\ x &= 5000 \end{aligned}$$
\$5000 was invested in bonds and $2(5000) = \$10,000$ was invested in mutual funds.

26. Let $x =$ profit at restaurant B
$x+15,000 =$ profit at restaurant A
$$\begin{aligned} x+(x+15,000) &= 75,000 \\ 2x+15,000 &= 75,000 \\ 2x+15,000-15,000 &= 75,000-15,000 \\ 2x &= 60,000 \\ x &= 30,000 \end{aligned}$$
The profit for restaurant B is \$30,000; the profit for restaurant A is $30,000+15,000 = \$45,000$.

27. $$\frac{2}{\frac{1}{3}}=\frac{3}{x}$$
$$2x=3\left(\frac{1}{3}\right)$$
$$2x=1$$
$$\frac{2x}{2}=\frac{1}{2}$$
$$x=\frac{1}{2}\text{ cup}$$

28. 1 hr 40 min = 60 min + 40 min = 100 min
$$\frac{120}{100}=\frac{300}{x}$$
$$120x=100(300)$$
$$120x=30,000$$
$$\frac{120x}{120}=\frac{30,000}{120}$$
$$x=250\text{ min, or 4 hr 10 min}$$

29. $$s=kt$$
$$60=k(10)$$
$$k=\frac{60}{10}=6$$
$$s=6t$$
$$s=6(12)$$
$$s=72$$

30. $$J=\frac{k}{A^2}$$
$$25=\frac{k}{2^2}$$
$$k=2^2(25)$$
$$k=100$$
$$J=\frac{100}{A^2}$$
$$J=\frac{100}{5^2}$$
$$J=\frac{100}{25}=4$$

31. $$W=\frac{kL}{A}$$
$$80=\frac{k(100)}{20}$$
$$100k=1600$$
$$\frac{100k}{100}=\frac{1600}{100}$$
$$k=16$$
$$W=\frac{16L}{A}$$
$$W=\frac{16(50)}{A}=\frac{800}{40}=20$$

32. $$z=\frac{kxy}{r^2}$$
$$12=\frac{k(20)(8)}{(8)^2}$$
$$160k=768$$
$$\frac{160k}{160}=\frac{768}{160}$$
$$k=4.8$$
$$z=\frac{4.8xy}{r^2}$$
$$z=\frac{4.8(10)(80)}{(3)^2}=\frac{3840}{9}=426.\overline{6}\approx 426.7$$

33. $$t=kv$$
$$2325=k(155,000)$$
$$k=\frac{2325}{155,000}$$
$$k=0.015$$
$$t=0.015v$$
$$t=0.015(210,000)=\$3150$$

34. $$\frac{1\ kWh}{\$0.162}=\frac{740\ kWh}{x}$$
$$x=\$119.88$$

35. $$8+9x\le 6x-4$$
$$8-8+9x\le 6x-4-8$$
$$9x\le 6x-12$$
$$9x-6x\le 6x-6x-12$$
$$3x\le -12$$
$$\frac{3x}{3}\le\frac{-12}{3}$$
$$x\le -4$$

−4

36.
$$3(x+9) \le 4x+11$$
$$3x+27 \le 4x+11$$
$$3x-3x+27 \le 4x-3x+11$$
$$27 \le x+11$$
$$27-11 \le x+11-11$$
$$16 \le x$$
$$x \ge 16$$

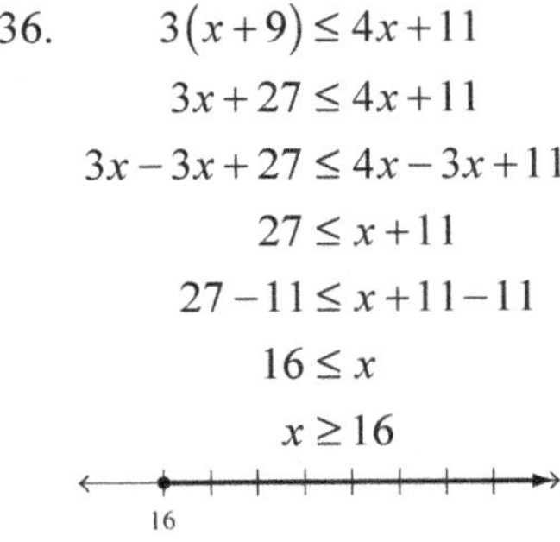

37.
$$5x+13 \ge -22$$
$$5x+13-13 \ge -22-13$$
$$5x \ge -35$$
$$\frac{5x}{5} \ge \frac{-35}{5}$$
$$x \ge -7$$

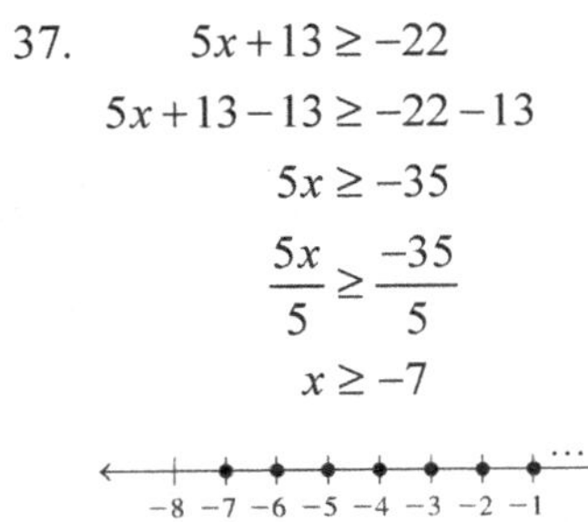

38.
$$-8 \le x+2 \le 7$$
$$-8-2 \le x+2-2 \le 7-2$$
$$-10 \le x \le 5$$

−10 −8 −6 −4 −2 0 2 4 6

39. – 42.

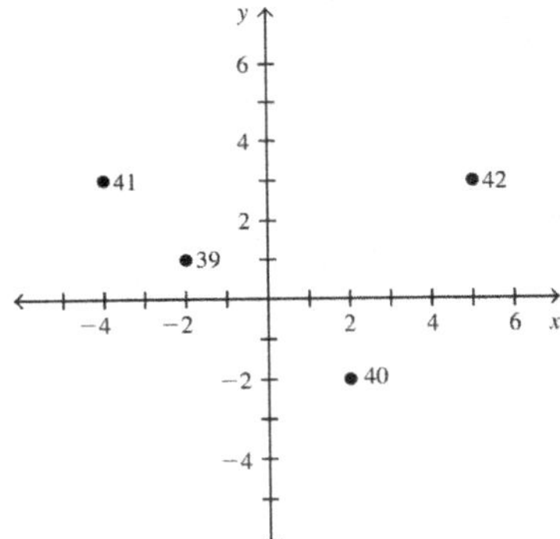

43.

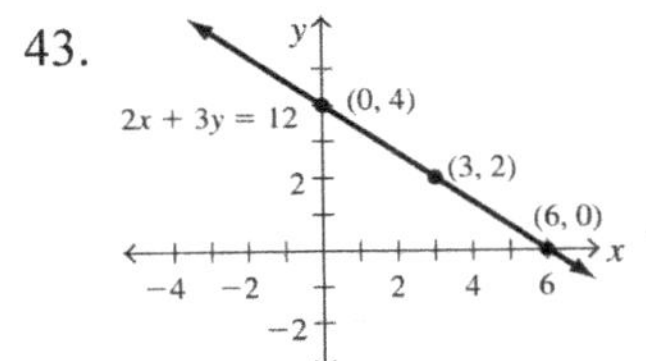

44.

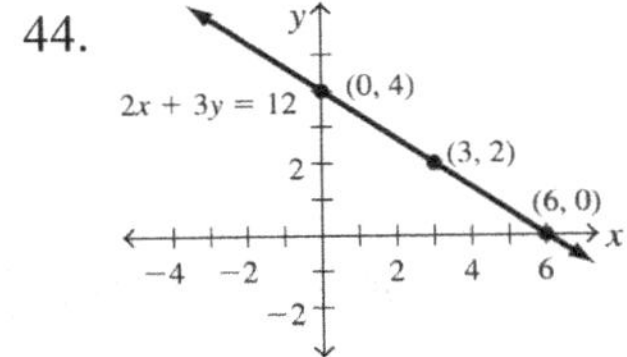

45.

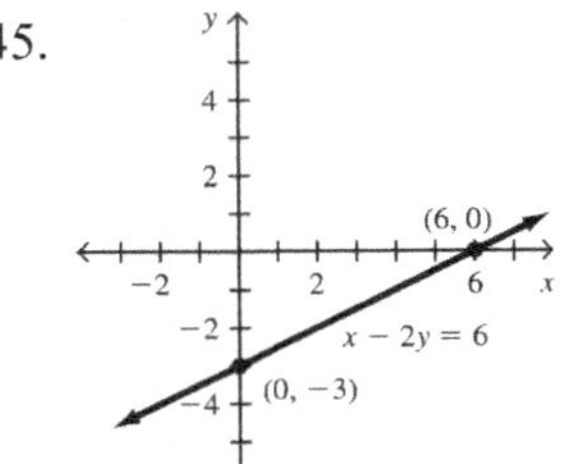

46.

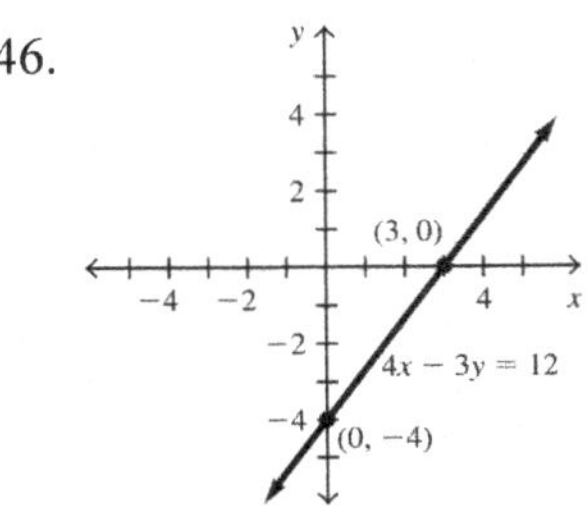

47. $m = \dfrac{y_2 - y_1}{x_2 - x_1} = \dfrac{7-3}{4-1} = \dfrac{4}{3}$

48. $m = \dfrac{y_2 - y_1}{x_2 - x_1} = \dfrac{-4-(-1)}{5-3} = \dfrac{-4+1}{5-3} = -\dfrac{3}{2}$

49. $m = \dfrac{y_2 - y_1}{x_2 - x_1} = \dfrac{3-(-4)}{2-(-1)} = \dfrac{3+4}{2+1} = \dfrac{7}{3}$

50. $m = \dfrac{y_2 - y_1}{x_2 - x_1} = \dfrac{-2-2}{6-6} = \dfrac{-4}{0}$; Undefined

51.

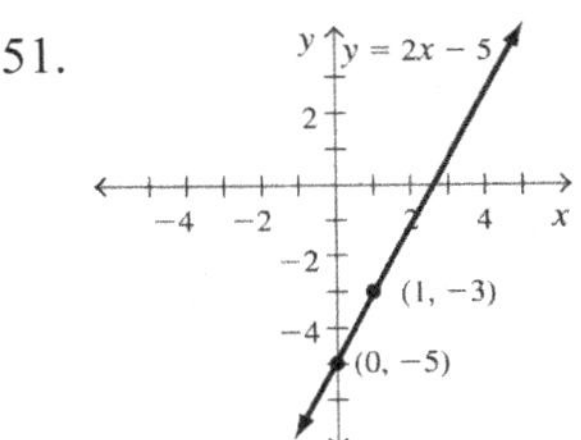

52.

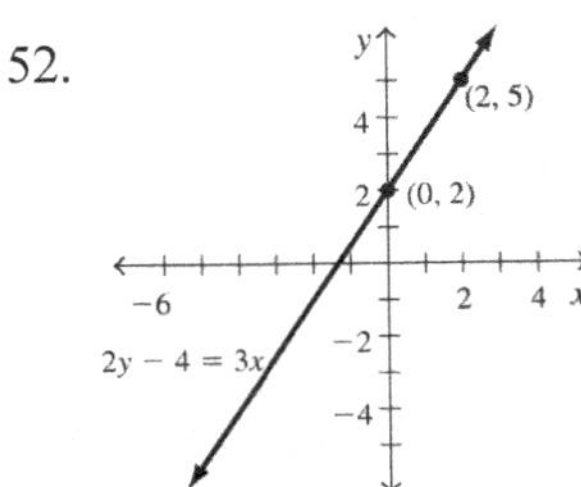

53. The y-intercept is 4, thus $b = 4$. Since the graph rises from left to right, the slope is positive. The change in y is 4 units while the change in x is 2. Thus, m, the slope is $\frac{4}{2}$ or 2. The equation is $y = 2x + 4$.

54. The y-intercept is 1, thus $b = 1$. Since the graph falls from left to right, the slope is negative. The change in y is 3 units while the change in x is 3. Thus, m, the slope is $\frac{-3}{3}$ or -1. The equation is $y = -x + 1$.

55. a)

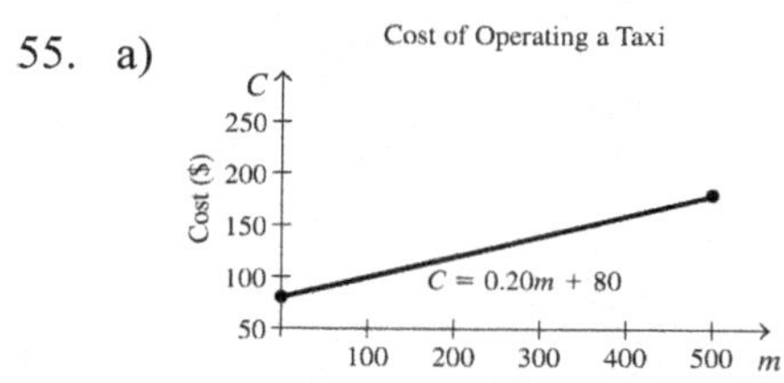

b) $C = 0.20(150) + 80 = \$110$

c)
$$104 = 0.20m + 80$$
$$104 - 80 = 0.20m$$
$$24 = 0.20m$$
$$m = 120 \text{ miles}$$

56.

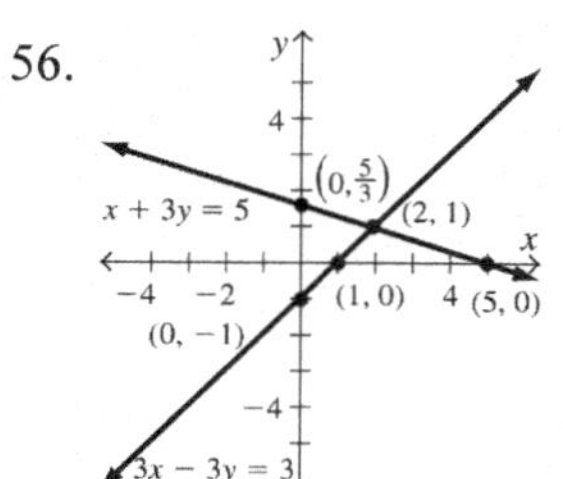

57.

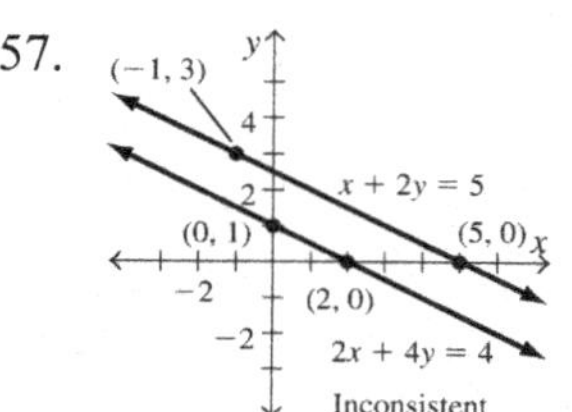

58. $-x + y = -2 \Rightarrow y = x - 2$

$x + 2y = 5$

Substitute $x - 2$ for y in the second equation and solve for x.

$$x + 2y = 5$$
$$x + 2(x - 2) = 5$$
$$x + 2x - 4 = 5$$
$$3x - 4 = 5$$
$$3x = 9$$
$$x = 5$$

Substitute 5 for x in $y = x - 2$.

$y = (5) - 2 = 3$

The solution is $(3, 1)$.

59. $x - 2y = 9$

$y = 2x - 3$

Substitute $2x - 3$ in place of y in the first equation and solve for x.

$$x - 2(2x - 3) = -9$$
$$x - 4x + 6 = -9$$
$$-3x + 6 = 9$$
$$-3x = 3$$
$$x = -1$$

Substitute (-1) for x in the second equation.

$y = 2(-1) - 3 = -5$

The solution is $(-1, -5)$.

60. $2x - y = 4 \Rightarrow y = 2x - 4$

$3x - y = 2$

Substitute $2x - 4$ for y in the second equation and solve for x.

$$3x - (2x - 4) = 2$$
$$3x - 2x + 4 = 2$$
$$x + 4 = 2$$
$$x = -2$$

Substitute -2 for x into $y = 2x - 4$.

$y = 2(-2) - 4 = -8$

The solution is $(-2, -8)$.

61. $3x + y = 1 \Rightarrow y = -3x + 1$

$3y = -9x - 4$

Substitute $-3x + 1$ for y in the second equation and solve for x.

$$3(-3x + 1) = -9x - 4$$
$$-9x + 3 = -9x -$$
$$3 \neq 4 \text{ False}$$

There is no solution to this system. The system is inconsistent.

62. $x + y = 2$

$x + 3y = -2$

Multiply the second equation by (–1).

$$\begin{aligned} -x - y &= -2 \\ x + 3y &= -2 \\ \hline 2y &= -4 \\ y &= -2 \end{aligned}$$

Substitute –2 for y in the first equation.

$$\begin{aligned} x + y &= 2 \\ x + (-2) &= 2 \\ x &= 4 \\ x &= -1 \end{aligned}$$

The solution is (4, –2).

63. $4x - 8y = 16$

$x - 2y = 4$

Multiply the second equation by –2.

$$\begin{aligned} 4x - 8y &= 16 \\ -4x + 8y &= -16 \\ \hline 0 &= 0 \text{ True} \end{aligned}$$

There are an infinite number of solutions. The system is dependent.

64. $3x - 4y = 10$

$5x + 3y = 7$

Multiply the first equation by 3 and the second equation by 4.

$$\begin{aligned} 9x - 12y &= 30 \\ 20x + 12y &= 28 \\ \hline 29x &= 58 \\ x &= 2 \end{aligned}$$

Substitute 2 for x in the second original equation.

$$\begin{aligned} 5(2) + 3y &= 7 \\ 10 + 3y &= 7 \\ 3y &= -3 \\ y &= -1 \end{aligned}$$

The solution is (2, –1).

65. $3x + 4y = 6$

$2x - 3y = 4$

Multiply the first equation by 2 and the second equation by –3.

$$\begin{aligned} 6x + 8y &= 12 \\ -6x + 9y &= -17 \\ \hline 17y &= 0 \\ y &= 0 \end{aligned}$$

Substitute 0 for y in the second original equation.

$$\begin{aligned} 2x - 3(0) &= 4 \\ 2x &= 4 \\ x &= 2 \end{aligned}$$

The solution is (2, 0).

66. Let $x =$ amount borrowed at 3%
$y =$ amount borrowed at 6%

$$x + y = 400{,}000 \Rightarrow x = 400{,}000 - y$$

$$0.03x + 0.06y = 16{,}500$$

Substitute $400{,}000 - y$ for x in the second equation and solve for y.

$$\begin{aligned} 0.03(400{,}000 - y) + 0.06y &= 16{,}500 \\ 12{,}000 - 0.03y + 0.06y &= 16{,}500 \\ 0\ .03y &= 4500 \\ y &= 150{,}000 \end{aligned}$$

\$150,000 was borrowed at 6% and $400{,}0000 - 150{,}000 = \$250{,}000$ was borrowed at 3%.

67. Let $s =$ liters of 80% acid solution
$w =$ liters of 50% acid solution

$$s + w = 100$$

$$0.80s + 0.50w = 100(0.75)$$

Rewrite equations.

$s = 100 - w$

$0.80s + 0.50w = 75$

Substitute $100 - w$ for s in the second equation and solve for w.

$$\begin{aligned} 0.80(100 - w) + 0.50w &= 75 \\ 80 - 0.80w + 0.50w &= 75 \\ -0.30w &= -5 \\ w &= 16.\overline{6} = 16\frac{2}{3} \end{aligned}$$

67. (continued)

Substitute $16\frac{2}{3}$ for w in $s = 100 - w$.

$s = 100 - 16\frac{2}{3} = 83\frac{1}{3}$

Mix $83\frac{1}{3}$ liters of 80% solution with $16\frac{2}{3}$ liters of 50% solution.

68. Let $c =$ total cost
$x =$ number of months to operate

a) Model 1600A: $c_A = 950 + 32x$

Model 6070B: $c_B = 1275 + 22x$

$$950 + 32x = 1275 + 22x$$
$$10x = 325$$
$$x = 32.5$$

After 32.5 months of operation the total cost of the units will be equal.

b) Model 1600A: $c_A = 950 + 32(10) = \$1270$

Model 6070B: $c_B = 1275 + 22(10) = \$1495$

The Model 6070B is more cost effective.

69. a) Let $C =$ total cost for parking
$x =$ number of additional hours

All-Day: $C = 5 + 0.50x$
Sav-A-Lot: $C = 4.25 + 0.75x$

$$5 + 0.50x = 4.25 + 0.75x$$
$$0.75 = 0.25x$$
$$x = 3$$

The total cost will be the same after 3 additional hours, or 4 hours total.

b) After 5 hours, or $x = 4$ additional hours:

All-Day: $C = 5 + 0.50(4) = \$7.00$

Sav-A-Lot: $C = 4.25 + 0.75(4) = \$7.25$

All-Day would be less expensive.

70. Graph $2x + 3y = 12$. Since the original inequality is less than or equal to, a solid line is drawn. Since the point (0, 0) satisfies the inequality $2x + 3y \leq 12$, all points on the line and in the half-plane below the line $2x + 3y = 12$ are in the solution set.

Graph for Exercise 70

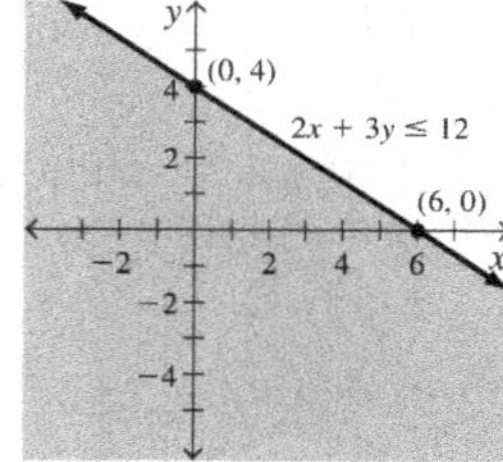

Graph for Exercise 71

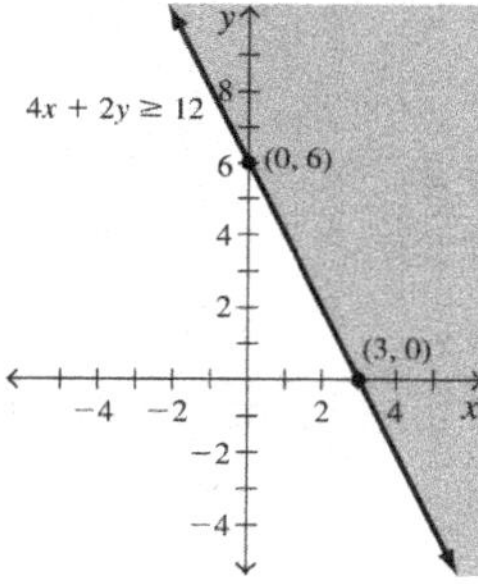

71. Graph $4x + 2y = 12$. Since the original inequality is greater than or equal to, a solid line is drawn. Since the point (0, 0) does not satisfy the inequality $4x + 2y \geq 12$, all points on the line and in the half plane above the line $4x + 2y = 12$ are in the solution set.

72. 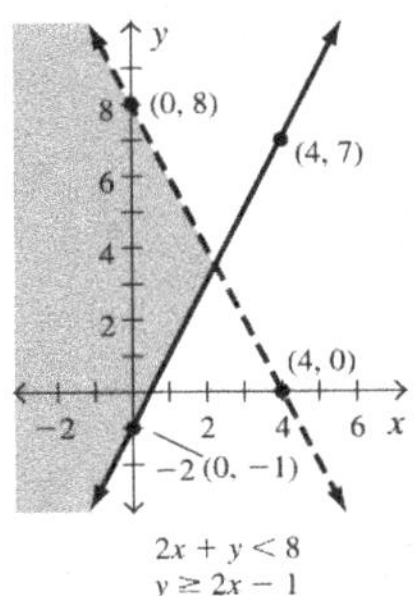

73. 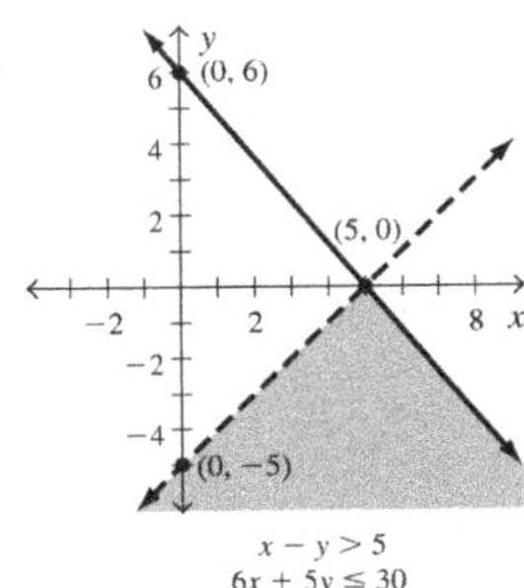

74. a) 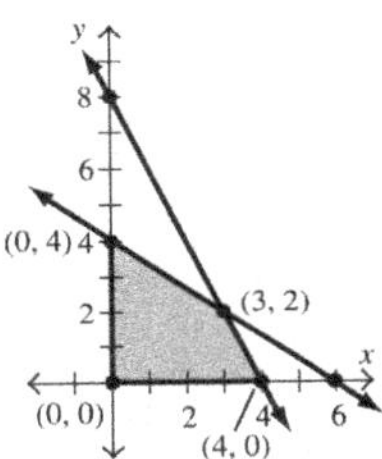

b) $P = 5x + 3(0)$

At (0, 0) $P = 5(0) + 3(0) = 0$

At (0, 4), $P = 5(0) + 3(4) = 12$

At (3, 2), $P = 5(3) + 3(2) = 21$

At (4, 0), $P = 5(4) + 3(0) = 20$

The maximum value is 21 at (3, 2); the minimum is value 0 at (0, 0).

75. $x^2 + 6x + 9 = (x+3)(x+3)$

76. $x^2 + 2x - 15 = (x+5)(x-3)$

77. $x^2 - 10x + 24 = (x-6)(x-4)$

78. $6x^2 + 7x - 3 = (3x-1)(2x+3)$

79.
$$x^2 + 9x + 20 = 0$$
$$(x+5)(x+4) = 0$$
$$x + 5 = 0 \quad \text{or} \quad x + 4 = 0$$
$$x = -5 \qquad\qquad x = -4$$

80.
$$x^2 + 3x = 10$$
$$x^2 + 3x - 10 = 0$$
$$(x+5)(x-2) = 0$$
$$x + 5 = 0 \quad \text{or} \quad x - 2 = 0$$
$$x = -5 \qquad\qquad x = 2$$

81.
$$3x^2 - 17x + 10 = 0$$
$$(3x-2)(x-5) = 0$$
$$3x - 2 = 0 \quad \text{or} \quad x - 5 = 0$$
$$3x = 2 \qquad\qquad x = 5$$
$$x = \frac{2}{3}$$

82.
$$3x^2 = -7x - 2$$
$$3x^2 + 7x + 2 = 0$$
$$(x+2)(3x+1) = 0$$
$$x + 2 = 0 \quad \text{or} \quad 3x + 1 = 0$$
$$x = -2 \qquad\qquad 3x = -1$$
$$x = -\frac{1}{3}$$

83. $x^2 - 6x - 16 = 0;\ a = 1,\ b = -6,\ c = -16$

$$x = \frac{-b \pm \sqrt{b^2 - 4ac}}{2a} = \frac{-(-6) \pm \sqrt{(-6)^2 - 4(1)(-16)}}{2(1)} = \frac{6 \pm \sqrt{36 + 64}}{2} = \frac{6 \pm \sqrt{100}}{2} = \frac{6 \pm 10}{2}$$

$$x = \frac{6-10}{2} = -2 \text{ or } x = \frac{6+10}{2} = 8$$

84. $2x^2 - x - 3 = 0;\ a = 2,\ b = -1,\ c = -3$

$$x = \frac{-b \pm \sqrt{b^2 - 4ac}}{2a} = \frac{-(-1) \pm \sqrt{(-1)^2 - 4(2)(-3)}}{2(2)} = \frac{1 \pm \sqrt{1 + 24}}{4} = \frac{1 \pm \sqrt{25}}{4} = \frac{1 \pm 5}{4}$$

$$x = \frac{1-5}{4} = -1 \text{ or } x = \frac{1+5}{4} = \frac{3}{2}$$

85. $2x^2 - 3x + 4 = 0;\ a = 2,\ b = -3,\ c = 2$

$$x = \frac{-b \pm \sqrt{b^2 - 4ac}}{2a} = \frac{-(-3) \pm \sqrt{(-3)^2 - 4(2)(4)}}{2(2)} = \frac{3 \pm \sqrt{9 - 32}}{4} = \frac{3 \pm \sqrt{-23}}{4};\ \text{No real solution}$$

86. $x^2 - 3x - 2 = 0;\ a = 1,\ b = -3,\ c = -2$

$$x = \frac{-b \pm \sqrt{b^2 - 4ac}}{2a} = \frac{-(-3) \pm \sqrt{(-3)^2 - 4(1)(-2)}}{2(1)} = \frac{3 \pm \sqrt{9 + 8}}{2} = \frac{3 \pm \sqrt{17}}{2}$$

87. Function since each value of x is paired with a unique value of y.
Domain: $x = -2, -1, 2, 3$; Range: $y = -1, 0, 2$

88. Not a function since it is possible to draw a vertical line that intersects the graph at more than one point.

89. Not a function since it is possible to draw a vertical line that intersects the graph at more than one point.

90. Function since each vertical line intersects the graph at only one point.
Domain: $\mathbb{R}$; Range: $\mathbb{R}$

91. $f(x) = 4x + 3$
$f(4) = 4x + 3 = 4(4) + 3 = 16 + 3 = 19$

92. $f(x) = -2x + 5$
$f(-3) = -2(-3) + 5 = 6 + 5 = 11$

93. $f(x) = 3x^2 - 2x + 1$
$f(5) = 3(5)^2 - 2(5) + 1 = 75 - 10 + 1 = 66$

94. $f(x) = -4x^2 + 7x + 9$
$f(-1) = -4(-1)^2 + 7(-1) + 9 = -4 - 7 + 9 = -2$

95. $y = -x^2 - 4x + 21$

a) $a = -1 < 0$, opens downward

b) $x = -2$

c) $(-2, 25)$

d) $(0, 21)$

e) $(-7, 0), (3, 0)$

f)

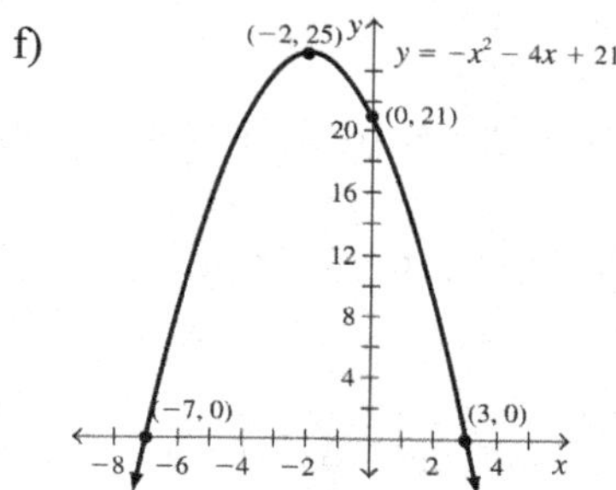

g) Domain: $\mathbb{R}$; Range: $y \le 25$

96. $f(x) = 2x^2 + 8x + 6$

a) $a = 2 > 0$, opens upward

b) $x = -\dfrac{8}{2(2)} = -2$

c) $(-2, -2)$

d) $(0, 6)$

e) $(-3, 0), (-1, 0)$

f)

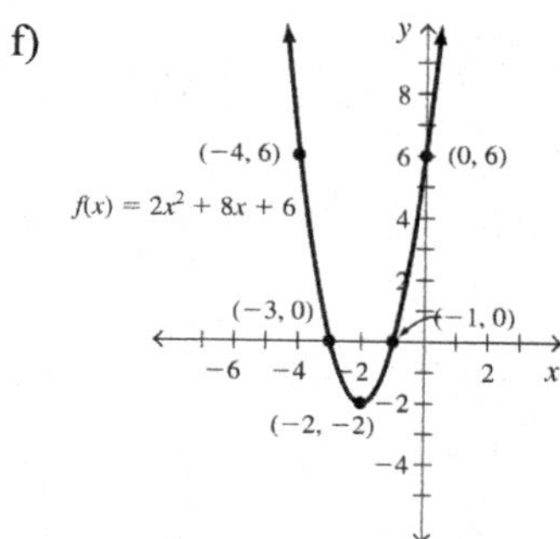

g) Domain: $\mathbb{R}$; Range: $y \ge -2$

97. Domain: $\mathbb{R}$; Range: $y > 0$

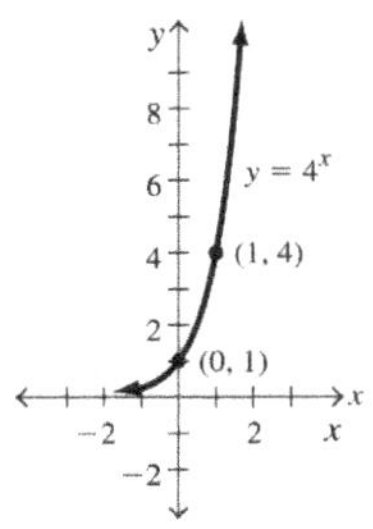

98. Domain: $\mathbb{R}$; Range: $y > 0$

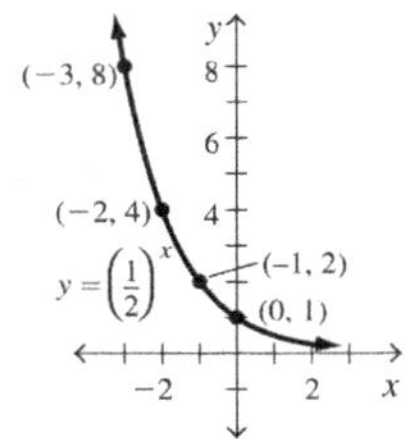

99. $m = 30 - 0.002n^2$

$$m = 30 - 0.002(60)^2 = 30 - 0.002(3600)$$
$$= 30 - 7.2 = 22.8 \text{ mpg}$$

100. $P = 100(0.92)^x$

$$P = 100(0.92)^{4.5} \approx 68.7\%$$

Chapter Test

1. $3x^2 + 6x - 1 = 3(-2)^2 + 6(-2) - 1$
$$= 12 - 12 - 1 = -1$$

2.
$$4x + 6 = 2(3x - 7)$$
$$4x + 6 = 6x - 14$$
$$4x - 4x + 6 = 6x - 4x - 14$$
$$6 = 2x - 14$$
$$6 + 14 = 2x - 14 + 14$$
$$20 = 2x$$
$$\frac{20}{2} = \frac{2x}{2}$$
$$10 = x$$

3.
$$-2(x-3) + 6x = 2x + 3(x-4)$$
$$-2x + 6 + 6x = 2x + 3x - 12$$
$$4x + 6 = 5x - 12$$
$$4x - 4x + 6 = 5x - 4x - 12$$
$$6 = x - 12$$
$$6 + 12 = x - 12 + 12$$
$$18 = x$$

4. Let $x =$ Mary's weekly sales
$0.06x =$ the amount of the commission
$$350 + 0.06x = 710$$
$$0.06x = 360$$
$$\frac{0.06x}{0.06} = \frac{360}{0.06}$$
$$x = \$6000$$

5. $L = ah + bh + ch = 2(7) + 5(7) + 4(7)$
$$= 14 + 35 + 28 = 77$$

6.
$$3x + 5y = 11$$
$$3x - 3x + 5y = -3x + 11$$
$$5y = -3x + 11$$
$$\frac{5y}{5} = \frac{-3x + 11}{5}$$
$$y = \frac{-3x + 11}{5} = -\frac{3}{5}x + \frac{11}{5}$$

7.
$$l = \frac{k}{w}$$
$$15 = \frac{k}{9}$$
$$k = 15(9) = 135$$
$$l = \frac{135}{w}$$
$$l = \frac{135}{20} = 6.75 \text{ ft}$$

8.
$$-5x + 14 \le 2x + 35$$
$$-5x - 2x + 14 \le 2x - 2x + 35$$
$$-7x + 14 \le 35$$
$$-7x + 14 - 14 \le 35 - 14$$
$$-7x \le 21$$
$$\frac{-7x}{-7} \ge \frac{21}{-7}$$
$$x \ge -3$$

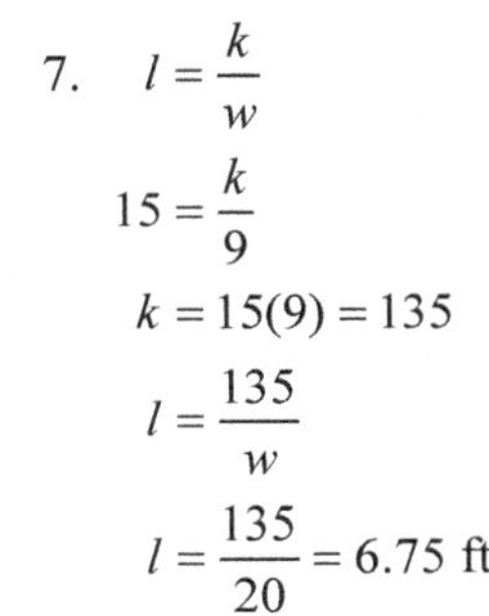

9. $m = \frac{y_2 - y_1}{x_2 - x_1} = \frac{14 - 8}{1 - (-2)} = \frac{6}{3} = 2$

10.

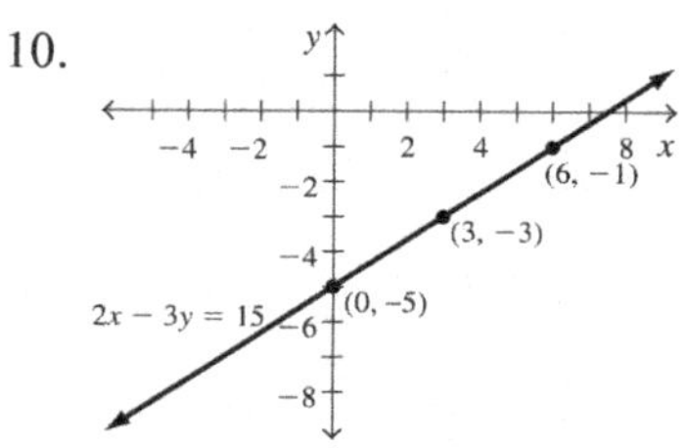

11. The solution is $(3,-6)$.

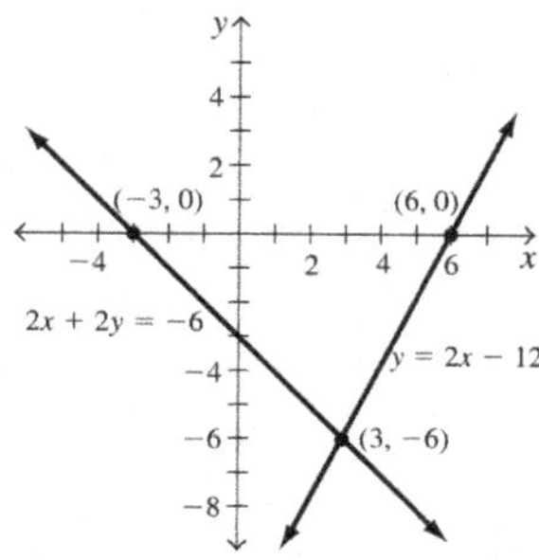

12. $x+\ y=-1 \Rightarrow x=-y-1$

$2x+3y=-5$

Substitute $-y-1$ for x in the second equation and solve for y.

$2(-y-1)+3y=-5$

$-2y-2+3y=-5$

$y-2+2=-5+2$

$y=-3$

Substitute -3 for y in the equation $x=-y-1$.

$x=-(-3)-1=2$

The solution is $(2,-3)$.

13. $4x+3y=5$

$2x+4y=10$

Multiply the second equation by (-2).

$4x+3y=5$

$-4x-8y=-20$

$5y=15$

$y=3$

Substitute 5 for y in the first equation.

$2x+4(3)=10$

$2x+12=10$

$2x=-2$

$x=-1$

The solution is $(-1, 3)$.

14. Let $x=$ daily fee
$y=$ mileage charge

$3x+150y=132$

$2x+400y=142 \Rightarrow x=71-200y$

Substitute $71-200y$ for x in the first equation and solve for y.

$3(71-200y)+150y=132$

$213-600y+150y=132$

$213-450y=132$

$-450y=-81$

$y=0.18$

Substitute 0.18 for y in the second equation and solve for x.

$3x+150(0.18)=132$

$3x+27=132$

$3x=105$

$x=35$

The daily fee is \$35 and the mileage charge is 18 cents, or \$0.18, per mile.

15. Graph $3y=5x-12$. Since the original statement is greater than or equal to, a solid line is drawn. Since the point (0, 0) satisfies the inequality $3y \geq 5x-12$, all points on the line and in the half-plane above the line $3y=5x-12$ are in the solution set.

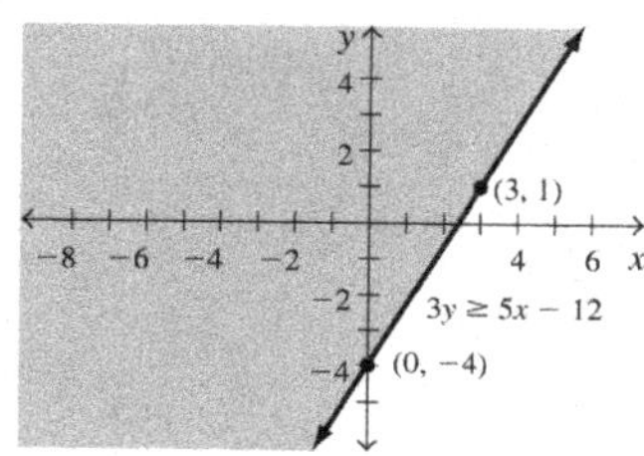

16. a)

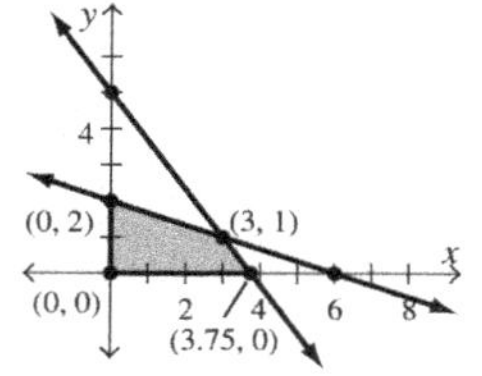

b) $P = 6x + 4y$

At (0, 0), $P = 6(0) + 4(0) = 0$

At (0, 2), $P = 6(0) + 4(2) = 8$

At (3, 1), $P = 6(3) + 4(1) = 22$

At (3.75, 0), $P = 6(3.75) + 4(0) = 22.5$

The maximum is 22.5 at (3.75, 0); the minimum is 0 at (0, 0).

17.

$$x^2 + 7x = -6$$
$$x^2 + 7x + 6 = 0$$
$$(x+6)(x+1) = 0$$
$$x + 6 = 0 \quad \text{or} \quad x + 1 = 0$$
$$x = -6 \qquad x = -1$$

18.

$$3x^2 + 2x = 8$$

$3x^2 + 2x - 8 = 0$; $a = 3$, $b = 2$, $c = -8$

$$x = \frac{-b \pm \sqrt{b^2 - 4ac}}{2a} = \frac{-(2) \pm \sqrt{(2)^2 - 4(3)(-8)}}{2(3)} = \frac{-2 \pm \sqrt{4 + 96}}{6} = \frac{-2 \pm \sqrt{100}}{6} = \frac{-2 \pm 10}{6}$$

$$x = \frac{-2 - 10}{6} = -2 \text{ or } x = \frac{-2 + 10}{6} = \frac{4}{3}$$

19.

$$f(x) = -2x^2 - 8x + 7$$
$$f(-2) = -2(-2)^2 - 8(-2) + 7$$
$$= -8 + 16 + 7 = 15$$

20. $y = x^2 - 2x + 4$

a) $a = 1 > 0$, opens upward

b) $x = 1$

c) $(1, 3)$

d) $(0, 4)$

e) no x-intercepts

f)

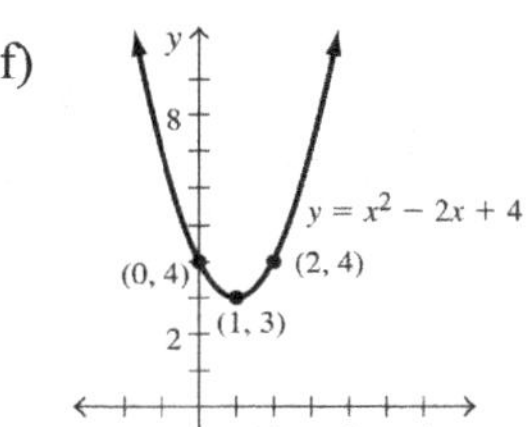

g) Domain: $\mathbb{R}$; Range: $y \geq 3$

Chapter Seven: The Metric System

Section 7.1: Basic Terms and Conversions Within the Metric System

1. Metric

3. a) Meter
   b) Kilogram
   c) Liter
   d) Celsius

5. a) Deka
   b) Deci

7. a) Kilo
   b) Milli

9. a) 0°C
   b) 100°C
   c) 37°C

11. b

13. c

15. f

17. a) 0.001 gram
    b) 100 grams
    c) 1000 grams
    d) 0.01 gram
    e) 10 grams
    f) 0.1 gram

19. cg; 0.01 g

21. dg; 0.1 g

23. kg; 1000 g

25. $603 \text{ km/h} = (603 \times 100) \text{ m/h} = 603\ 000 \text{ m/h}$

27. $9.2 \text{ m} = (9.2 \times 1000) \text{ mm} = 9200 \text{ mm}$

29. $0.057 \text{ m} = (0.057)(0.001) \text{ km} = 0.000\ 057 \text{ km}$

31. $186.2 \text{ cm} = (186.2)(0.01) \text{ m} = 1.862 \text{ m}$

33. $6.5 \text{ km} = (6.5)(1\ 000\ 000) \text{ mm} = 6\ 500\ 000 \text{ mm}$

35. $24 \text{ hm} = (25)(0.1) \text{ km} = 2.4 \text{ km}$

37. $40\ 302 \text{ m}\ell = (40\ 302)(0.000\ 1) \text{ da}\ell = 4.030\ 2 \text{ da}\ell$

39. $5.8 \text{ m} = (5.8)(100) \text{ cm} = 580 \text{ cm}$

41. $50 \text{ cm} = (50)(0.01) \text{ m} = 0.5 \text{ m}$

43. $1270 \text{ kg} = (1270)(1000) \text{ g} = 1\ 270\ 000 \text{ g}$

45. $56.3 \text{ km/h} = (56.3)(1\ 000) \text{ m/h} = 56\ 300 \text{ m/h}$

47. 620 cm, 4.4 dam, 0.52 km

49. 1.4 kg, 1600 g, 16 300 dg

51. 105 000 mm, 2.6 km, 52.6 hm

53. 1 hectometer in 10 min is faster, since 1 hm > 1 dam.

55. The pump that removes 1 $\text{da}\ell$ of water per min is faster, since $1 \text{ da}\ell > 1 \text{ d}\ell$.

57. a) $4302 \text{ m} - 3776 \text{ m} = 526 \text{ m}$
    b) $526 \text{ m} = (526)(100) \text{ cm} = 52\ 600 \text{ cm}$
    c) $526 \text{ m} = (526)(0.01) \text{ hm} = 5.26 \text{ hm}$

59. a) $(2)(250)(7) = 3500 \text{ mg}$
    b) $3500 \text{ mg/week} = 3.5 \text{ g}$

61. a) $6(360) = 2160 \text{ m}\ell$
    b) $2160 \text{ m}\ell = (2160)(0.001)\ \ell = 2.16\ \ell$
    c) $\frac{2.45}{2.16} \approx \$1.13$ per liter

63. a) $(4)(27) = 108 \text{ m}$
    b) $108 \text{ m} = 0.108 \text{ km}$
    c) $108 \text{ m} = 108\ 000 \text{ km}$

65. 1 gigameter = 1000 megameters (Refer to the chart on page 403.)

67. 1 teraliter $= 1 \times 10^{24}$ picoliters = 1 000 000 000 000 000 000 000 000 picoliters

69. 9000 cm = 9 dam

71. 0.000 06 hg = 6 mg

73. $0.02 \text{ k}\ell = 2 \text{ da}\ell$

Section 7.2: Length, Area, and Volume

1. Length
3. Area
5. Length
7. Volume
9. Volume
11. Centimeters
13. Meters
15. Centimeters
17. Kilometers
19. c; 120 m
21. c; 2070 km
23. b; 168 m
25. Answers will vary.
27. Answers will vary.
29. Answers will vary.
31. Square meters
33. Square kilometers or hectares
35. Square centimeters or square millimeters
37. a; 50 cm2
39. b; $\frac{1}{8}$ ha
41. a; 100 cm^2
43. Answers will vary.
45. Answers will vary.
47. Answers will vary.
49. Milliliters
51. Kiloliters
53. Cubic meters
55. Cubic meters
57. c; 7780 cm^3
58. a; 2500 $k\ell$
59. b; 45 m^3
61. b; 120 $\text{m}\ell$
63. b; 355 $\text{m}\ell$
65. a) Answers will vary.
    b) $V = lwh = (46)(45)(5) = 10\,580 \text{ cm}^3$
67. a) Answers will vary.
    b) $V = \pi r^2 h = \pi\left(\frac{0.5}{2}\right)^2 (1) = \pi(0.25)^2(1)$
    $\approx 0.20 \text{ m}^3$
69. A cubic decimeter
71. A cubic centimeter
73. Longer side ≈ 4 cm, shorter side ≈ 2.2 cm,
    $A = lw \approx (4)(2.2) = 8.8 \text{ cm}^2$
75. $A = \pi r^2 = \pi\left(\frac{32}{2}\right)^2 = \pi(16)^2 = 256\pi$
    $\approx 804.25 \text{ m}^2$
77. a) $(4)(0.8) = 3.2 \text{ km}^2$
    b) $(3.2)(100) = 320$ ha
79. a) $A = (42.4)(32.5) = 1378 \text{ m}^2$
    b) $\left(1378 \text{ m}^2\right)\left(\frac{0.0001 \text{ ha}}{1 \text{ m}^2}\right) = 0.1378 \text{ ha}$
81. $A = (37)(28) = 1036 \text{ cm}^2$
    $A_{\text{matting}} = 2540 - 1036 = 1504 \text{ cm}^2$
83. a) $V = lwh = (70)(40)(20) = 56\,000 \text{ cm}^3$
    b) $56\,000 \text{ cm}^3 = 56\,000 \text{ m}\ell$
    c) $(56\,000 \text{ m}\ell)\left(\frac{1\ \ell}{1000 \text{ m}\ell}\right) = 56\ \ell$
85. $10^2 = 100$ times larger
87. $10^3 = 1000$ times larger
89. $\left(1 \text{ cm}^2\right)\left(\frac{100 \text{ mm}}{1 \text{ cm}}\right)^2 = 100 \text{ mm}^2$
91. $\left(1 \text{ cm}^2\right)\left(\frac{1 \text{ m}}{100 \text{ cm}}\right)^2 = 0.0001 \text{ m}^2$
93. $\left(1 \text{ m}^3\right)\left(\frac{100 \text{ m}}{1 \text{ m}}\right)^3 = 1\,000\,000 \text{ cm}^3$
95. $724 \text{ cm}^3 = 724 \text{ m}\ell$
97. $189 \text{ k}\ell = 189 \text{ m}^3$
99. $60 \text{ m}^3 = 60 \text{ k}\ell$

101. $5.3\text{ k}\ell = 5300\ \ell = (5300\ \ell)\left(\frac{1\text{ dm}^3}{1\ \ell}\right) = 5300\text{ dm}^3$

103. a) $1\text{ mi}^2 = (1\text{ mi}^2)\left(\frac{5280\text{ ft}}{1\text{ mi}}\right)^2 = 27,878,400\text{ ft}^2$

$27,878,400\text{ ft}^2 = (27,878,400\text{ ft}^2)\left(\frac{12\text{ in}}{1\text{ ft}}\right)^2 = 4,014,489,600\text{ in}^2$

b) Answers will vary. It is easier to convert in the metric system because it is a base 10 system.

Section 7.3: Mass and Temperature

1. Mass
3. Kilogram
5. Celsius
7. 0°C
9. Kilograms
11. Grams or kilograms
13. Grams
15. Kilograms
17. Kilograms or metric tonnes
19. c; 1.4 kg
21. a; 5.3 g
23. b; 190 kg
25. Answers will vary.
27. Answers will vary.
29. c; 0°C
31. b; Dress warmly and walk.
33. c; 1200°C
35. b; −5°C

37. $F = \frac{9}{5}(25) + 32 = 45 + 32 = 77°\text{F}$

39. $C = \frac{5}{9}(-25 - 32) = \frac{5}{9}(-57) \approx -31.7°\text{C}$

41. $C = \frac{5}{9}(0 - 32) = \frac{5}{9}(-32) \approx -17.8°\text{C}$

43. $F = \frac{9}{5}(37) + 32 \approx 66.6 + 32 = 98.6°\text{F}$

45. $C = \frac{5}{9}(10 - 32) = \frac{5}{9}(-22) \approx -12.2°\text{C}$

47. $F = \frac{9}{5}(0) + 32 = 0 + 32 = 32°\text{F}$

49. $C = \frac{5}{9}(-40 - 32) = \frac{5}{9}(-72) = -40.0°\text{C}$

51. $F = \frac{9}{5}(22.0) + 32 = 39.6 + 32 = 71.6°\text{F}$

53. $F = \frac{9}{5}(19.7) + 32 = 35.46 + 32 = 67.46°\text{F}$

55. Minimum: $F = \frac{9}{5}(17.8) + 32 = 32.04 + 32 = 64.04°\text{F}$; Maximum: $F = \frac{9}{5}(23.5) + 32 = 42.3 + 32 = 74.3°\text{F}$

57. $560\text{ g} + 62\text{ g} + 130\text{ m}\ell = 560\text{ g} + 62\text{ g} + 130\text{ g} = 752\text{ g}$

59. a) $(1\text{ kg})(1000\text{ g})\left(\frac{7\text{ euros}}{100\text{ g}}\right) = 70\text{ euros}$

b) $(500\text{ g})\left(\frac{7\text{ euros}}{100\text{ g}}\right) = 35\text{ euros}$

c) $(21\text{ euros})\left(\frac{100\text{ g}}{7\text{ euros}}\right) = 300\text{ g}$

61. a) $V = (20)(10)(8) = 1600\text{ m}^3$

b) $1600\text{ m}^3 = 1600\text{ k}\ell$

c) $1600\text{ k}\ell = 1600\text{ t}$

63. $3.5\text{ kg} = (3.5\text{ kg})\left(\frac{1\text{ t}}{1000\text{ kg}}\right) = 0.003\ 5\text{ t}$

65. 1 460 000 mg

$= (1\ 460\ 000\text{ mg})\left(\frac{1\text{ kg}}{1000\text{ mg}}\right)\left(\frac{1\text{ t}}{1000\text{ kg}}\right)$

$= 0.001\ 46\text{ t}$

67. a) Yes; mass is a measure of the amount of matter in an object.
    b) No; weight is a measure of gravitational force.

69. a) $1.2\ \ell = 1200\ \text{m}\ell$, with a mass of 1200 g.
    b) $1200\ \text{m}\ell = 1200\ \text{cm}^3$

71. $(3\,\text{kg})(2\,\text{m}) = (400\ \text{cm})(?\,\text{g})$
$(3000\ \text{g})(200\ \text{cm}) = (400\ \text{cm})(?\,\text{g})$
$? = 1500\ \text{g}$

Section 7.4: Dimensional Analysis and Conversions to and from the Metric System

1. Dimensional

3. $\dfrac{60\ \text{sec}}{1\ \text{min}}$ and $\dfrac{1\ \text{min}}{60\ \text{sec}}$

5. $\dfrac{100\ \text{cm}}{1\ \text{m}}$ and $\dfrac{1\ \text{m}}{100\ \text{cm}}$

7. a) $\dfrac{1\ \text{lb}}{0.45\ \text{kg}}$; Since we need to eliminate kilograms, kg must appear in the denominator. Since we need to convert to pounds, lb must appear in the numerator.
   b) $\dfrac{0.45\ \text{kg}}{1\ \text{lb}}$; Since we need to eliminate pounds; lb must appear in the denominator. Since we need to convert to kilograms, kg must appear in the numerator.

9. a) $\dfrac{3.8\ \ell}{1\ \text{gal}}$; Since we need to eliminate gallons, gal must appear in the denominator. Since we need to convert to liters, ℓ must appear in the numerator.
   b) $\dfrac{1\ \text{gal}}{3.8\ \ell}$; Since we need to eliminate liters, ℓ must appear in the denominator. Since we need to convert to gallons, gal must appear in the numerator.

11. $18\ \text{lb} = (18\ \text{lb})\left(\dfrac{0.45\ \text{kg}}{1\ \text{lb}}\right) = 8.1\ \text{kg}$

13. $425\ \text{g} = (425\ \text{g})\left(\dfrac{1\ \text{oz}}{28\ \text{g}}\right) \approx 15.18\ \text{oz}$

15. $18 = 75\ \text{kg} = (175\ \text{kg})\left(\dfrac{1\ \text{lb}}{0.45\ \text{kg}}\right) \approx 388.89\ \text{lb}$

17. $39\ \text{mi} = (39\ \text{mi})\left(\dfrac{1.6\ \text{km}}{1\ \text{mi}}\right) = 62.4\ \text{km}$

19. $675\ \text{ha} = (675\ \text{ha})\left(\dfrac{1\ \text{acre}}{0.4\ \text{ha}}\right) = 1687.5\ \text{acres}$

21. $25.2\ \ell = (25.2\ \ell)\left(\dfrac{1\ \text{pt}}{0.47\ \ell}\right) \approx 53.62\ \text{pints}$

23. $3.8\ \text{km}^2 = (3.8\ \text{km}^2)\left(\dfrac{1\ \text{mi}^2}{2.6\ \text{km}^2}\right) \approx 1.46\ \text{mi}^2$

25. $120\ \text{lb} = (120\ \text{lb})\left(\dfrac{0.45\ \text{kg}}{1\ \text{lb}}\right) = 54\ \text{kg}$

27. $414\ \text{m} = (414\ \text{m})\left(\dfrac{1\ \text{yd}}{0.9\ \text{m}}\right) \approx 460\ \text{yd}$

29. $357\ \text{m} = (357\ \text{m})\left(\dfrac{100\ \text{cm}}{1\ \text{m}}\right)\left(\dfrac{1\ \text{ft}}{30.5\ \text{cm}}\right)$
$\approx 1170.49\ \text{ft}$

31. $80\ \text{kph} = (80\ \text{kph})\left(\dfrac{1\ \text{mi}}{1.6\ \text{km}}\right) = 50\ \text{mph}$

33. $(15\ \text{yd})(10\ \text{yd}) = 150\ \text{yd}^2 = (150\ \text{yd}^2)\left(\dfrac{0.8\ \text{m}^2}{1\ \text{yd}^2}\right)$
$= 120\ \text{m}^2$

35. $6\ \text{g} = (6\ \text{g})\left(\dfrac{1\ \text{oz}}{28\ \text{g}}\right) \approx 0.21\ \text{oz}$

37. $(91{,}696\ \text{acres})\left(\dfrac{0.4\ \text{ha}}{1\ \text{acre}}\right) = 36\ 678.4\ \text{ha}$

39. a) $1{,}000{,}000\ \text{ft} = (1{,}000{,}000\ \text{ft})\left(\dfrac{0.305\ \text{ft}}{1\ \text{m}}\right)$
$= 305\ 000\ \text{m}$
    b) $305\ 000\ \text{m}\left(\dfrac{1\ \text{km}}{1000\ \text{m}}\right) = 305\ \text{km}$

41. a) $3.4\text{ fl oz} = (3.4\text{ fl oz})\left(\frac{30\text{ m}\ell}{1\text{ fl oz}}\right) = 102\text{ m}\ell$

b) 3.4 fluid ounces is greater.

43. a) $15\text{ mi} + 1800\text{ mi} + 1400\text{ mi} + 760\text{ mi}$
$= 3975\text{ mi}$

b) $3975\text{ mi} = (3975\text{ mi})\left(\frac{1.6\text{ km}}{1\text{ mi}}\right) = 6360\text{ km}$

c) $12.2\text{ km} = (12.2\text{ km})\left(\frac{1\text{ mi}}{1.6\text{ km}}\right) \approx 7.63\text{ mi}$

d) $24\text{ km} = (24\text{ km})\left(\frac{1\text{ mi}}{1.6\text{ km}}\right) = 15\text{ mi}$

45. a) $18.5\text{ acres} = (18.5\text{ acres})\left(\frac{0.4\text{ ha}}{1\text{ acre}}\right) = 7.4\text{ ha}$

b) $7.4\text{ ha} = (7.4\text{ ha})\left(\frac{10\ 000\text{ m}^2}{1\text{ ha}}\right)$
$= 74\ 000\text{ m}^2$

47. $120\text{ lb} = (120\text{ lb})\left(\frac{0.45\text{ kg}}{1\text{ lb}}\right)\left(\frac{1.5\text{ mg}}{1\text{ kg}}\right) = 81\text{ mg}$

49. $76\text{ lb} = (76\text{ lb})\left(\frac{0.45\text{ kg}}{1\text{ lb}}\right)\left(\frac{200\text{ mg}}{1\text{ kg}}\right) = 6840\text{ mg} = (6840\text{ mg})\left(\frac{1\text{ g}}{1000\text{ mg}}\right) = 6.84\text{ g}$

51. a) $2\text{ tablespoons} = (2\text{ tablespoons})\left(\frac{236\text{ mg}}{1\text{ tablespoon}}\right) = 472\text{ mg}$

b) $8\text{ fl oz} = (8\text{ fl oz})\left(\frac{30\text{ m}\ell}{1\text{ fl oz}}\right)\left(\frac{1\text{ tablespoon}}{15\text{ m}\ell}\right)\left(\frac{236\text{ mg}}{1\text{ tablespoon}}\right) = 3776\text{ mg}$

53. a) $32\text{ kph} = (32\text{ kph})\left(\frac{1\text{ mi}}{1.6\text{ km}}\right) = 20\text{ mph}$ for the roadrunner's speed. Usain Bolt ran faster.

b) $42\text{ kph} = (42\text{ kph})\left(\frac{1\text{ mi}}{1.6\text{ km}}\right) = 26.25\text{ mph}$ for the roadrunner's speed. The roadrunner ran faster.

55. a) $9730\text{ m} = (9730\text{ m})\left(\frac{1\text{ km}}{1000\text{ m}}\right)\left(\frac{1\text{ mi}}{1.6\text{ km}}\right)$
$\approx 6.08\text{ mi}$

b) $709\text{ kph} = (709\text{ kph})\left(\frac{1\text{ mi}}{1.6\text{ km}}\right)$
$\approx 443.13\text{ mph}$

c) $181\text{ kph} = (181\text{ kph})\left(\frac{1\text{ mi}}{1.6\text{ km}}\right)$
$\approx 113.13\text{ mph}$

d) $-45°\text{C} = \frac{9}{5}(-45) + 32 = -49°\text{F}$

57. a) $\frac{2.40\text{ €}}{1\text{ kg}} = \left(\frac{2.40\text{ €}}{1\text{ kg}}\right)\left(\frac{0.45\text{ kg}}{1\text{ lb}}\right) = \frac{1.08\text{ €}}{1\text{ lb}}$

b) $\frac{1.08\text{ €}}{1\text{ lb}} = \left(\frac{1.08\text{ €}}{1\text{ lb}}\right)\left(\frac{\$1.14}{1\text{ €}}\right) = \frac{\$1.23}{1\text{ lb}}$

59. a) $\frac{745\text{ kph}}{402\text{ knots}} \approx \frac{1.85\text{ kph}}{1\text{ knot}}$

b) $\frac{402\text{ knots}}{463\text{ mph}} \approx \frac{0.87\text{ knot}}{1\text{ mph}}$

c) $\frac{463\text{ mph}}{745\text{ kph}} \approx \frac{0.62\text{ mph}}{1\text{ kph}}$

61. b; $(0.2\text{ mg})\left(\frac{1\text{ grain}}{60\text{ mg}}\right)\left(\frac{1\text{ grain}}{300\text{ m}\ell}\right) = 1\text{ m}\ell = 1.0\text{ cc}$

63. a) $(6.2\ \ell)\left(\frac{1000\text{ m}\ell}{1\ \ell}\right)\left(\frac{1\text{ cm}^3}{1\text{ m}\ell}\right) = 6200\text{ cc}$

b) $\left(6200\text{ cm}^3\right)\left(\frac{1\text{ in.}}{2.54\text{ cm}}\right)^3 \approx 378.35\text{ in.}^3$

Review Exercises

1. $\frac{1}{100}$ of basic unit

2. $1000 \times$ basic unit

3. $100 \times$ basic unit

4. $\frac{1}{1000}$ of basic unit

5. $10 \times$ basic unit

6. $\frac{1}{10}$ of basic unit

7. $(80\,\text{mg})\left(\dfrac{1\,\text{g}}{1000\,\text{mg}}\right) = 0.08\,\text{g}$

8. $(0.52\text{ km})\left(\dfrac{10{,}000\text{ dm}}{1\text{ km}}\right) = 5200\text{ dm}$

9. $(5700\,\text{cm})\left(\dfrac{0.0001\,\text{hm}}{1\,\text{cm}}\right) = 0.57\,\text{hm}$

10. $(1\ 000\ 000\,\text{mg})\left(\dfrac{0.000\ 001\,\text{kg}}{1\,\text{mg}}\right) = 1\,\text{kg}$

11. $(4.6\,\text{k}\ell)\left(\dfrac{1000\ell}{\text{k}\ell}\right) = 4620\ell$

12. $(192.6\text{ dag})\left(\dfrac{100\text{ dm}}{1\text{ dag}}\right) = 19\ 260\,\text{dg}$

13. $2.67\text{ k}\ell = 2\ 670\ 000\text{ m}\ell$
$14\ 630\text{ c}\ell = 146\ 300\text{ m}\ell$
$3000\text{ m}\ell,\ 14\ 630\text{ c}\ell,\ 2.67\text{ k}\ell$

14. $0.047\text{ km} = 47\text{ m}$
$47000\text{ cm} = 470\text{ m}$
$0.047\text{ km},\ 47\ 000\text{ cm},\ 4700\text{ m}$

15. Degrees Celsius

16. Centimeters

17. Square meters

18. Milliliters or liters

19. Kilograms or tonnes

20. Kilometers

21. a) Answers will vary.
b) Answers will vary.

22. a) Answers will vary.
b) Answers will vary.

23. b; 2000 km

24. a; 1300 kg

25. c; 4 ℓ

26. a; 200 m^2

27. a; 34°C

28. b; 30 m

29. $1800\text{ kg} = (1800\text{ kg})\left(\dfrac{1\text{ t}}{1000\text{ kg}}\right) = 1.8\text{ t}$

30. $9.2\text{ t} = (9.2\text{ t})\left(\dfrac{1000\text{ kg}}{1\text{ t}}\right)\left(\dfrac{1000\text{ g}}{1\text{ kg}}\right)$
$= 9\ 200\ 000\text{ g}$

31. $F = \frac{9}{5}(24) + 32 = 75.2°\text{F}$

32. $C = \frac{5}{9}(68 - 32) = 20°\text{C}$

33. $F = \frac{5}{9}(-6 - 32) = -21.\overline{1} \approx -21.1°\text{C}$

34. $C = \frac{9}{5}(39) + 32 = 102.2°\text{F}$

35. $l \approx 4$ cm, $w \approx 1.6$ cm; $A = lw \approx 4(1.6)$
$= 6.4\text{ cm}^2$

36. $r \approx 1.5$ cm; $A = \pi r^2 \approx \pi(1.5)^2 = 2.25\pi$
$\approx 7.07\text{ cm}^2$

37. a) $V = \pi r^2 h = \pi\left(\frac{2}{1}\right)^2(1) = \pi \approx 3.14\text{ m}^3$
b) $(3.14\text{ m}^3)\left(\dfrac{1000\text{ kg}}{1\text{ m}^3}\right) \approx 3140\text{ kg}$

38. a) $A = lw = (33.7)(26.7) = 899.79\text{ cm}^2$
b) $899.79\text{ cm}^2 = (899.79\text{ cm}^2)\left(\dfrac{1\text{ m}^2}{10\ 000\text{ cm}^2}\right)$
$= 0.089\ 979\text{ m}^2$

39. a) $V = lwh = (90)(50)(40) = 180\ 000\text{ cm}^3$
b) $180\ 000\text{ cm}^3 = (180\ 000\text{ cm}^3)\left(\dfrac{1\text{ m}^3}{(100)^3\text{ cm}^3}\right) = 0.18\text{ m}^3$

39. (continued)

c) $180\ 000\ \text{cm}^3 = \left(180\ 000\ \text{cm}^3\right)\left(\frac{1\ \text{m}\ell}{1\ \text{cm}^3}\right) = 180\ 000\ \text{m}\ell$

d) $0.18\ \text{m}^3 = \left(0.18\ \text{m}^3\right)\left(\frac{1\ \text{k}\ell}{1\ \text{m}^3}\right) = 0.18\ \text{k}\ell$

40. Since $1\ \text{km} = 100 \times 1\ \text{dam}$, $1\ \text{km}^2 = 100^2 \times 1\ \text{dam}^2 = 10\ 000\ \text{dam}^2$. Thus, 1 square kilometer is 10,000 times larger than a square dekameter.

41. $(41\ \text{kg})\left(\frac{1\ \text{lb}}{0.45\ \text{kg}}\right) \approx 91.11\ \text{lb}$

42. $(18\ \text{cm})\left(\frac{1\ \text{in.}}{2.54\ \text{cm}}\right) \approx 7.09\ \text{in.}$

43. $(37\ \text{yd})\left(\frac{0.9\ \text{m}}{1\ \text{yd}}\right) = 33.3\ \text{m}$

44. $(100\ \text{m})\left(\frac{1\ \text{yd}}{0.9\ \text{m}}\right) \approx 111.11\ \text{yd}$

45. $(52\ \text{mi})\left(\frac{1.6\ \text{km}}{1\ \text{mi}}\right) = 83.2\ \text{km}$

46. $(27\ \ell)\left(\frac{1\ \text{qt}}{0.95\ \ell}\right) \approx 28.42\ \text{qt}$

47. $(20\ \text{gal})\left(\frac{3.8\ \ell}{1\ \text{gal}}\right) = 76\ \ell$

48. $\left(60\ \text{m}^3\right)\left(\frac{1\ \text{yd}^3}{0.76\ \text{m}^3}\right) \approx 78.95\ \text{yd}^3$

49. $\left(96\ \text{cm}^2\right)\left(\frac{1\ \text{in.}^2}{6.5\ \text{cm}^2}\right) \approx 14.77\ \text{in.}^2$

50. $(4\ \text{qt})\left(\frac{0.95\ \ell}{1\ \text{qt}}\right) = 3.8\ \ell$

51. a) $(137\ \text{cm})\left(\frac{1\ \text{in.}}{2.54\ \text{cm}}\right) \approx 53.94\ \text{in.}$

b) $(44.2\ \text{kg})\left(\frac{1\ \text{lb}}{0.45\ \text{kg}}\right) \approx 98.22\ \text{lb}$

52. $(241\ \text{mi})\left(\frac{1.6\ \text{km}}{1\ \text{mi}}\right) = 385.6\ \text{km}$

53. $A = lw = (24)(15) = 360\ \text{ft}^2 = \left(360\ \text{ft}^2\right)\left(\frac{0.09\ \text{m}^2}{1\ \text{ft}^2}\right) = 32.4\ \text{m}^2$

54. a) $(50{,}000\ \text{gal})\left(\frac{3.8\ \ell}{1\ \text{gal}}\right)\left(\frac{1\ \text{k}\ell}{1000\ \ell}\right) = 190\ \text{k}\ell$

b) $(190\ \text{k}\ell)\left(\frac{1000\ \ell}{1\ \text{k}\ell}\right)\left(\frac{1\ \text{kg}}{1\ \ell}\right) = 190\ 000\ \text{kg}$

55. a) $70\ \text{km/hr} = (70\ \text{km/hr})\left(\frac{1\ \text{m}}{1.6\ \text{km}}\right)$

$= 43.75\ \text{mi/hr}$

b) $70\ \text{km/hr} = (70\ \text{km/hr})\left(\frac{1000\ \text{m}}{1\ \text{km}}\right)$

$= 70\ 000\ \text{m/hr}$

56. a) $V = lwh = (90)(70)(40) = 252\ 000\ \text{cm}^3 = \left(252\ 000\ \text{cm}^3\right)\left(\frac{1\ \text{m}\ell}{1\ \text{cm}^3}\right)\left(\frac{1\ \ell}{1000\ \text{m}\ell}\right) = 252\ \ell$

b) $252\ \ell = (252\ \ell)\left(\frac{1\ \text{kg}}{1\ \ell}\right) = 252\ \text{kg}$

57. $1\ \text{kg} = (1\ \text{kg})\left(\frac{1\ \text{lb}}{0.45\ \text{kg}}\right) = 2.\overline{2}\ \text{lb}; \frac{\$3.50}{2.\overline{2}} = \$1.575 \approx \$1.58$ per pound

Chapter Test

1. $2400\ \ell = (2400)(0.001) = 2.4\ \text{k}\ell$

2. $46.2\ \text{cm} = (46.2)(0.0001) = 0.004\ 62\ \text{hm}$

3. $1\text{ km} = (1\text{ km})\left(\frac{100\text{ dam}}{1\text{ km}}\right) = 100\text{ dam}$ or 100 times greater

4. $400(6) = 2400\text{ m}; (2400\text{ m})\left(\frac{1\text{ km}}{1000\text{ m}}\right)$
$= 2.4\text{ km}$

5. b; 20 cm

6. a; 2 m^2

7. c; 75 ℓ

8. b; 500 km

9. c; 24°C

10. $1\text{ m}^2 = (1\text{ m}^2)\left(\frac{100\text{ cm}}{1\text{ m}}\right)^2 = 10\ 000\text{ cm}^2$ or 10,000 times greater

11. $1\text{ m}^3 = (1\text{ m}^3)\left(\frac{1000^2\text{ mm}^3}{1\text{ m}^3}\right) = 1\ 000\ 000\text{ cm}^3$ or 1,000,000 times greater

12. $88\text{ lb} = (88\text{ lb})\left(\frac{0.45\text{ kg}}{1\text{ lb}}\right) = 39.6\text{ kg}$

13. $169.29\text{ m} = (169.29\text{ m})\left(\frac{1\text{ ft}}{0.305\text{ m}}\right) \approx 555.05\text{ ft}$

14. $53\ 321\text{ km}^2 = (53\ 321\text{ km}^2)\left(\frac{1\text{ mi}^2}{2.6\text{ km}^2}\right) \approx 20{,}508.1\text{ mi}^2$

15. $F = \frac{5}{9}(40-32) = \frac{5}{9}(8) \approx 4.44°\text{C}$

16. $C = \frac{9}{5}(-5)+32 = -9+32 = 23°\text{F}$

17. a) $300\text{ kg} = (300\text{ kg})\left(\frac{1000\text{ g}}{1\text{ kg}}\right) = 300\ 000\text{ g}$

b) $300\text{ kg} = (300\text{ kg})\left(\frac{1\text{ lb}}{0.45\text{ kg}}\right) \approx 666.67\text{ lb}$

18. a) $V = lwh = 20(20)(8) = 3200\text{ m}^3$

b) $3200\text{ m}^3 = (3200\text{ m}^3)\left(\frac{1000\text{ k}\ell}{1\text{ m}^3}\right) = 3\ 200\ 000\ell$ (or 3200 $k\ell$)

c) $3200\text{ k}\ell = (3200\text{ k}\ell)\left(\frac{1000\ \ell}{1\text{ k}\ell}\right)\left(\frac{1\text{ kg}}{1\ \ell}\right) = 3\ 200\ 000\text{ kg}$

19. Total surface area: $2lh + 2wh = 2(20)(6) + 2(15)(6) = 420\text{ m}^2$

Liters needed for first coat: $(420\text{ m}^2)\left(\frac{1\ \ell}{10\text{ m}^2}\right) = 42\ \ell$

Liters needed for second coat: $(420\text{ m}^2)\left(\frac{1\ \ell}{15\text{ m}^2}\right) = 28\ \ell$

Total liters needed: $42 + 28 = 70\ \ell$

Total cost: $(70\ \ell)\left(\frac{\$3.50}{1\ \ell}\right) = \$245$

20. a) $\left(\frac{1.61\text{ euros}}{1\ \ell}\right)\left(\frac{3.8\ \ell}{1\text{ gal}}\right) \approx 6.12$ euros per gallon

b) $6.12\text{ euros per gallon} = (6.12\text{ euros per gallon})\left(\frac{\$1}{0.87\text{ euros}}\right) \approx \$7.03$ per gallon

Chapter Eight: Geometry

Section 8.1: Points, Lines, Planes, and Angles

1. Parallel
3. Angle
5. Supplementary
7. Right
9. Acute
11. Line segment, $\overline{AB}$
13. Ray, $\overrightarrow{AB}$
15. Ray, $\overrightarrow{BA}$
17. Line, $\overleftrightarrow{AB}$
19. $\overline{AD}$
21. $\overset{\circ\!\!-\!\!\circ}{BD}$
23. $\overleftrightarrow{BC}$
25. $\overrightarrow{IC}$
27. ΔBCF
29. $\{F\}$
31. $\measuredangle CBG$ or $\measuredangle GBC$
33. $\overrightarrow{BC}$
35. $\{B\}$
37. $\overline{BF}$
39. $\measuredangle ABE$ or $\measuredangle EBA$
41. $\measuredangle CBG$ or $\measuredangle GBC$
43. Acute
45. Obtuse
47. None of these angles
49. $90° - 13° = 77°$
51. $90° - 32\frac{3}{4}° = 57\frac{1}{4}°$
53. $90° - 64.7° = 25.3°$
55. $180° - 29° = 151°$
57. $180° - 20.5° = 159.5°$
59. $180° - 43\frac{5}{7}° = 136\frac{2}{7}°$
61. c
63. b
65. a
67. Let $x =$ measure of $\measuredangle 1$
    $5x =$ measure of $\measuredangle 2$
    $$x + 5x = 90$$
    $$6x = 90$$
    $$x = \frac{90}{6} = 15°,\ m\measuredangle 1$$
    $$5x = (5)(15) = 75°,\ m\measuredangle 2$$
69. Let $x =$ measure of $\measuredangle 1$
    $180 - x =$ measure of $\measuredangle 2$
    $$x - (180 - x) = 102$$
    $$x - 180 + x = 102$$
    $$2x - 180 = 102$$
    $$2x = 282$$
    $$x = \frac{282}{2} = 141°, m\measuredangle 1$$
    $$180 - x = 180 - 141 = 39°, m\measuredangle 2$$
71. $m\measuredangle 1 + 130° = 180°$
    $m\measuredangle 1 = 50°$
    $m\measuredangle 2 = m\measuredangle 1$ (vertical angles)
    $m\measuredangle 3 = 130°$ (vertical angles)
    $m\measuredangle 5 = m\measuredangle 2$ (alternate interior angles)
    $m\measuredangle 4 = m\measuredangle 3$ (alternate interior angles)
    $m\measuredangle 7 = m\measuredangle 4$ (vertical angles)
    $m\measuredangle 6 = m\measuredangle 5$ (vertical angles)
    Measures of angles 3, 4, and 7 are each 130°.
    Measures of angles 1, 2, 5, and 6 are each 50°.
73. $m\measuredangle 3 + 20° = 180°$
    $m\measuredangle 3 = 160°$
    $m\measuredangle 1 = 20°$ (vertical angles)
    $m\measuredangle 2 = m\measuredangle 3$ (vertical angles)
    $m\measuredangle 4 = m\measuredangle 1$ (corresponding angles)
    $m\measuredangle 7 = m\measuredangle 4$ (vertical angles)
    $m\measuredangle 6 = m\measuredangle 3$ (alternate interior angles)
    $m\measuredangle 5 = m\measuredangle 6$ (vertical angles)
    Measures of angles 1, 4, and 7 are each 20°.
    Measures of angles 2, 3, 5, and 6 are each 160°.
75. $$x + 2x + 12 = 90$$
    $$3x + 12 = 90$$
    $$3x = 78$$
    $$x = 26°,\ m\measuredangle 2$$
    $$90 - x = 90 - 26 = 64°,\ m\measuredangle 1$$

77. $x + 2x - 9 = 90$

$3x - 9 = 90$

$3x = 99$

$x = 33°,\ m\measuredangle 1$

$90 - x = 90 - 33 = 57°,\ m\measuredangle 2$

79. $x + 3x - 4 = 180$

$4x - 4 = 180$

$4x = 184$

$x = 46°,\ m\measuredangle 2$

$180 - x = 180 - 46 = 134°,\ m\measuredangle 1$

81. $x + 5x + 6 = 180$

$6x + 6 = 180$

$6x = 174$

$x = 29°,\ m\measuredangle 1$

$180 - x = 180 - 29 = 151°,\ m\measuredangle 2$

For Exercises 83–89, the answers given are one of many possible answers.

83. $\overleftrightarrow{EF}$ and $\overleftrightarrow{DG}$

85. Plane *ABG* and plane *JCD*

87. Plane $ABG \cap$ plane $ABC \cap$ plane $BCD = \{B\}$

89. $\overrightarrow{BC} \cap$ plane $ABG = \{B\}$

91. Always true. If any two lines are parallel to a third line, then they must be parallel to each other.

93. Sometimes true. Vertical angles are only complementary when each is equal to $45°$.

95. Sometimes true. Alternate interior angles are only complementary when each is equal to $45°$.

97. a) An infinite number of lines can be drawn through a given point.
    b) An infinite number of planes can be drawn through a given point.

99. An infinite number of planes can be drawn through a given line.

101. a) Answers will vary.
     b) Answers will vary.
     c) Answers will vary.
     d) Answers will vary.

103. No. Line *l* and line *n* may be parallel or skew.

105. a) Other answers are possible.

b) Let $m\measuredangle ABC = x$ and $m\measuredangle CBD = y$.

$x + y = 90°$ and $y = 2x$

$x + y = 90°$

$x + (2x) = 90°$

$3x = 90°$

$x = 30° = m\measuredangle ABC$

c) $m\measuredangle CBD = y$

$y = 2x = 2(30°) = 60°$

d) $m\measuredangle ABD + m\measuredangle DBE = 180°$

$m\measuredangle ABD = x + y = 30° + 60° = 90°.$

$90° + m\measuredangle DBE = 180°$

$m\measuredangle DBE = 180° - 90° = 90°.$

Section 8.2: Polygons

1. Polygon

3. Proportion

5. Congruent

7. a) Pentagon
   b) Regular

9. a) Heptagon
   b) Regular

11. a) Heptagon
    b) Not regular

13. a) Decagon
    b) Regular

15. a) Isosceles
    b) Acute

17. a) Isosceles
    b) Right

19. a) Scalene
    b) Acute

21. a) Equilateral
    b) Acute

23. Square

25. Trapezoid

27. Rhombus

29. The measures of the other two angles of the triangle are 37° (by vertical angles) and $180° - 132°$ (supplementary angles). The measure of the third angle of the triangle is $180° - (37°) - (180° - 132°) = 95°$. Since angle x is a vertical angle with the $95°$ angle, the measure of angle x is $95°$.

31. The measures of two angles of the triangle are $180° - 105°$ and $180° - 133°$ (supplementary angles). The measure of the third angle of the triangle is $180° - (180° - 105°) - (180° - 133°) = 58°$. Since angle x is a vertical angle with the 58° angle, the measure of angle x is 58°.

33.

Angle	Measure	Reason
1	90°	∡ 1 and ∡ 7 are vertical angles
2	50°	∡ 2 and ∡ 4 are corresponding angles
3	130°	∡ 3 and ∡ 4 form a straight angle
4	50°	Vertical angle with the given 50° angle
5	50°	∡ 2 and ∡ 5 are vertical angles
6	40°	Vertical angle with the given 40° angle
7	90°	∡ 2, ∡ 6, and ∡ 7 form a straight angle
8	130°	∡ 3 and ∡ 8 are vertical angles
9	140°	∡ 9 and ∡ 10 form a straight angle
10	40°	∡ 10 and ∡ 12 are vertical angles
11	140°	∡ 9 and ∡ 11 are vertical angles
12	40°	∡ 6 and ∡ 12 are corresponding angles

35. $n = 5$; $(5-2)\times 180° = 3\times 180° = 540°$

37. $n = 8$; $(8-2)\times 180° = 6\times 180° = 1080°$

39. $n = 12$; $(12-2)\times 180° = 10\times 180° = 1800°$

41. a) The sum of the measures of the interior angles of a triangle is 180°. Dividing by 3, the number of angles, each interior angle measures 60°.
    b) Each exterior angle measures $180° - 60° = 120°$.

43. a) The sum of the measures of the interior angles of a hexagon is $(6-2)\times 180° = 4\times 180° = 720°$. Dividing by 6, the number of angles, each interior angle measures 120°.
    b) Each exterior angle measures $180° - 120° = 60°$.

45. a) The sum of the measures of the interior angles of a nonagon is $(9-2)\times 180° = 7\times 180° = 1260°$. Dividing by 9, the number of angles, each interior angle measures 140°.
    b) Each exterior angle measures $180° - 140° = 40°$.

47. $\frac{A'C'}{AC} = \frac{A'B'}{AB}$ $\quad$ $\frac{B'C'}{BC} = \frac{A'B'}{AB}$

$\frac{x}{50} = \frac{10}{25}$ $\quad$ $\frac{y}{40} = \frac{10}{25}$

$25x = 500$ $\quad$ $25y = 400$

$x = 20$ $\quad$ $y = 16$

49. $\frac{DC}{D'C'} = \frac{AB}{A'B'}$ $\quad$ $\frac{B'C'}{BC} = \frac{A'B'}{AB'}$

$\frac{x}{10} = \frac{9}{15}$ $\quad$ $\frac{y}{6} = \frac{15}{9}$

$\frac{x}{10} = \frac{3}{5}$ $\quad$ $\frac{y}{6} = \frac{5}{3}$

$5x = 30$ $\quad$ $3y = 30$

$x = 6$ $\quad$ $y = 10$

51. $\frac{D'C'}{DC} = \frac{A'D'}{AD}$ $\quad$ $\frac{A'B'}{AB} = \frac{A'D'}{AD}$

$\frac{x}{16} = \frac{22.5}{18}$ $\quad$ $\frac{y}{17} = \frac{22.5}{18}$

$18x = 360$ $\quad$ $18y = 382.5$

$x = 20$ $\quad$ $y = 21.25$

53. $\frac{BC}{EC} = \frac{AB}{DE}$

$\frac{BC}{1} = \frac{3}{1}$

$BC = 3$

55. $AD = AC - DC = 5 - \frac{5}{3} = \frac{15}{3} - \frac{5}{3} = 3\frac{1}{3}$

57. $AB = A'B' = 33$

59. $B'C' = BC = 18$

61. $m\measuredangle ACB = m\measuredangle A'C'B'$

$= 180° - 31° - 78°$

$= 71°$

63. $AD = A'D' = 9$

65. $A'B' = AB = 10$

67. $m\measuredangle DAB = m\measuredangle D'A'B'$

$= 360° - 130° - 70° - 50°$

$= 110°$

69. a) $\frac{160 \text{ mi}}{2 \text{ in.}} = \frac{\text{A-SA}}{1 \text{ in.}}$

$\text{A-SA} = \frac{(160)(1)}{2} \text{ mi} = 80 \text{ mi}$

b) $\frac{160 \text{ mi}}{2 \text{ in.}} = \frac{\text{SA-H}}{2.5 \text{ in.}}$

$\text{SA-H} = \frac{(160)(2.5)}{2} \text{ mi} = 200 \text{ mi}$

71. a) $\frac{44 \text{ mi}}{0.875 \text{ in.}} = \frac{\text{SP-A}}{2.25 \text{ in.}}$

$\text{SP-A} = \frac{(44)(2.25)}{0.875} \text{ mi} = 113.1 \text{ mi}$

b) $\frac{44 \text{ mi}}{0.875 \text{ in.}} = \frac{\text{SP-R}}{1.5 \text{ in.}}$

$\text{SP-R} = \frac{(44)(1.5)}{0.875} \text{ mi} = 75.4 \text{ mi}$

73. Let $x =$ height of silo

$\frac{x}{6} = \frac{105}{9}$

$9x = 630$

$x = 70$ ft

75. $m\measuredangle GBC = 180° - 125° = 55°$

77. $m\measuredangle DFE = 180° - 90° - 55° = 35°$

79. a) $m\measuredangle HMF = m\measuredangle TMB$, $m\measuredangle HFM = m\measuredangle TMB$, $m\measuredangle MHF = m\measuredangle MTB$

b) Let $x =$ height of the wall

$\frac{x}{20} = \frac{5.5}{2.5}$

$2.5x = 110$

$x = \frac{110}{2.5} = 44$ ft

Section 8.3: Perimeter and Area

Throughout this section, in exercises involving π, we used the π key on a scientific calculator to determine the answer. If you use 3.14 for π, your answers may vary slightly.

1. a) Perimeter
   b) Area

3. Circle

5. $A = \frac{1}{2}bh = \frac{1}{2}(1)(3) = 1.5 \text{ in.}^2$

7. $A = \frac{1}{2}bh = \frac{1}{2}(100)(50) = 2500 \text{ cm}^2$

9. a) $A = lw = 8(4) = 32 \text{ ft}^2$
   b) $P = 2l + 2w = 2(8) + 2(4) = 24 \text{ ft}$

11. $0.25 \text{ m} = 0.25(100) = 25 \text{ cm}$
    a) $A = bh = 25(24) = 6000 \text{ cm}^2$
    b) $P = 2l + 2w = 2(25) + 2(25) = 100 \text{ cm}$

13. $2 \text{ ft} = 2(12) = 24 \text{ in.}$
    a) $A = \frac{1}{2}h(b_1 + b_2) = \frac{1}{2}(24)(5 + 19) = 288 \text{ in.}^2$
    b) $P = s_1 + s_2 + b_1 + b_2 = 25 + 25 + 5 + 19$
    $= 74 \text{ in.}$

15. a) $A = \pi r^2 = \pi(8)^2 = 64\pi \approx 201.06 \text{ in.}^2$
    b) $C = 2\pi r = 2\pi(8) = 16\pi \approx 50.27 \text{ in.}$

17. $r = \frac{13}{2} = 6.5 \text{ ft}$
    a) $A = \pi r^2 = \pi(6.5)^2 = 42.25\pi \approx 132.73 \text{ ft}^2$
    b) $C = \pi d = \pi(13) \approx 40.84 \text{ ft}$

19. a) $c^2 = 15^2 + 8^2$
    $c^2 = 225 + 64$
    $c^2 = 289$
    $c = \sqrt{289} = 17 \text{ yd}$
    b) $P = s_1 + s_2 + s_3 = 8 + 15 + 17 = 40 \text{ yd}$
    c) $A = \frac{1}{2}bh = \frac{1}{2}(8)(15) = 60 \text{ yd}^2$

21. a) $b^2 + 5^2 = 13^2$
    $b^2 + 25 = 169$
    $b^2 = 144$
    $b = \sqrt{144} = 12 \text{ km}$
    b) $P = s_1 + s_2 + s_3 = 5 + 12 + 13 = 30 \text{ km}$
    c) $A = \frac{1}{2}bh = \frac{1}{2}(5)(12) = 30 \text{ km}^2$

23. Area of square: $(6)^2 = 36 \text{ ft}^2$
    Area of circle: $\pi(3)^2 = 9\pi \text{ ft}^2$
    Shaded area: $36 - 9\pi \approx 7.73 \text{ ft}^2$

25. Find the length of a side of the shaded square.
    $x^2 = 2^2 + 2^2$
    $x^2 = 4 + 4$
    $x^2 = 8$
    $x = \sqrt{8}$
    Shaded area: $\sqrt{8} \cdot \sqrt{8} = 8 \text{ in.}^2$

27. Area of Trapezoid: $18\left(\frac{9+11}{2}\right) = 180$
    Area of Triangle: $\frac{1}{2}(18)(10) = 90$
    Shaded Area: $180 - 90 = 90 \text{ yd}^2$

29. Area of Parallelogram: $bh = 10(4) = 40 \text{ ft}^2$
    Area of circle: $\pi(2)^2 = 4\pi \text{ ft}^2$
    Shaded area: $40 - 4\pi \approx 27.43 \text{ ft}^2$

31. Radius of larger circle: $\frac{12}{2} = 6 \text{ mm}$
    Area of large circle: $\pi(6)^2 = 36\pi \text{ mm}^2$
    Radius of each smaller circle: $\frac{6}{2} = 3 \text{ mm}$
    Area of each smaller circle: $\pi(3)^2 = 9\pi \text{ mm}^2$
    Shaded area: $36\pi - 9\pi - 9\pi = 18\pi$
    $\approx 56.55 \text{ mm}^2$

33. $9x = 63$

$$\frac{1}{x} = \frac{9}{63}$$

$$x = \frac{63}{9} = 7 \text{ yd}^2$$

35. $\frac{1}{13.5} = \frac{9}{x}$

$x = 13.5(9) = 121.5 \text{ ft}^2$

37. $\frac{1}{7} = \frac{10,000}{x}$

$x = 7(10,000) = 70,000 \text{ cm}^2$

39. $\frac{1}{x} = \frac{10,000}{4072}$

$10,000x = 4072$

$$x = \frac{4072}{10,000} = 0.4072 \text{ m}^2$$

41. Area of living/dining room: $25(22) = 550 \text{ ft}^2$

a) $550(11.99) = \$6594.50$

b) $550(15.99) = \$8794.50$

43. Area of kitchen: $12(14) = 168 \text{ ft}^2$

Area of first floor bathroom: $6(10) = 60 \text{ ft}^2$

Area of second floor bathroom: $8(14) = 112 \text{ ft}^2$

Area of kitchen and both bathrooms: 340 ft^2

Cost: $340(\$10.49) = \$3566.60$

45. Area of bedroom 1: $10(14) = 140 \text{ ft}^2$

Area of bedroom 2: $10(20) = 200 \text{ ft}^2$

Area of bedroom 3: $10(14) = 140 \text{ ft}^2$

Total area: $140 + 200 + 140 = 480 \text{ ft}^2$

Cost: $480(\$9.99) = \$4795.20$

47. Area of entire lawn if all grass: $400(300) = 120,000 \text{ ft}^2$

Area of house: $\frac{1}{2}(50)(100+150) = 6250 \text{ ft}^2$

Area of goldfish pond: $\pi(20)^2 = 400\pi \text{ ft}^2$

Area of privacy hedge: $200(20) = 4000 \text{ ft}^2$

Area of garage: $70(30) = 2100 \text{ ft}^2$

Area of driveway: $40(25) = 1000 \text{ ft}^2$

Area of lawn: $120,000 - 6250 - 400\pi - 4000 - 2100 - 1000 = 105,393.3629 \text{ ft}^2$

$$\frac{105,393.3629}{9} = 11,710.37366 \text{ yd}^2$$

Cost to cut lawn: $11,710.37366(\$0.03) = \$351.3112098 \approx \$351.31$

49. a) Perimeter: $2(94) + 2(50) = 288$ ft

b) Area: $(94)(50) = 4700 = 4700$ tiles

51. Let $h =$ height on the wall of the ladder.

$h^2 + 20^2 = 29^2$

$h^2 + 400 = 841$

$h^2 = 441$

$h = \sqrt{441} = 21$ ft

53. Let $d =$ the distance

$d^2 = 37^2 + 310^2$

$d^2 = 97,469$

$d = \sqrt{97,469} \approx 312$ ft

55. $s = \frac{1}{2}(a+b+c) = \frac{1}{2}(8+6+10) = 12;$ $A = \sqrt{12(12-8)(12-6)(12-10)} = \sqrt{12(4)(6)(2)} = \sqrt{576} = 24 \text{ cm}^2$

$A = \frac{1}{2}(6)(8) = 24 \text{ cm}^2$; The area is the same for both formulas.

57. Answers will vary.

Section 8.4: Volume and Surface Area

In this section, we use the π key on the calculator to determine answers in calculations involving π. If you use 3.14 for π, your answers may vary slightly.

1. Volume

3. Platonic

5. Right

7. a) $V = lwh = (8)(2)(4) = 64 \text{ ft}^3$

   b) $SA = 2lw + 2wh + 2lh$

   $SA = 2(8)(2) + 2(2)(4) + 2(8)(4) = 112 \text{ ft}^2$

9. a) $V = \pi r^2 h = \pi(2^2)(12) = 48\pi \approx 150.80 \text{ in.}^3$

   b) $SA = 2\pi r^2 + 2\pi rh$

   $SA = 2\pi(2^2) + 2\pi(2)(12) = 56\pi$

   $SA \approx 175.93 \text{ in}^2$

11. a) $V = \frac{1}{3}\pi r^2 h = \frac{1}{3}\pi(3^2)(14) = 42\pi$

    $\approx 131.95 \text{ cm}^3$

    b) $SA = \pi r^2 + \pi r\sqrt{r^2 + h^2}$

    $SA = \pi\left(3^2 + 3\sqrt{3^2 + 14^2}\right) = \pi\left(9 + 3\sqrt{205}\right)$

    $SA \approx 163.22 \text{ cm}^2$

13. a) $V = \frac{4}{3}\pi r^3 = \frac{4}{3}\pi(9^3) = \frac{4}{3}\pi(729)$

    $\approx 3053.63 \text{ mi}^3$

    b) $SA = 4\pi r^2 = 4\pi(9^2) = 4\pi(81) = 324\pi$

    $\approx 1017.88 \text{ mi}^2$

15. Area of the base: $B = \frac{1}{2}bh = \frac{1}{2}(6)(6) = 18 \text{ m}^2$

    $V = Bh = 18(9) = 162 \text{ m}^3$

17. Area of the base: $B = s^2 = 12^2 = 144 \text{ cm}^2$

    $V = \frac{1}{3}Bh = \frac{1}{3}(144)(15) = 720 \text{ cm}^3$

19. Volume of large prism: $(10)(10)(20) = 2000 \text{ ft}^3$

    Volume of small prism: $(5)(5)(20) = 500 \text{ ft}^3$

    Shaded volume: $2000 - 500 = 1500 \text{ ft}^3$

21. Volume of cylinder: $\pi r^2 h = \pi(3.5)^2(20.8)$

    $= 254.8\pi \text{ cm}^3$

    Volume of one sphere: $\frac{4}{3}\pi r^3 = \frac{4}{3}\pi(3.45)^3$

    $= 54.7515\pi$

    Shaded volume: $254.8\pi - 3(54.7515\pi)$

    $\approx 284.46 \text{ cm}^3$

23. Volume of trapezoidal solid: $7\left(\frac{4+8}{2}\right)(14)$

    $= 588 \text{ ft}^3$

    Volume of triangular solid: $\frac{1}{2}(4)(7)(14)$

    $= 196 \text{ ft}^3$

    Shaded volume: $588 - 196 = 392 \text{ ft}^3$

25. Volume of rectangular solid: $lwh = (3)(3)(4)$

    $= 36 \text{ ft}^3$

    Volume of pyramid: $\frac{1}{3}Bh = \frac{1}{3}(9)(4) = 12 \text{ ft}^3$

    Shaded volume: $36 - 12 = 24 \text{ ft}^3$

27. $7 \text{ yd}^3 = 7(27) = 189 \text{ ft}^3$

29. $453.6 \text{ ft}^3 = \frac{453.6}{27} = 16.8 \text{ yd}^3$

31. $1.5 \text{ m}^3 = 1.5(1,000,000) = 1,500,000 \text{ cm}^3$

33. $500,000 \text{ cm}^3 = \frac{500,000}{1,000,000} = 0.5 \text{ m}^3$

35. a) $V = lwh = (20)(15)\left(\frac{9}{12}\right) = 225 \text{ ft}^3$

b) Cost = $(\$11)(225) = \$2475$

37. a) $V = 80(50)(30) = 120{,}000 \text{ cm}^3$

b) $120{,}000 \text{ m}\ell$

c) $120{,}000 \text{ m}\ell = \frac{120{,}000}{1000} = 120\ \ell$

39. $V_{air} = V_{box} - V_{ball}$

$$= (7.5)^3 - \frac{4}{3}\pi\left(\frac{7.5}{2}\right)^3$$

$$= (7.5)^3 - \frac{4}{3}\pi(3.75)^3$$

$$\approx 200.98 \text{ cm}^3$$

41. a) $V = \frac{4}{3}\pi r^3 = \frac{4}{3}\pi\left(\frac{9.15}{2}\right)^3 = \frac{4}{3}\pi(4.575)^3$

$\approx 401.11 \text{ in.}^3$

b) $SA = 4\pi r^2 = 4\pi\left(\frac{9.15}{2}\right)^2 = 4\pi(4.575)^2$

$\approx 263.02 \text{ in.}^2$

43. a) Circular pan: $A = \pi r^2 = \pi\left(\frac{9}{2}\right)^2 = 20.25\pi$

$\approx 63.62 \text{ in.}^2$

Rectangular pan: $A = lw = 7(9) = 63 \text{ in.}^2$

b) Circular pan: $V = \pi r^2 h \approx 63.62(2)$

$= 127.24 \text{ in.}^3$

Rectangular pan: $V = lwh = 7(9)(2)$

$= 126 \text{ in.}^3$

c) The circular pan has the larger volume.

45. $V = \frac{1}{3}\pi r^2 h = \frac{1}{3}\pi\left(\frac{3}{2}\right)^2(6) = 4.5\pi \approx 14.14 \text{ in.}^3$

47. Volume of each cylinder: $\pi r^2 h = \pi\left(\frac{3.875}{2}\right)^2(3) = \pi(1.9375)^2(3) = 11.26171875\pi$

Total volume: $8(11.26171875\pi) = 283.0378631 \approx 283.04 \text{ in.}^3$

49. $8 - 12 + x = 2$

$-4 + x = 2$

$x = 6$ faces

51. $x - 30 + 12 = 2$

$x - 4 = 2$

$x = 6$ vertices

53. $12 - x + 20 = 2$

$32 - x = 2$

$x = 30$ edges

55. $r_E = \frac{12{,}756.3}{2} = 6378.15 \text{ km}$

$r_M = \frac{3474.8}{2} = 1737.4 \text{ km}$

a) $SA_E = 4\pi(6378.15^2) \approx 5.11\times10^8 \text{ km}^2$

b) $SA_M = 4\pi(1737.4^2) \approx 3.79\times10^7 \text{ km}^2$

c) $\frac{SA_E}{SA_M} = \frac{5.11\times10^8}{3.79\times10^7} \approx 13$ times larger

d) $V_E = \frac{4}{3}\pi(6378.15^3) \approx 1.09\times10^{12} \text{ km}^3$

e) $V_M = \frac{4}{3}\pi(1737.4^3) \approx 2.20\times10^{10} \text{ km}^3$

f) $\frac{V_E}{V_M} = \frac{1.09\times10^{12}}{2.20\times10^{10}} \approx 50$ times larger

57. a) $V_1 = (a)(a)(a) = a^3$ $\quad V_2 = (a)(a)(b) = a^2b$ $\quad V_3 = (a)(a)(b) = a^2b$ $\quad V_4 = (a)(b)(b) = ab^2$

$V_5 = (a)(a)(b) = a^2b$ $\quad V_6 = (a)(b)(b) = ab^2$ $\quad V_7 = (b)(b)(b) = b^3$

b) The volume of the piece not shown is $(a)(b)(b) = ab^2$.

c) Answers will vary. One possible solution: Find the volume of each numbered region. Since the length of each side is $a+b$, the sum of the volumes of each region will equal $(a+b)^3$.

59. If we double the radius of a sphere, the new volume will be eight times the original volume.

Section 8.5: Transformational Geometry, Symmetry, and Tessellations

1. Reflection

3. Vector

5. Rotation

7. Symmetry

9.

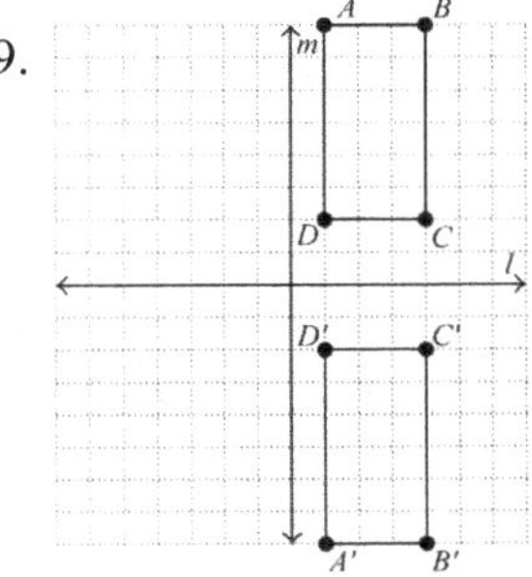

11.

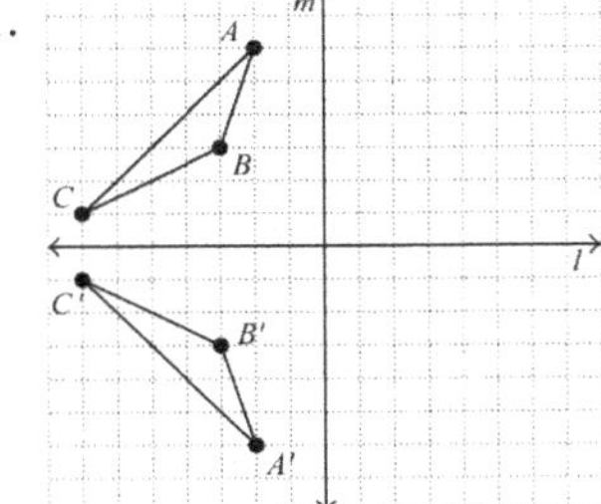

13.

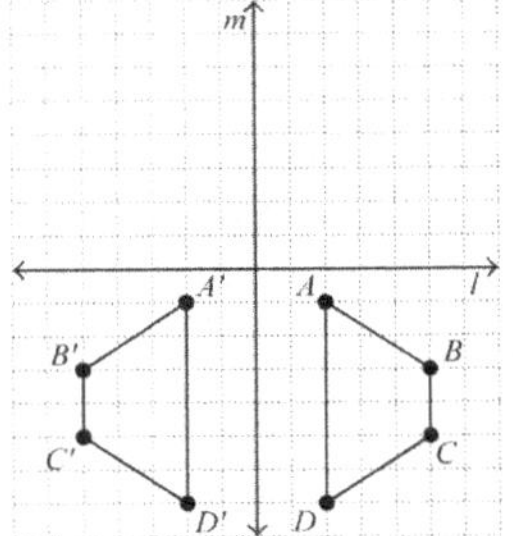

15.

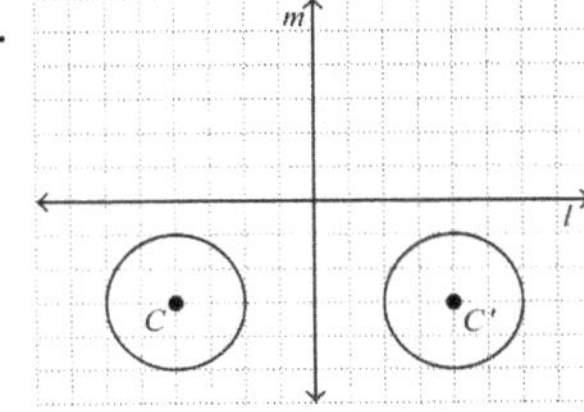

17.

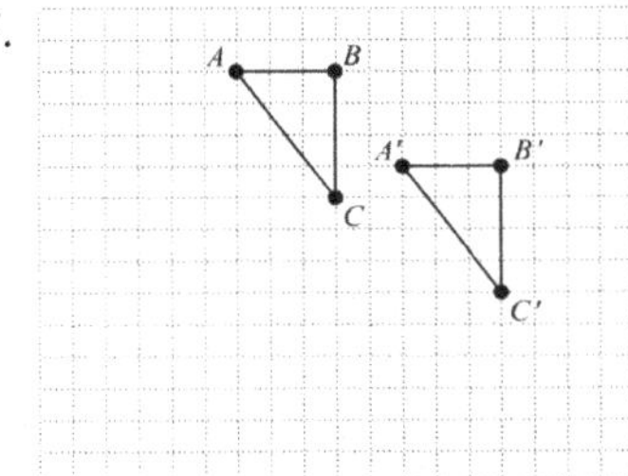

19.

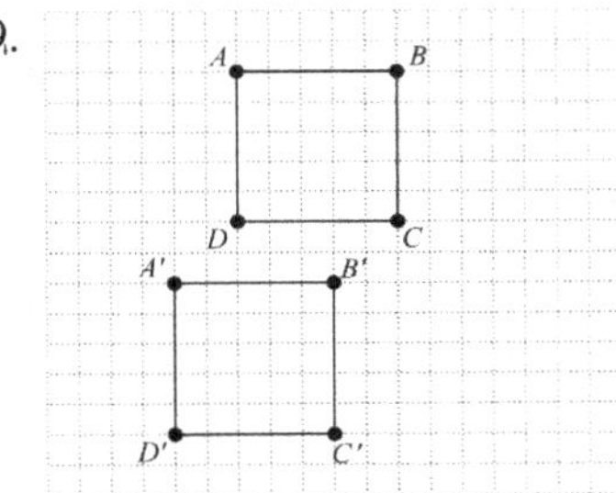

21.

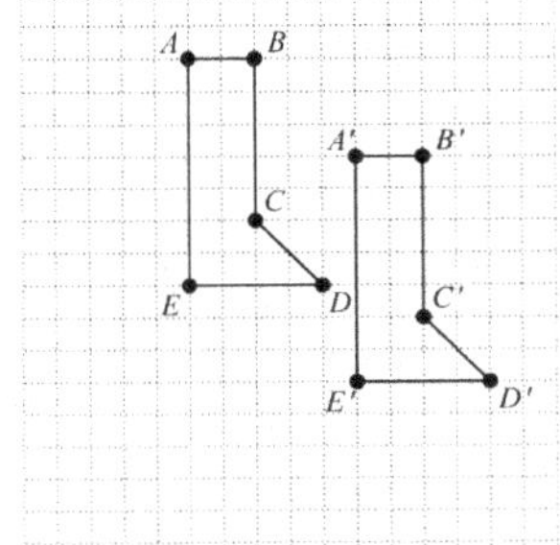

23.

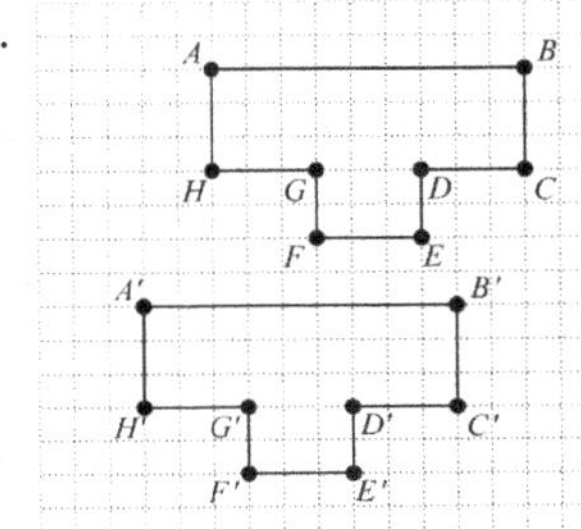

25.

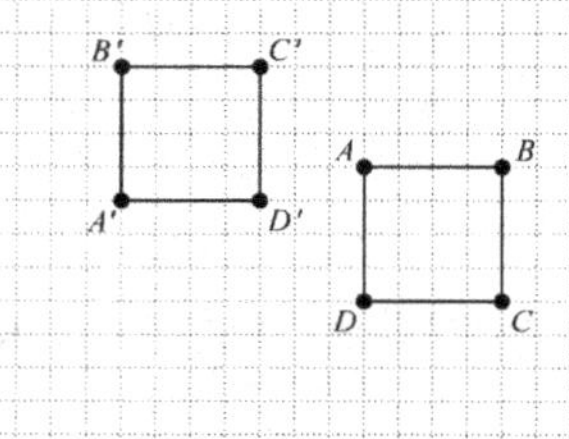

27.

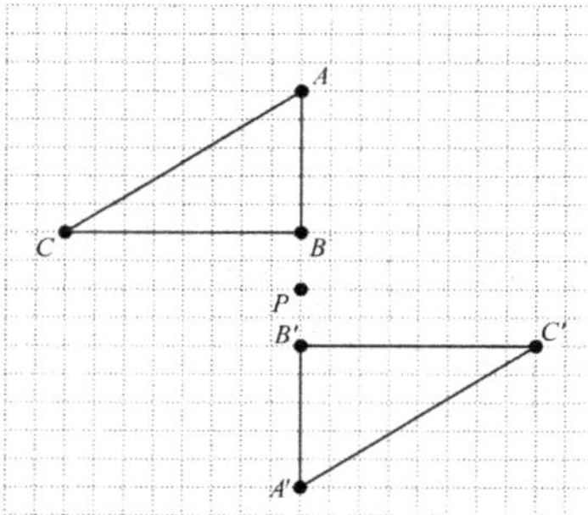

29.

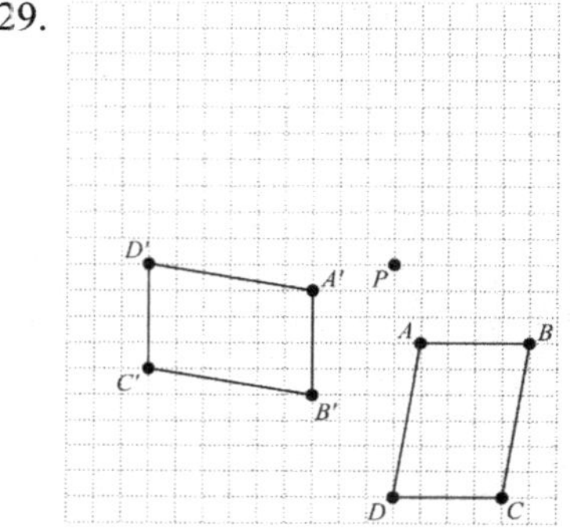

31.

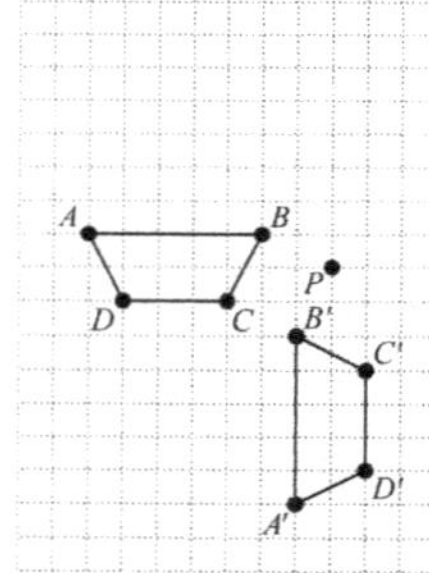

33.

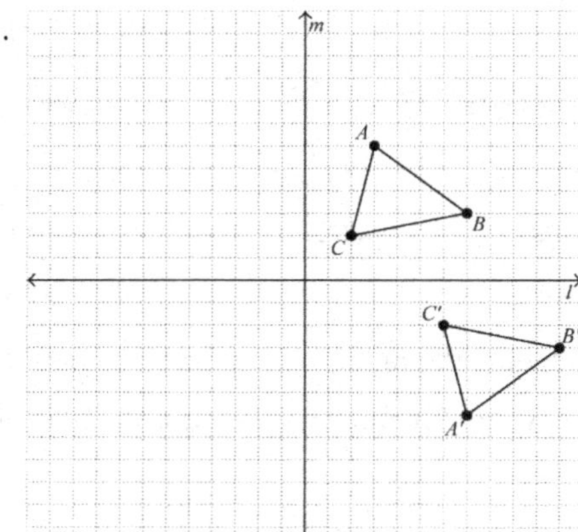

35.

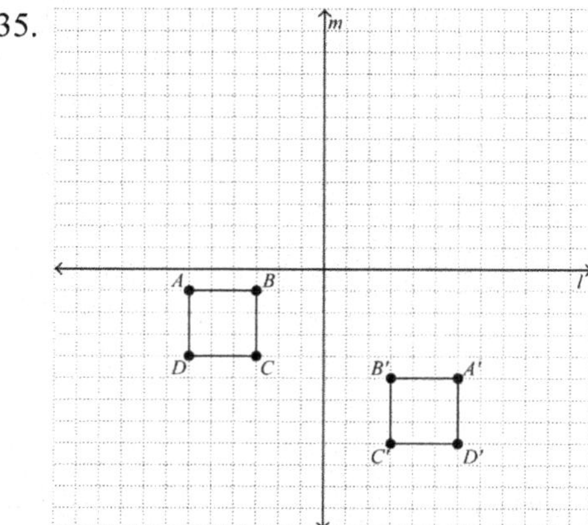

37.

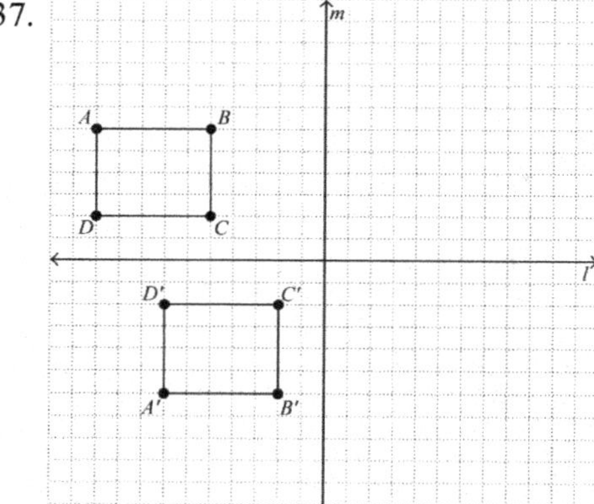

39.

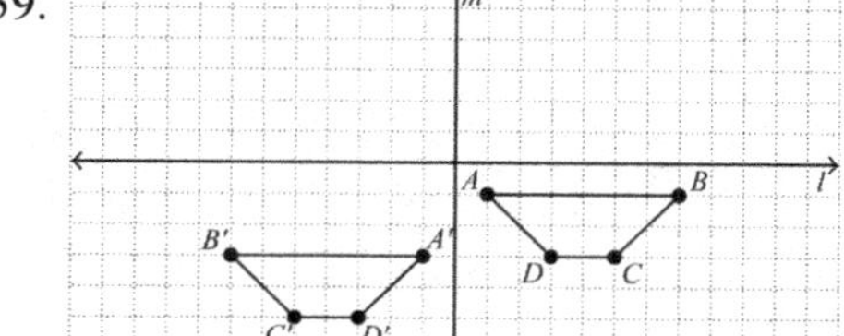

41. a) 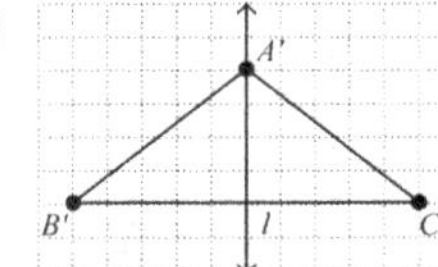

b) Yes c) Yes

43. a) 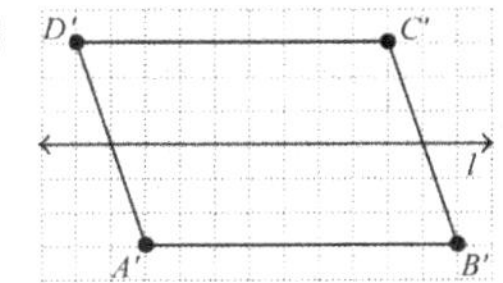

b) No c) No

45. a) 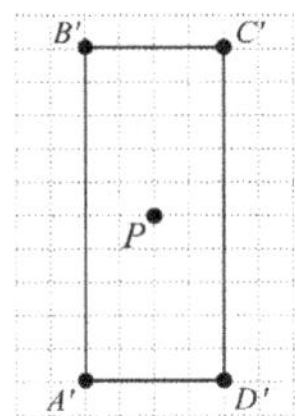

b) No c) No

d) 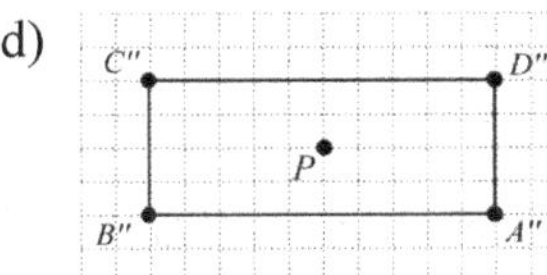

e) Yes f) Yes

47. a) – c) 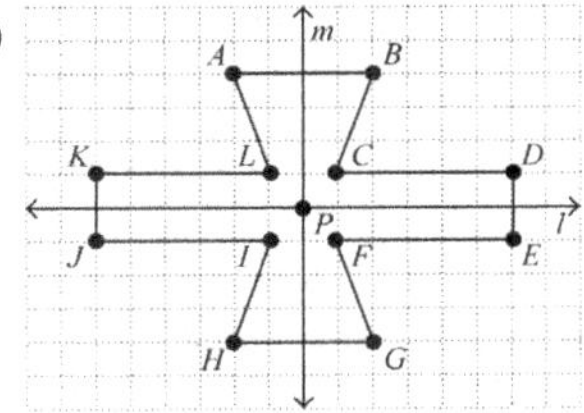

d) No. Any 90° rotation will result in the figure being in a different position than the starting position.

49. Answers will vary.

51. A regular octagon cannot be used as a tessellating shape.

53. Although answers will vary depending on the font, the following capital letters have reflective symmetry about a horizontal line drawn through the center of the letter: B, C, D, E, H, I, K, O, X.

55. Although answers will vary depending on the font, the following capital letters have 180° rotational symmetry about a point in the center of the letter: H, I, N, O, S, X, Z.

Section 8.6: Topology

1. Rubber

3. Klein

5. Jordan

7. Answers will vary.

   Green: 7
   Yellow: 2, 4, 6
   Red: 1, 3, 5

9. Answers will vary.

   Red: 1, 7
   Green: 4, 6, 8
   Blue: 2, 3, 5

11. Answers will vary.

   Red: CA, WA, MT, UT
   Green: OR, WY, AZ
   Blue: ID, NM
   Yellow: NV, CO

13. Answers will vary.

   Red: YT, NU, AB, ON
   Green: BC, SK
   Blue: NT, QC
   Yellow: MB

15. Outside; a straight line from point A to a point clearly outside the curve crosses the curve an even number of times.

17. Inside; a straight line from point C to a point clearly outside the curve crosses the curve an odd number of times.

19. Inside; a straight line from point E to a point clearly outside the curve crosses the curve an odd number of times.

21. 0

23. 1

25. 5

27. 3

29. Larger than 5

31. 4

33. a) Answers will vary.
 b) Answers will vary.
 c) Answers will vary.
 d) Answers will vary.

35. One

37. Two

39. a) No, it has an inside and an outside.
 b) Two
 c) Two
 d) Two strips, one inside the other

41. Answers will vary.

43. Answers will vary.

Section 8.7: Non-Euclidean Geometry and Fractal Geometry

1. Parallel

3. Two

5. Sphere

7. Geodesic

9.

11.

13. a)

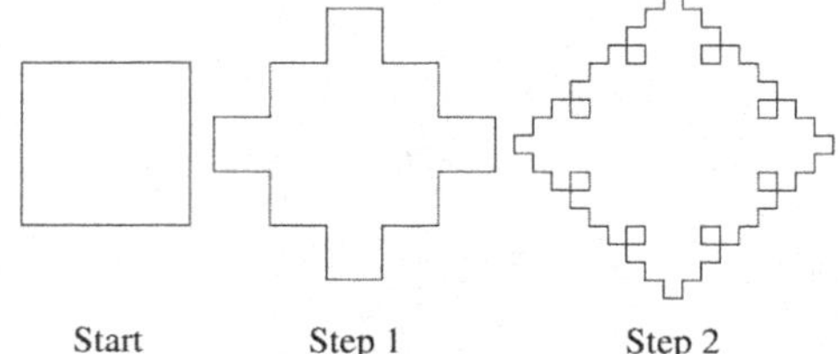

 b) Infinite since it consists of an infinite number of pieces.
 c) Finite since it covers a finite or closed area.

15. Each type of geometry can be used in its own frame of reference.

17. Coastlines, trees, mountains, galaxies, rivers, weather patterns, brains, lungs, blood supply

Review Exercises

In the Review Exercises, the π key on the calculator is used to determine answers in calculations involving π. If you use 3.14 for π, your answers may vary slightly.

1. $\measuredangle CBF$ or $\measuredangle FBC$

2. $\overline{BC}$

3. $\triangle BFC$

4. $\overleftrightarrow{BH}$

5. $\{F\}$

6. $\{\ \}$

7. $90° - 27.6° = 62.4°$

8. $180° - 100.5° = 79.5°$

9. $\dfrac{BC}{B'C} = \dfrac{AC}{A'C}$

$\dfrac{BC}{3.4} = \dfrac{12}{4}$

$\dfrac{BC}{3.4} = \dfrac{3}{1}$

$BC = 10.2$ in.

10. $\dfrac{A'B'}{AB} = \dfrac{A'C}{AC}$

$\dfrac{A'B'}{6} = \dfrac{4}{12}$

$\dfrac{A'B'}{6} = \dfrac{1}{3}$

$3(A'B') = 6$

$A'B' = 2$ in.

11. $m\measuredangle ABC = m\measuredangle A'B'C$

$m\measuredangle A'B'C = 180° - 88° = 92°$

$m\measuredangle ABC = 92°$

$m\measuredangle BAC = 180° - 30° - 92° = 58°$

12. $m\measuredangle ABC = m\measuredangle A'B'C$

$m\measuredangle A'B'C = 180° - 88° = 92°$

$m\measuredangle ABC = 92°$

13. $m\measuredangle 1 = 43°$

$m\measuredangle 6 = 180° - 117° = 63°$

$m\measuredangle 2 = m\measuredangle 1 + m\measuredangle 6 = 106°$

$m\measuredangle 3 = 74°$

$m\measuredangle 4 = m\measuredangle 6 = 63°$

$m\measuredangle 5 = 180° - m\measuredangle 4 = 180° - 63° = 117°$

14. $(n-2)180° = (8-2)180° = 6(180°) = 1080°$

15. a) $A = lw = 11(9) = 99 \text{ mi}^2$

b) $P = 2l + 2w = 2(11) + 2(9) = 40 \text{ mi}$

16. a) $A = \frac{1}{2}h(b_1 + b_2) = \frac{1}{2}(4)(6+12) = 36 \text{ m}^2$

b) $P = 5+5+3+3+6+6 = 28 \text{ m}$

17. a) $A = bh = 12(7) = 84 \text{ in.}^2$

b) $P = 2(9) + 2(12) = 42 \text{ in.}$

18. a) $A = \frac{1}{2}bh = \frac{1}{2}(3)(4) = 6 \text{ km}^2$

b) $P = 3 + 4 + \sqrt{3^2 + 4^2} = 7 + \sqrt{25} = 12 \text{ km}$

19. a) $A = \pi r^2 = \pi(7)^2 = 49\pi \approx 153.94 \text{ ft}^2$

b) $C = 2\pi r = 2\pi(7) = 14\pi \approx 43.98 \text{ ft}$

20. Area of rectangle:
$lw = (10+10+10)(10) = (10)(30) = 300 \text{ m}^2$

Area of one circle: $\pi r^2 = \pi(5)^2 = 25\pi \text{ m}^2$

Area of circle: $\pi r^2 = \pi(3)^2 = 9\pi \text{ m}^2$

Shaded area:
$144 - 4(9) - 9\pi = 108 - 9\pi \approx 79.73 \text{ yd}^3$

21. Area of outer square: $lw = (12)(12) = 144 \text{ yd}^2$

Area of one missing corner: $lw = (3)(3) = 9 \text{ yd}^2$

Shaded area: $300 - 3(25\pi) = 300 - 75\pi$
$\approx 64.38 \text{ yd}^3$

22. $A = lw = 14(16) = 224 \text{ ft}^2$

Cost: $224(\$9.75) = \$2184$

23. a) $V = lwh = 11(3)(5) = 165 \text{ cm}^3$

b) $SA = 2lw + 2wh + 2lh$
$= 2(11)(3) + 2(3)(5) + 2(11)(5)$
$= 206 \text{ cm}^2$

24. a) $V = \pi r^2 h = \pi\left(3^2\right)(9) = 81\pi$
$\approx 254.47 \text{ in}^3$

b) $SA = 2\pi r^2 + 2\pi rh$
$SA = 2\pi\left(3^2\right) + 2\pi(3)(9) = 72\pi$
$SA \approx 226.19 \text{ in}^2$

25. a) $r = \frac{12}{2} = 6 \text{ mm}$
$V = \frac{1}{3}\pi r^2 h = \frac{1}{3}\pi\left(6^2\right)(16) = 192\pi$
$\approx 603.19 \text{ mm}^3$

b) $SA = \pi r^2 + \pi r\sqrt{r^2 + h^2}$
$= \pi\left(6^2 + 6\sqrt{6^2 + 16^2}\right)$
$= \pi\left(36 + 6\sqrt{292}\right)$
$\approx 435.20 \text{ mm}^2$

26. a) $V = \frac{4}{3}\pi r^3 = \frac{4}{3}\pi\left(5^3\right) = \frac{4}{3}\pi(125)$
$= \frac{625\pi}{3} \approx 523.60 \text{ yd}^3$

b) $SA = 4\pi r^2 = 4\pi\left(5^2\right) = 4\pi(25)$
$\approx 314.16 \text{ yd}^2$

27. $B = \frac{1}{2}bh = \frac{1}{2}(9)(12) = 54 \text{ m}^2$

$V = Bh = 54(8) = 432 \text{ m}^3$

28. Let $h =$ height of triangular base of pyramid

$h^2 + 3^2 = 5^2$

$h^2 + 9 = 25$

$h^2 = 16$

$h = \sqrt{16} = 4 \text{ ft}$

$B = \frac{1}{2}bh = \frac{1}{2}(6)(4) = 12 \text{ ft}^2$

$V = \frac{1}{3}Bh = \frac{1}{3}(12)(7) = 28 \text{ ft}^3$

29. Volume of cylinder: $\pi r^2 h = \pi(2)^2 9 = 36\pi \text{ cm}^3$

Volume of cone:

$\frac{1}{3}\pi r^2 h = \frac{1}{3}\pi(2)^2(9) = 12\pi \text{ cm}^3$

Shaded volume: $36\pi - 12\pi = 24\pi \approx 75.40 \text{ cm}^3$

30. Volume of large sphere: $\frac{4}{3}\pi r^3 = \frac{4}{3}\pi(7)^3$

$= \frac{1372\pi}{3} \text{ in.}^3$

Volume of small sphere: $\frac{4}{3}\pi r^3 = \frac{4}{3}\pi\left(\frac{7}{2}\right)^3$

$= \frac{343\pi}{6} \text{ in.}^3$

Shaded volume: $\frac{1372\pi}{3} - \frac{343\pi}{6} = \frac{2401\pi}{6}$

$\approx 1257.16 \text{ in.}^3$

31. Let $h =$ the height of the trough

$h^2 + 1^2 = 3^2$

$h^2 + 1 = 9$

$h^2 = 8$

$h = \sqrt{8}$ ft

$B = \frac{1}{2}h(b_1 + b_2) = \frac{1}{2}(\sqrt{8})(2+4) = 3\sqrt{8} \text{ ft}^2$

a) $V = Bh = 3\sqrt{8}(8) = 24\sqrt{8} \approx 67.88 \text{ ft}^3$

b) Weight of water: $67.88(62.4) + 375$

≈ 4610.71 lb

Yes, it will support the trough filled with water.

c) $(4610.71 - 375) = 4235.71$ lb of water

$\frac{4235.7}{8.3} \approx 510.33$ gal

32.

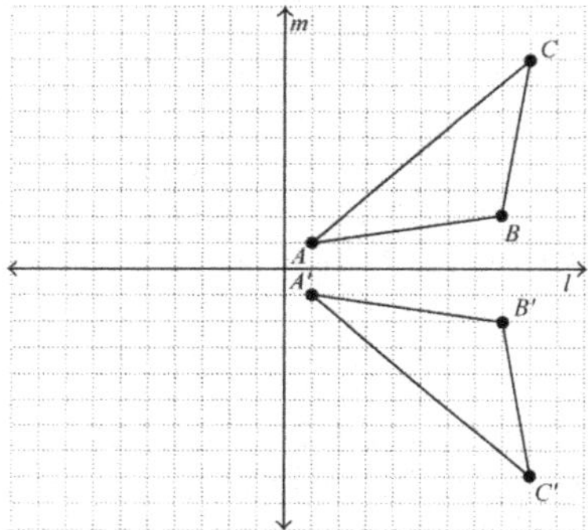

33.

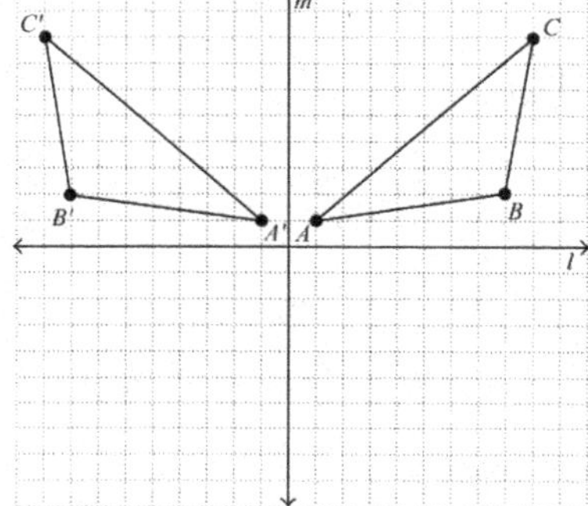

34.

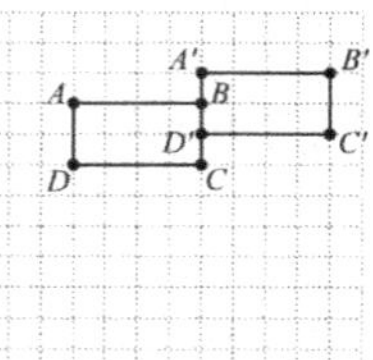

35.

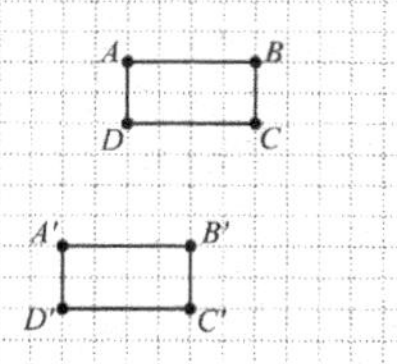

36.

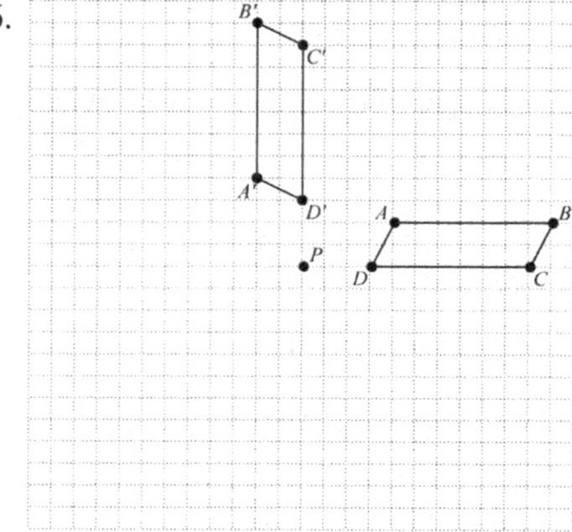

37.

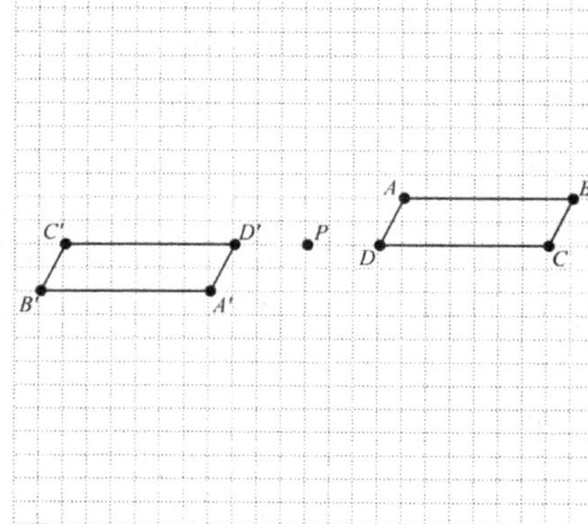

38.

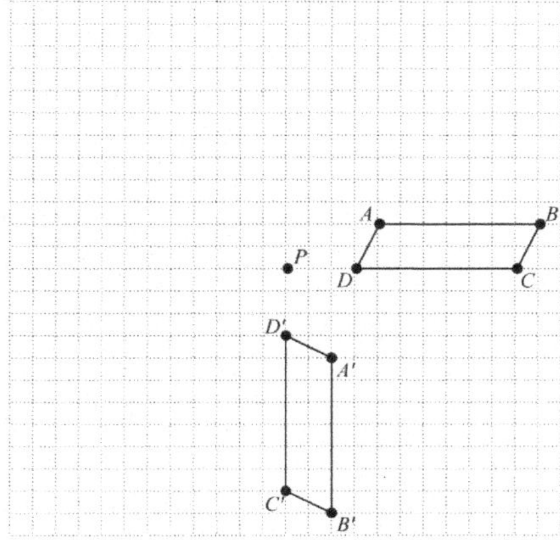

39.

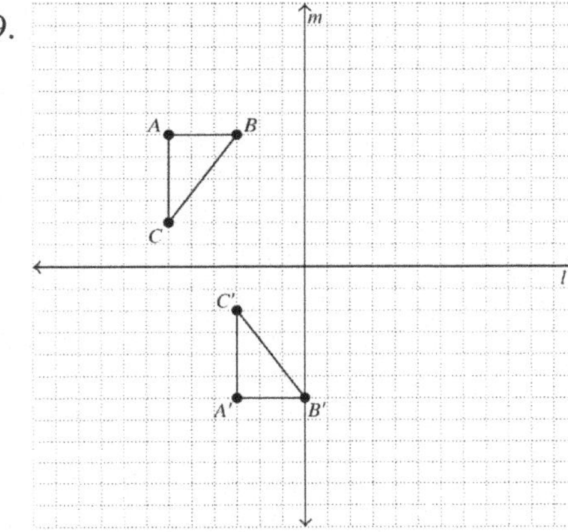

40.

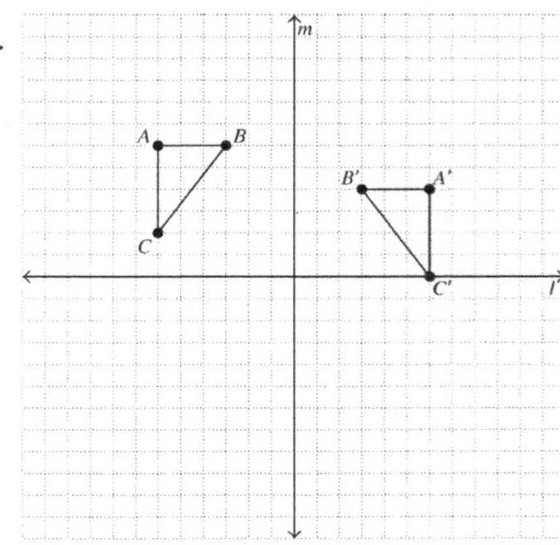

41. Yes. A reflection about the line *l* will result in the figure being in the same position as the starting position.

42. No. Any reflection about any horizontal line will result in the figure being in a different position than the starting position.

43. No. Any 90° rotation will result in the figure being in a different position than the starting position.

44. Yes. Any 180° rotation will result in the figure being in the same position as the starting position.

45. a) Answers will vary.
    b) Answers will vary.
    c) Answers will vary.
    d) Answers will vary.

46. Answers will vary.
    Red: 1, 3, 9
    Green:2, 8,10
    Blue: 4, 5, 6, 7

47. Outside; a straight line from point *A* to a point clearly outside the curve crosses the curve an even number of times.

48. Euclidean: Given a line and a point not on the line, one and only one line can be drawn parallel to the given line through the given point.

    Elliptical: Given a line and a point not on the line, no line can be drawn through the given point parallel to the given line.

    Hyperbolic: Given a line and a point not on the line, two or more lines can be drawn through the given point parallel to the given line.

49.

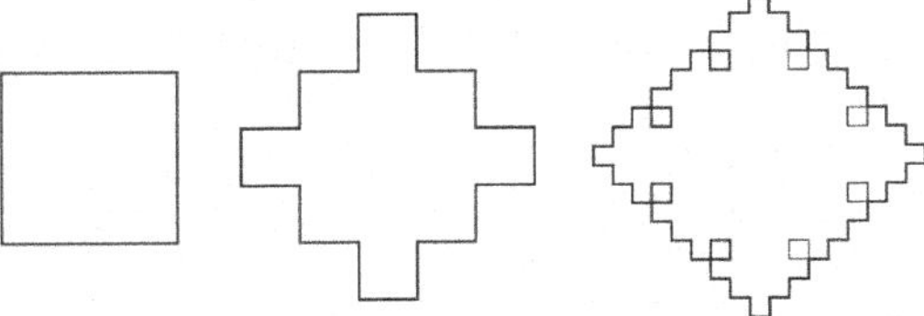

50.

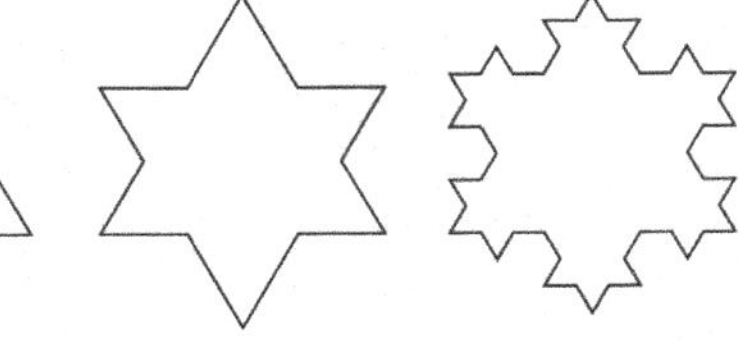

Chapter Test

In the Chapter Test, the π key on the calculator is used to determine answers in calculations involving π. If you use 3.14 for π, your answers may vary slightly.

1. $\measuredangle BAE$ or $\measuredangle EAB$
2. ΔBCD
3. $\{D\}$
4. $\overleftrightarrow{AC}$

5. $90° - 41.8° = 48.2°$

6. $180° - 73.5° = 106.5°$

7. One angle of the triangle is 48° (by vertical angles) and $180° - 116° = 64°$. Thus, the measure of angle x is $180° - 48° - 64° = 68°$.

8. $(n-2)180° = (6-2)180° = 4(180°) = 720°$

9. Let $x = B'C'$

$$\frac{B'C'}{BC} = \frac{A'C'}{AC}$$
$$\frac{x}{7} = \frac{5}{13}$$
$$13x = 35$$
$$x = \frac{35}{13} \approx 2.69 \text{ cm}$$

10. a) Let $x = AB$

$$x^2 + 12^2 = 13^2$$
$$x^2 + 144 = 169$$
$$x^2 = 25$$
$$x = \sqrt{25}$$
$$x = 5 \text{ in.}$$

b) $P = 5 + 13 + 12 = 30$ in.

c) $A = \frac{1}{2}bh = \frac{1}{2}(5)(12) = 30 \text{ in.}^2$

11. $r = \frac{14}{2} = 7$ cm

a) $V = \frac{4}{3}\pi r^3 = \frac{4}{3}\pi\left(7^3\right) \approx 1436.76 \text{ cm}^3$

b) $SA = 4\pi r^2 = 4\pi\left(7^2\right) = 4\pi(49) \approx 615.75 \text{ cm}^2$

12. Shaded volume = volume of prism − volume of cylinder

Volume of prism: $V = lwh = (6)(4)(3) = 72 \text{ m}^3$

Volume of cylinder: $V = \pi r^2 h = \pi(1)^2(4) = 4\pi \text{ m}^3$

Shaded volume: $72 - 4\pi \approx 59.43 \text{ m}^3$

13. $B = lw = 4(7) = 28 \text{ ft}^2$

$V = \frac{1}{3}Bh = \frac{1}{3}(28)(12) = 112 \text{ ft}^3$

14. 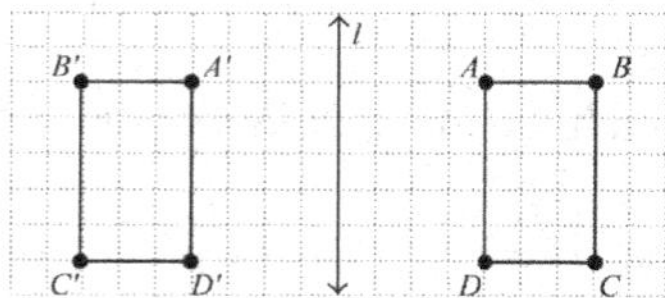

15. 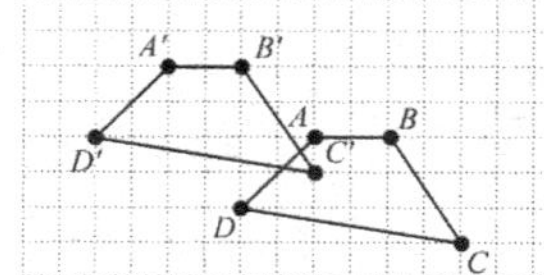

16. 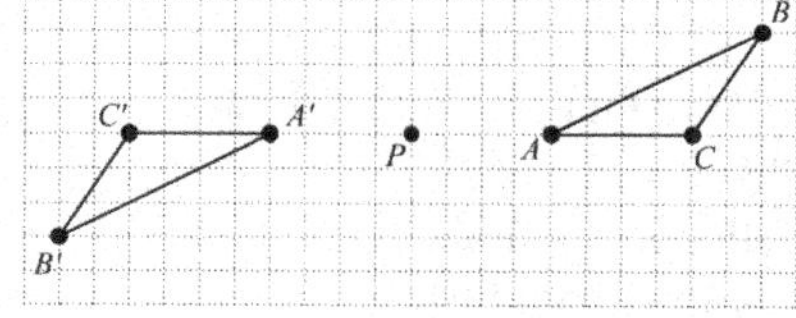

17. 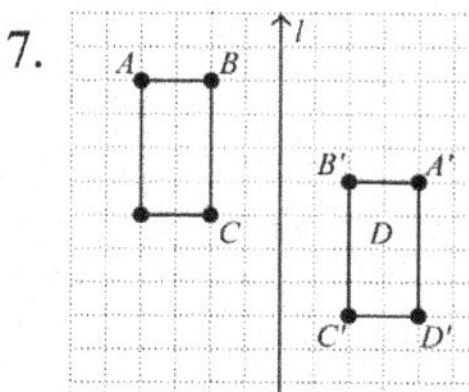

18. a) No. Any 90° rotation will result in the figure being in a different position than the starting position.
b) Yes. Any 180° rotation will result in the figure being in the same position as the starting position.

19. A Möbius strip is a surface with one side and one edge.

20. a) Answers will vary.

b) Answers will vary.

Chapter Nine: Mathematical Systems

Section 9.1: Groups

1. Binary
3. Closed
5. Identity
7. Inverse
9. Commutative
11. The commutative property of addition stated that $a+b=b+a$, for any elements a and b.
    Example: $3+4=4+3$
13. The associative property of multiplication states that $(a\cdot b)\cdot c=a\cdot(b\cdot c)$, for any real numbers a, b, and c.
    Example: $(2\cdot 3)\cdot 4=2\cdot(3\cdot 4)$
15. $7-3=4$, but $3-7=-4$
17. $(8\div 4)\div 2=2\div 2=1$, but $8\div(4\div 2)=8\div 2=4$
19. a) Yes; the sum of any two integers is an integer.
    b) Yes, the identity element is 0.
    c) Yes, each element has an inverse, its opposite.
    d) Answer will vary. One possible example is $(1+2)+3=1+(2+3)$.
    e) Answer will vary. One possible example is $2+4=4+2$.
    f) Yes. It satisfies the five properties needed.
21. a) Yes; the sum of any two positive integers is a positive integer.
    b) No
    c) Since there is no identity element, each element does not have an inverse.
    d) Answer will vary. One possible example is $(2+3)+4=2+(3+4)$.
    e) Answer will vary. One possible example is $1+2=2+1$.
    f) No; there is no identity element.
23. Yes; it satisfies the four properties needed.
25. No; the system is not closed.
27. Yes; it satisfies the five properties needed.
29. No; there is no identity element.
31. No; the system is not closed. For example, $\frac{1}{0}$ is undefined
33. No; the system is not closed, $\pi+(-\pi)=0$, which is rational. There is no identity element.
35. Answers will vary.

Section 9.2: Finite Mathematical Systems

1. $\{1, 2, 3, 4, 5, 6, 7, 8, 9, 10, 11, 12\}$
3. Identity
5. Associative
7. Commutative
9. Commutative
11. $7+6=1$
13. $8+7=3$
15. $4+12=4$
17. $3+(8+9)=3+5=8$
19. $10-7=3$
21. $4-10=6$
23. $6-10=8$
25. $5-5=12$

27.

+	1	2	3	4	5	6
1	2	3	4	5	6	1
2	3	4	5	6	1	2
3	4	5	6	1	2	3
4	5	6	1	2	3	4
5	6	1	2	3	4	5
6	1	2	3	4	5	6

29. $6+2=2$

31. $5-2=3$

33. $(3-5)-6=4-6=4$

35.

+	1	2	3	4	5	6	7
1	2	3	4	5	6	7	1
2	3	4	5	6	7	1	2
3	4	5	6	7	1	2	3
4	5	6	7	1	2	3	4
5	6	7	1	2	3	4	5
6	7	1	2	3	4	5	6
6	1	2	3	4	5	6	7

37. $4+4=1$

39. $5+5=3$

41. $(4-5)-6=6-6=7$

43. Yes. It satisfies the five required properties.

45. No. The system is not closed. It contains a symbol other than x, y, and z.

47. The identity element is C. The top row is the same as the third row and the left column is the same as the third column.

49. a) The inverse of A is A because A ╬ $A = C$.

    b) The inverse of B is B because B ╬ $B = C$.

    c) The inverse of C is C because C ╬ $C=C$.

51. No; the elements are not symmetric about the main diagonal.

53. a) {2, 4, 6, 8}
    b) Q
    c) Yes; all elements in the table are in the original set.
    d) Yes; the identity element is 0.
    e) Yes; $0\ Q\ 0=0$, $2\ Q\ 6=0$, $4\ Q\ 4=0$, and $6\ Q\ 2=0$
    f) $(2\ Q\ 4)\ Q\ 6=6\ Q\ 6=4$ and $2\ Q\ (4\ Q\ 6)=2\ Q\ 2=4$
    g) Yes; $2\ Q\ 6=0$ and $6\ Q\ 2=0$
    h) Yes; the system satisfies the five properties needed.

55. a) {4, 5, L}
    b) \$
    c) Yes; all elements in the table are in the original set.
    d) Yes; the identity element is L.
    e) Yes; $4\ \$\ 5=L$, $5\ \$\ 4=L$, $L\ \$\ L=L$
    f) $(4\ \$\ 5)\ \$\ 5=L\ \$\ 5=5$ and $4\ \$\ (5\ \$\ 5)=4\ \$\ 4=5$
    g) Yes; $L\ \$\ 4=4$ and $4\ \$\ L=4$
    h) Yes; the system satisfies the five properties needed.

57. a) Yes; all elements in the table are in the original set.
    b) Yes; the identity element is O.
    c) $G☆D = O$, $O☆O = O$, $D☆G = O$; L does not have an inverse.
    d) $(L☆ L) ☆ D = D☆D = L$ and $L☆(L☆D) = L☆G = D$
    e) No
    f) Yes
    g) No; not every element has an inverse and the associative property does not hold.

59. Not closed, Π is not an element. Not associative: $(\Gamma\ \beta\ \Phi)\ \beta\ \Delta \neq \Gamma\ \beta\ (\Phi\ \beta\ \Delta)$.

61. There is no inverse for ⊙ and C.
    Not associative: $(P$ ⊡ $C)$ ⊡ ⊙ = T ⊡ ⊙ = C and P ⊡ $(C$ ⊡ ⊙$)$ = P ⊡ P = ⊙

63. There is no identity element and therefore no inverses.
    Not associative: $(d\ F\ e)\ F\ d = d\ F\ d = e$ and $d\ F\ (e\ F\ d) = d\ F\ e = d$
    Not communitive: $e\ F\ d = e$ and $d\ F\ e = d$

65. a)

+	E	O
E	E	O
O	O	E

    b) The system is closed, the identity element is E, each element is its own inverse, and the system is commutative since the table is symmetric about the main diagonal. Since the system has fewer than 6 elements and satisfies the above properties, it is a commutative group.

67. a) All elements in the table are in the set {1, 2, 3, 4, 5, 6} so the system is closed. The identity is 6. 5 and 1 are inverses of each other, and 2, 3, 4, and 6 are their own inverses. Thus, if the associative property is assumed, the system is a group.

    Examples of associativity:
    (2 ? 3) ? 4 = 5 ? 4 = 3 and 2 ? (3 ? 4) = 2 ? 5 = 3
    (1 ? 3) ? 5 = 4 ? 5= 2 and 1 ? (3 ? 5) = 1 ? 4 = 2
    b) 3 ? 1 = 2, but 1 ? 3 = 4

69. a)

*	R	S	T	U	V	I
R	V	T	U	S	I	R
S	U	I	V	R	T	S
T	S	R	I	V	U	T
U	T	V	R	I	S	U
V	I	U	S	T	R	V
I	R	S	T	U	V	I

    b) Yes. It is closed and the identity element is I. The inverse of R is V, of S is S, of T is T, of U is U, of V is R, and of I is I. The associative property will hold: $R*(T*V) = R*U = S$ and $(R*T)*V = U*V = S$.
    c) No, it is not commutative: $R*S = T$, but $U = S*R$.

Section 9.3: Modular Arithmetic

1. $m-1$
3. Remainder
5. Congruent
7. $17 \div 7 = 2$ remainder 3, so 17 days is 3 days later than today (Thursday), or Sunday.
9. $105 \div 7 = 15$ remainder 0, so 105 days is 0 days later than today (Thursday), or Thursday.
11. 3 years $= (3 \cdot 365)$ days $= 1095$ days; $1095 \div 7 = 156$ remainder 3, so 3 years is 3 days later than today (Thursday), or Sunday.

13. $400 \div 7 = 57$ remainder 1, so 400 days is one day after today (Thursday), or Friday.

15. $22 \div 12 = 1$ remainder 10, so 22 months is 10 months after October, which is August.

17. Since 2 years is 24 months and $24 \equiv 0 \pmod{12}$, 2 years 10 months is the same as 10 months and 10 months after October is August.

19. $83 \div 12 = 6$ remainder 11; So, 83 months is 11 months after October, which is September.

21. $105 \div 12 = 8$ remainder 9; So, 105 months is 9 months after October, which is July.

23. $4+1=5;\ 5 \equiv 0 \pmod 5$

25. $10-1=9;\ 9 \equiv 4 \pmod 5$

27. $8 \cdot 9 = 72;\ 72 \equiv 2 \pmod 5$

29. $$\begin{aligned} 4-8 &\equiv (5+4)-8 \pmod 5 \\ &\equiv 9-8 \pmod 5 \\ &\equiv 1 \pmod 5 \end{aligned}$$

31. $(15 \cdot 4) - 8 = 60 - 8 = 52;\ 52 \equiv 2 \pmod 5$

33. $1 \equiv 15 \pmod 2$

35. $5 \equiv 77 \pmod 8$

37. $5 \equiv 41 \pmod 9$

39. $6 \equiv -1 \pmod 7$

41. $5 \equiv -27 \pmod 8$

43. $4+5 \equiv 3 \pmod 6$

45. $6+3 \equiv 2 \pmod 7$

47. $4-5 \equiv 5 \pmod 6$

49. $5 \cdot 5 \equiv 7 \pmod 9$

51. No solution or { }

53. $4 \cdot 1 \equiv 4 \pmod{10}$
    $4 \cdot 6 \equiv 4 \pmod{10}$

55. $5-8 \equiv 9 \pmod{12}$

57. a) 1792, 1796, 1800, 1804, 1808
    b) 2024, 2028, 2032, 2036, 2040
    c) 2552, 2556, 2560, 2564, 2568, 2572

59. His schedule is: L L L L M M M R R, a nine-day cycle.
    a) $30 \equiv 3 \pmod 9$; 3 days after today he works at Longboat Key.
    b) $100 \equiv 1 \pmod 9$; 1 day after today he rests.
    c) $366 \equiv 6 \pmod 9$; 6 days after today he works at The Meadows.

61. The manager's schedule is repeated every seven weeks. If this is week two of his schedule, then this is his second weekend that he works, or week 1 in a mod 7 system. His schedule in mod 7 on any given weekend is shown in the following table.

Weekend (mod 7)	0	1	2	3	4	5	6
Work/Off	W	W	W	W	W	W	O

    a) If this is weekend 1, then in 5 more weeks $(1+5=6)$ he will have the weekend off.
    b) $25 \div 7 = 3$, remainder 4; Thus $25 \equiv 4 \pmod 7$, and 4 weeks from weekend 1 will be weekend 5. He will not have the weekend off.
    c) $50 \div 7 = 7$, remainder 1; One week from weekend 1 will be weekend 2. It will be 4 more weeks before he has the weekend off. Thus, in 54 weeks he will have the weekend off.

63. The shift cycle is 5 days.

Week	0	1	2	3	4
Shift	Early	Early	Early	Late	Late

    a) $6 \equiv 1 \pmod 5$; 1 week after the third week will be a late shift.
    b) $7 \equiv 2 \pmod 5$; 2 weeks after the fourth week will be an early shift.
    c) $11 \equiv 1 \pmod 5$; 1 week after the first week will be an early shift.

65. a)

+	0	1	2
0	0	1	2
1	1	2	0
2	2	0	1

b) Yes. All the numbers in the table are from the set $\{0, 1, 2\}$.

c) The identity element is 0.

d) Yes, each element has an inverse: $0+0=0$, $1+2=0$, and $2+1=0$.

e) $(1+2)+2=1+(2+2)$

f) Yes, the table is symmetric about the main diagonal: $1+2=0$ and $2+1=0$.

g) Yes. All five properties are satisfied.

h) Yes. The modulo system behaves the same no matter how many elements are in the system.

67. a)

×	0	1	2	3
0	0	0	0	0
1	0	1	2	3
2	0	2	0	2
3	0	3	2	1

b) Yes. All the elements in the table are from the set $\{0, 1, 2, 3\}$.

c) Yes. The identity element is 1.

d) $1\times 1=1$, $3\times 3=1$; 0 and 2 do not have inverses.

e) $(1\times 2)\times 3=2\times 3=2$ and $1\times(2\times 3)=1\times 2=2$

f) Yes, the commutative property holds: $2\times 3=2$ and $3\times 2=2$.

g) No. Not all elements have inverses.

69. $4\cdot 3=12$ and $12\equiv 2 \pmod 5$, so $2\div 3\equiv 4 \pmod 5$

71. $1\div 1\equiv 1 \pmod 4$, $2\div 2\equiv 1 \pmod 4$, $3\div 3\equiv 1 \pmod 4$, so $?=\{1, 2, 3\}$ (Note: $0\div 0$ is undefined.)

73. $8k\equiv x \pmod 8$

$$\begin{array}{r} k \\ 8\overline{)8k} \\ \underline{8k} \\ 0 \end{array}$$

$0\equiv x \pmod 8$

$x=0$

75.

$$4k-2\equiv x \pmod 4$$
$$4k-2+4\equiv x \pmod 4$$
$$4k+2\equiv x \pmod 4$$

$$\begin{array}{r} k \\ 4\overline{)4k+2} \\ \underline{4k} \\ 2 \end{array}$$

$2\equiv x \pmod 4$

$x=2$

77. a) $365\equiv 1 \pmod 7$, so his birthday will be one day later in the week next year and will fall on Tuesday.

b) $366\equiv 2 \pmod 7$, so if next year is a leap year his birthday will fall on Wednesday.

79. Subtract 5 from each number.

Original numbers	16	19	20	1	17	10	9	12	10	5	14	24	5	21	20	1	10	23
Decoded numbers	11	14	15	23	12	5	4	7	5	0	9	19	0	16	15	23	5	18
Message	K	N	O	W	L	E	D	G	E		I	S		P	O	W	E	R

Section 9.4: Matrices

1. Matrix
3. 2
5. Dimensions
7. Columns, rows

9. $A+B=\begin{bmatrix}3 & 8\\ 2 & -6\end{bmatrix}+\begin{bmatrix}-4 & 1\\ 5 & -3\end{bmatrix}=\begin{bmatrix}3+(-4) & 8+1\\ 2+5 & -6+(-3)\end{bmatrix}=\begin{bmatrix}-1 & 9\\ 7 & -9\end{bmatrix}$

11. $A+B=\begin{bmatrix}5 & 2\\ -1 & 4\\ 7 & 0\end{bmatrix}+\begin{bmatrix}-3 & 3\\ -4 & 0\\ 1 & 6\end{bmatrix}=\begin{bmatrix}5+(-3) & 2+3\\ -1+(-4) & 4+0\\ 7+1 & 0+6\end{bmatrix}=\begin{bmatrix}2 & 5\\ -5 & 4\\ 8 & 6\end{bmatrix}$

13. $A-B=\begin{bmatrix}4 & 5\\ 1 & -3\end{bmatrix}-\begin{bmatrix}-2 & 1\\ 6 & 7\end{bmatrix}=\begin{bmatrix}4-(-2) & 5-1\\ 1-6 & -3-7\end{bmatrix}=\begin{bmatrix}6 & 4\\ -5 & -10\end{bmatrix}$

15. $A-B=\begin{bmatrix}-5 & 1\\ 8 & 6\\ 1 & -5\end{bmatrix}-\begin{bmatrix}-6 & -8\\ -10 & -11\\ 3 & -7\end{bmatrix}=\begin{bmatrix}-5-(-6) & 1-(-8)\\ 8-(-10) & 6-(-11)\\ 1-3 & -5-(-7)\end{bmatrix}=\begin{bmatrix}1 & 9\\ 18 & 17\\ -2 & 2\end{bmatrix}$

17. $2B=2\begin{bmatrix}3 & 2\\ 5 & 0\end{bmatrix}=\begin{bmatrix}2(3) & 2(2)\\ 2(5) & 2(0)\end{bmatrix}=\begin{bmatrix}6 & 4\\ 10 & 0\end{bmatrix}$

19. $2B+3C=2\begin{bmatrix}3 & 2\\ 5 & 0\end{bmatrix}+3\begin{bmatrix}-2 & 3\\ 4 & 0\end{bmatrix}=\begin{bmatrix}6 & 4\\ 10 & 0\end{bmatrix}+\begin{bmatrix}-6 & 9\\ 12 & 0\end{bmatrix}=\begin{bmatrix}0 & 13\\ 22 & 0\end{bmatrix}$

21. $A\times B=\begin{bmatrix}2 & 7\\ 1 & 3\end{bmatrix}\begin{bmatrix}1 & 0\\ 3 & 8\end{bmatrix}=\begin{bmatrix}2(1)+7(3) & 2(0)+7(8)\\ 1(1)+3(3) & 1(0)+3(8)\end{bmatrix}=\begin{bmatrix}23 & 56\\ 10 & 24\end{bmatrix}$

23. $A\times B=\begin{bmatrix}2 & 3 & -1\\ 0 & 4 & 6\end{bmatrix}\begin{bmatrix}2\\ 4\\ 1\end{bmatrix}=\begin{bmatrix}2(2)+3(4)+(-1)(1)\\ 0(2)+4(4)+6(1)\end{bmatrix}=\begin{bmatrix}15\\ 22\end{bmatrix}$

25. $A\times B=\begin{bmatrix}2 & -5 & 0\end{bmatrix}\begin{bmatrix}4 & 1\\ -1 & 0\\ -2 & 6\end{bmatrix}=\begin{bmatrix}2(4)+(-5)(-1)+0(-2) & 2(1)+(-5)(0)+0(6)\end{bmatrix}=\begin{bmatrix}13 & 2\end{bmatrix}$

27. $A+B=\begin{bmatrix}-2 & 5\\ 0 & 4\end{bmatrix}+\begin{bmatrix}1 & 0\\ -6 & 3\end{bmatrix}=\begin{bmatrix}(-2)+1 & 5+0\\ 0+(-6) & 4+3\end{bmatrix}=\begin{bmatrix}-1 & 5\\ -6 & 7\end{bmatrix}$

$A\times B=\begin{bmatrix}-2 & 5\\ 0 & 4\end{bmatrix}\begin{bmatrix}1 & 0\\ -6 & 3\end{bmatrix}=\begin{bmatrix}(-2)(1)+5(-6) & (-2)(0)+5(3)\\ 0(1)+4(-6) & 0(0)+4(3)\end{bmatrix}=\begin{bmatrix}-32 & 15\\ -24 & 12\end{bmatrix}$

29. $A+B=\begin{bmatrix}1 & 3 & 0\\ 2 & 4 & -1\end{bmatrix}+\begin{bmatrix}7 & -2 & 3\\ 2 & -1 & 1\end{bmatrix}=\begin{bmatrix}1+7 & 3+(-2) & 0+3\\ 2+2 & 4+(-1) & -1+1\end{bmatrix}=\begin{bmatrix}8 & 1 & 3\\ 4 & 3 & 0\end{bmatrix}$

A and B cannot be multiplied because number of columns of A is not equal to number of rows of B.

31. A and B cannot be added because they do not have the same dimensions.

$A\times B=\begin{bmatrix}2 & 5 & 1\\ 8 & 3 & 6\end{bmatrix}\times\begin{bmatrix}3 & 2\\ 4 & 6\\ -2 & 0\end{bmatrix}=\begin{bmatrix}2(3)+5(4)+1(-2) & 2(2)+5(6)+1(0)\\ 8(3)+3(4)+6(-2) & 8(2)+3(6)+6(0)\end{bmatrix}=\begin{bmatrix}24 & 34\\ 24 & 34\end{bmatrix}$

33. A and B cannot be added because they do not have the same dimensions.
A and B cannot be multiplied because the number of columns in A is not equal to the number of rows in B.

35. $A+B=\begin{bmatrix} 3 & 1 \\ -1 & 4 \end{bmatrix}+\begin{bmatrix} 4 & 1 \\ 7 & -3 \end{bmatrix}=\begin{bmatrix} 3+4 & 1+1 \\ -1+7 & 4+(-3) \end{bmatrix}=\begin{bmatrix} 7 & 2 \\ 6 & 1 \end{bmatrix}$

$B+A=\begin{bmatrix} 4 & 1 \\ 7 & -3 \end{bmatrix}+\begin{bmatrix} 3 & 1 \\ -1 & 4 \end{bmatrix}=\begin{bmatrix} 4+3 & 1+1 \\ 7+(-1) & -3+4 \end{bmatrix}=\begin{bmatrix} 7 & 2 \\ 6 & 1 \end{bmatrix}$

Thus, $A+B=B+A$.

37. $A+B=\begin{bmatrix} 0 & -1 \\ 3 & -4 \end{bmatrix}+\begin{bmatrix} 8 & 1 \\ 3 & -4 \end{bmatrix}=\begin{bmatrix} 0+8 & -1+1 \\ 3+3 & -4+(-4) \end{bmatrix}=\begin{bmatrix} 8 & 0 \\ 6 & -8 \end{bmatrix}$

$B+A=\begin{bmatrix} 8 & 1 \\ 3 & -4 \end{bmatrix}+\begin{bmatrix} 0 & -1 \\ 3 & -4 \end{bmatrix}=\begin{bmatrix} 8+0 & 1+(-1) \\ 3+3 & -4+(-4) \end{bmatrix}=\begin{bmatrix} 8 & 0 \\ 6 & -8 \end{bmatrix}$

Thus, $A+B=B+A$.

39. $(A+B)+C=\left(\begin{bmatrix} 6 & 5 \\ -1 & 3 \end{bmatrix}+\begin{bmatrix} 3 & 4 \\ -2 & 7 \end{bmatrix}\right)+\begin{bmatrix} -2 & 4 \\ 5 & 0 \end{bmatrix}=\begin{bmatrix} 9 & 9 \\ -3 & 10 \end{bmatrix}+\begin{bmatrix} -2 & 4 \\ 5 & 0 \end{bmatrix}=\begin{bmatrix} 7 & 13 \\ 2 & 10 \end{bmatrix}$

$A+(B+C)=\begin{bmatrix} 6 & 5 \\ -1 & 3 \end{bmatrix}+\left(\begin{bmatrix} 3 & 4 \\ -2 & 7 \end{bmatrix}+\begin{bmatrix} -2 & 4 \\ 5 & 0 \end{bmatrix}\right)=\begin{bmatrix} 6 & 5 \\ -1 & 3 \end{bmatrix}+\begin{bmatrix} 1 & 8 \\ 3 & 7 \end{bmatrix}=\begin{bmatrix} 7 & 13 \\ 2 & 10 \end{bmatrix}$

Thus, $(A+B)+C=A+(B+C)$.

41. $(A+B)+C=\left(\begin{bmatrix} 3 & 2 \\ 2 & 0 \\ -5 & 9 \end{bmatrix}+\begin{bmatrix} 1 & 8 \\ 0 & 6 \\ 4 & -4 \end{bmatrix}\right)+\begin{bmatrix} 0 & 4 \\ 1 & 9 \\ 9 & -6 \end{bmatrix}=\begin{bmatrix} 4 & 10 \\ 2 & 6 \\ -1 & 5 \end{bmatrix}+\begin{bmatrix} 0 & 4 \\ 1 & 9 \\ 9 & -6 \end{bmatrix}=\begin{bmatrix} 4 & 14 \\ 3 & 15 \\ 8 & -1 \end{bmatrix}$

$A+(B+C)=\begin{bmatrix} 3 & 2 \\ 2 & 0 \\ -5 & 9 \end{bmatrix}+\left(\begin{bmatrix} 1 & 8 \\ 0 & 6 \\ 4 & -4 \end{bmatrix}+\begin{bmatrix} 0 & 4 \\ 1 & 9 \\ 9 & -6 \end{bmatrix}\right)=\begin{bmatrix} 3 & 2 \\ 2 & 0 \\ -5 & 9 \end{bmatrix}+\begin{bmatrix} 1 & 12 \\ 1 & 15 \\ 13 & -10 \end{bmatrix}=\begin{bmatrix} 4 & 14 \\ 3 & 15 \\ 8 & -1 \end{bmatrix}$

Thus, $(A+B)+C=A+(B+C)$.

43. $A\times B=\begin{bmatrix} -2 & 1 \\ -3 & 4 \end{bmatrix}\begin{bmatrix} -3 & 1 \\ 4 & 2 \end{bmatrix}=\begin{bmatrix} -2(-3)+1(4) & -2(1)+1(2) \\ -3(-3)+4(4) & -3(1)+4(2) \end{bmatrix}=\begin{bmatrix} 10 & 0 \\ 25 & 5 \end{bmatrix}$

$B\times A=\begin{bmatrix} -3 & 1 \\ 4 & 2 \end{bmatrix}\begin{bmatrix} -2 & 1 \\ -3 & 4 \end{bmatrix}=\begin{bmatrix} -3(-2)+1(-3) & -3(1)+1(4) \\ 4(-2)+2(-3) & 4(1)+2(4) \end{bmatrix}=\begin{bmatrix} 6 & 1 \\ -14 & 12 \end{bmatrix}$

Thus, $A\times B\neq B\times A$.

45. $A\times B=\begin{bmatrix} 2 & 0 \\ -4 & 1 \end{bmatrix}\begin{bmatrix} 2 & 0 \\ 8 & 4 \end{bmatrix}=\begin{bmatrix} 2(2)+0(8) & 2(0)+0(4) \\ -4(2)+1(8) & -4(0)+1(4) \end{bmatrix}=\begin{bmatrix} 4 & 0 \\ 0 & 4 \end{bmatrix}$

$B\times A=\begin{bmatrix} 2 & 0 \\ 8 & 4 \end{bmatrix}\begin{bmatrix} 2 & 0 \\ -4 & 1 \end{bmatrix}=\begin{bmatrix} 2(2)+0(-4) & 2(0)+0(1) \\ 8(2)+4(-4) & 8(0)+4(1) \end{bmatrix}=\begin{bmatrix} 4 & 0 \\ 0 & 4 \end{bmatrix}$

Thus, $A\times B=B\times A$.

47. $(A\times B)\times C=\left(\begin{bmatrix}1&3\\4&0\end{bmatrix}\begin{bmatrix}4&2\\3&1\end{bmatrix}\right)\begin{bmatrix}2&1\\3&0\end{bmatrix}=\begin{bmatrix}13&5\\16&8\end{bmatrix}\begin{bmatrix}2&1\\3&0\end{bmatrix}=\begin{bmatrix}41&13\\56&16\end{bmatrix}$

$A\times(B\times C)=\begin{bmatrix}1&3\\4&0\end{bmatrix}\left(\begin{bmatrix}4&2\\3&1\end{bmatrix}\begin{bmatrix}2&1\\3&0\end{bmatrix}\right)=\begin{bmatrix}1&3\\4&0\end{bmatrix}\begin{bmatrix}14&4\\9&3\end{bmatrix}=\begin{bmatrix}41&13\\56&16\end{bmatrix}$

Thus, $(A\times B)\times C=A\times(B\times C)$.

49. $(A\times B)\times C=\left(\begin{bmatrix}4&3\\-6&2\end{bmatrix}\begin{bmatrix}1&2\\0&1\end{bmatrix}\right)\begin{bmatrix}4&3\\0&-2\end{bmatrix}=\begin{bmatrix}4&11\\-6&-10\end{bmatrix}\begin{bmatrix}4&3\\0&-2\end{bmatrix}=\begin{bmatrix}16&-10\\-24&2\end{bmatrix}$

$A\times(B\times C)=\begin{bmatrix}4&3\\-6&2\end{bmatrix}\left(\begin{bmatrix}1&2\\0&1\end{bmatrix}\begin{bmatrix}4&3\\0&-2\end{bmatrix}\right)=\begin{bmatrix}4&3\\-6&2\end{bmatrix}\begin{bmatrix}4&-1\\0&-2\end{bmatrix}=\begin{bmatrix}16&-10\\-24&2\end{bmatrix}$

Thus, $(A\times B)\times C=A\times(B\times C)$.

51. Let S = student, A = adults, SC = senior citizens

$A+B=\begin{bmatrix}85&150&50\\95&162&41\end{bmatrix}+\begin{bmatrix}73&130&45\\120&200&53\end{bmatrix}=\begin{bmatrix}85+73&150+130&50+45\\95+120&162+200&41+53\end{bmatrix}=\begin{bmatrix}158&280&95\\215&362&94\end{bmatrix}$

Total tickets: $\begin{matrix} & S & A & SC & \\ & \begin{bmatrix}158&280&95\\215&362&94\end{bmatrix} & & & \begin{matrix}\text{Matinee}\\\text{Evening}\end{matrix}\end{matrix}$

53. $A\times B=\begin{bmatrix}2&2&\frac{1}{2}&1\\3&2&1&2\\0&1&0&3\\\frac{1}{2}&1&0&0\end{bmatrix}\begin{bmatrix}10&12\\5&8\\8&8\\4&6\end{bmatrix}=\begin{bmatrix}2\cdot10+2\cdot5+\frac{1}{2}\cdot8+1\cdot4 & 2\cdot12+2\cdot8+\frac{1}{2}\cdot8+1\cdot6\\3\cdot10+2\cdot5+1\cdot8+2\cdot4 & 3\cdot12+2\cdot8+1\cdot8+2\cdot6\\0\cdot10+1\cdot5+0\cdot8+3\cdot4 & 0\cdot12+1\cdot8+0\cdot8+3\cdot6\\\frac{1}{2}\cdot10+1\cdot5+0\cdot8+0\cdot4 & \frac{1}{2}\cdot12+1\cdot8+0\cdot8+0\cdot6\end{bmatrix}=\begin{matrix}\text{L}\quad\text{S}\\\begin{bmatrix}38&50\\56&72\\17&26\\10&14\end{bmatrix}\end{matrix}\begin{matrix}\text{Sugar}\\\text{Flour}\\\text{Milk}\\\text{Eggs}\end{matrix}$

55. $\begin{bmatrix}180&172&50&136\end{bmatrix}\begin{bmatrix}10&12\\5&8\\5&5\\4&6\end{bmatrix}=\begin{bmatrix}3604&4753\end{bmatrix}$; Small: \$36.04, Large: \$47.52

57. a) Yes

b) Yes, $\begin{bmatrix}0&0\\0&0\\0&0\end{bmatrix}$

c) Yes

d) Answers will vary.

e) Yes

f) Yes

59. Answers will vary

61. Answers will vary.

63. $A\times I=\begin{bmatrix}a&b\\c&d\end{bmatrix}\begin{bmatrix}1&0\\0&1\end{bmatrix}=\begin{bmatrix}a&b\\c&d\end{bmatrix}$

$I\times A=\begin{bmatrix}1&0\\0&1\end{bmatrix}\begin{bmatrix}a&b\\c&d\end{bmatrix}=\begin{bmatrix}a&b\\c&d\end{bmatrix}$

Thus, $A\times I=I\times A$.

Review Exercises

1. A binary operation is an operation that can be performed on two and only two elements of a set. The result is a single element.

2. A mathematical system consists of a set of elements and at least one binary operation.

3. No; for example, $2\div3=\frac{2}{3}$, and $\frac{2}{3}$ is not a whole number.

4. Yes. The difference of any two real numbers is always a real number.

5. $9+8=5$

6. $10+8=6$

7. $8-10=8+12-10=10$

8. $7-4+6=3+6=9$

9. The system is closed.
   There exists an identity element in the set.
   Every element in the set has a unique inverse.
   The set is associative under the operation.

10 Yes; the integers satisfy the four properties needed.

11. The set of integers under the operation of multiplication does not form a group since not all elements have an inverse. For example, $4\cdot\frac{1}{4}=1$, but $\frac{1}{4}$ is not an integer. Only 1 and –1 have inverses.

12. No, the set of natural numbers does not have an identity element.

13. The set of rational numbers under the operation of multiplication does not form a group since zero does not have an inverse.

14. There is no identity element and no inverses.

15. Not every element has an inverse. The system is not associative: $(P\ ?\ P)\ ?\ 4 = L\ ?\ 4 = \#$, but $P\ ?\ (P\ ?\ 4) = P\ ?\ L = 4$.

16. The system is not associative: $(!\ \square\ p)\ \square\ ? = p\ \square\ ? = !$, but $!\ \square\ (p\ \square\ ?) = !\ \square\ ! = ?$.

17. a) {☺, ●, ♀, ♂}
    b) ⌂
    c) Yes; all elements in the table are in the original set.
    d) The identity element is ☺.
    e) Every element has an inverse: ☺ ⌂ ☺ = ☺, ● ⌂ ♂ = ☺, ♀ ⌂ ♀ = ☺, and ♂ ⌂ ● = ☺
    f) The system is associative: (☺ ⌂ ♀) ⌂ ♂ = ♀ ⌂ ♂ = ● and ☺ ⌂ (♀ ⌂ ♂) = ☺ ⌂ ● = ●.
    g) Yes, the elements in the table are symmetric around the diagonal: ♂ ⌂ ♀ = ● and ♀ ⌂ ♂ = ●.
    h) Yes, all five properties are satisfied.

18. $19\div 7=2$, remainder 5; $19\equiv 5 \pmod 2$

19. $24\div 6=4$, remainder 0; $24\equiv 0 \pmod 6$

20. $47\div 9=5$, remainder 2; $47\equiv 2 \pmod 9$

21. $34\div 5=6$, remainder 4; $34\equiv 4 \pmod 5$

22. $71\div 12=5$, remainder 11; $71\equiv 11 \pmod{12}$

23. $54\div 14=3$, remainder 12; $54\equiv 12 \pmod{14}$

24. 1; $2+5=7\equiv 1 \pmod 6$

25. 6; $2-6=(2+7)-6=3\equiv 3 \pmod 7$

26. 3; $7\cdot 3=21\equiv 5 \pmod 8$

27. 0, 2, 4, 6
    $0\cdot 4=0\equiv 0 \pmod 8$
    $1\cdot 4=4\equiv 4 \pmod 8$
    $2\cdot 4=8\equiv 0 \pmod 8$
    $3\cdot 4=12\equiv 4 \pmod 8$
    $4\cdot 4=16\equiv 0 \pmod 8$
    $5\cdot 4=20\equiv 4 \pmod 8$
    $6\cdot 4=24\equiv 0 \pmod 8$
    $7\cdot 4=28\equiv 4 \pmod 8$

28. 4; $10\cdot 7=70\equiv 4 \pmod{11}$

29. 9; $9\cdot 7=63\equiv 3 \pmod{10}$

30. No solution or { }

$0 \cdot 3 = 0 \equiv 0 \pmod 6$
$1 \cdot 3 = 3 \equiv 3 \pmod 6$
$2 \cdot 3 = 6 \equiv 0 \pmod 6$
$3 \cdot 3 = 9 \equiv 3 \pmod 6$
$4 \cdot 3 = 12 \equiv 0 \pmod 6$
$5 \cdot 3 = 15 \equiv 3 \pmod 6$

31. 8; $7 \cdot 8 = 56 \equiv 2 \pmod 9$

32.

+	0	1	2	3	4	5
0	0	1	2	3	4	5
1	1	2	3	4	5	0
2	2	3	4	5	0	1
3	3	4	5	0	1	2
4	4	5	0	1	2	3
5	5	0	1	2	3	4

Yes, it is a commutative group.

33.

×	0	1	2	3
0	0	0	0	0
1	0	1	2	3
2	0	2	0	2
3	0	3	2	1

34. In the following table, W represents a work day, O an off day.

Day	0	1	2	3	4	5	6	7	8	9	10	11	12	13
Work/Off	W	W	W	O	O	O	O	W	W	W	W	O	O	O

a) Since $30 \equiv 2 \pmod{14}$, Julie will be working 30 days after today.

b) Since $45 \equiv 3 \pmod{14}$, Julie will have the day off in 45 days.

35. $A+B=\begin{bmatrix} 2 & -1 \\ -3 & 0 \end{bmatrix}+\begin{bmatrix} 3 & 5 \\ 1 & 2 \end{bmatrix}=\begin{bmatrix} 2+3 & (-1)+5 \\ -3+1 & 0+2 \end{bmatrix}=\begin{bmatrix} 5 & 4 \\ -2 & 2 \end{bmatrix}$

36. $A-B=\begin{bmatrix} 2 & -1 \\ -3 & 0 \end{bmatrix}-\begin{bmatrix} 3 & 5 \\ 1 & 2 \end{bmatrix}=\begin{bmatrix} 2-3 & (-1)-5 \\ -3-1 & 0-2 \end{bmatrix}=\begin{bmatrix} -1 & -6 \\ -4 & -2 \end{bmatrix}$

37. $3A-2B=3\begin{bmatrix} 2 & -1 \\ -3 & 0 \end{bmatrix}-2\begin{bmatrix} 3 & 5 \\ 1 & 2 \end{bmatrix}=\begin{bmatrix} 6 & -3 \\ -9 & 0 \end{bmatrix}-\begin{bmatrix} 6 & 10 \\ 2 & 4 \end{bmatrix}=\begin{bmatrix} 6-6 & (-3)-10 \\ (-9)-2 & 0-4 \end{bmatrix}=\begin{bmatrix} 0 & -13 \\ -11 & -4 \end{bmatrix}$

38. $A\times B=\begin{bmatrix} 2 & -1 \\ -3 & 0 \end{bmatrix}\begin{bmatrix} 3 & 5 \\ 1 & 2 \end{bmatrix}=\begin{bmatrix} 2(3)+(-1)(1) & 2(5)+(-1)(2) \\ -3(3)+0(1) & -3(5)+0(2) \end{bmatrix}=\begin{bmatrix} 5 & 8 \\ -9 & -15 \end{bmatrix}$

39. $B\times A=\begin{bmatrix} 3 & 5 \\ 1 & 2 \end{bmatrix}\begin{bmatrix} 2 & -1 \\ -3 & 0 \end{bmatrix}=\begin{bmatrix} 3(2)+5(-3) & 3(-1)+5(0) \\ 1(2)+2(-3) & 1(-1)+2(0) \end{bmatrix}=\begin{bmatrix} -9 & -3 \\ -4 & -1 \end{bmatrix}$

Chapter Test

1. A mathematical system consists of a set of elements and at least one binary operation.
2. Closure, identity element, inverses, associative property, and commutative property.
3. Yes; the integers satisfy the five properties needed.
4. No, the set of natural numbers is not a commutative group under the operation of subtraction because it is not closed, not associative, and not commutative.
5.

+	1	2	3	4	5
1	2	3	4	5	1
2	3	4	5	1	2
3	4	5	1	2	3
4	5	1	2	3	4
5	1	2	3	4	5

6. Yes. It is closed since the only elements in the table are from the original set. The identity element is 5. Every element has an inverse: $1+4=5$, $2+3=5$, $3+2=5$, $4+1=5$, and $5+5=5$. The system is associative. The system is commutative since the table is symmetric about the main diagonal. Thus, all five properties are satisfied.

7. a) The binary operation is □.

   b) Yes. All elements in the table are from the original set.

   c) The identity element is T, since $T \square x = x = x \square T$, where x is any member of the set $\{W, S, T, R\}$.

   d) The inverse of R is S, since $R \square S = T$ and $S \square R = T$.

   e) $(T \square R) \square W = R \square W = S$

8. The system is not a group. It does not have the closure property since $c * c = d$, and d is not a member of $\{a, b, c\}$. It is also not associative.

9. Since all the numbers in the table are elements of $\{1, 2, 3\}$, the system is closed. The commutative property holds since the elements are symmetric about the main diagonal. The identity element is 2. Every element has an inverse: $1 ? 3 = 2$, $2 ? 2 = 2$, and $3 ? 1 = 2$. It is assumed the associative property holds as illustrated by the example: $(1 ? 2) ? 3 = 1 ? 3 = 2$ and $1 ? (2 ? 3) = 1 ? 3 = 2$, then the system is a commutative group.

10. Since all the numbers in the table are elements of {@, $, &, %}, the system is closed. The commutative property holds since the elements are symmetric about the main diagonal. The identity element is $. Every element has an inverse: @ ○ & = $, $ ○ $ = $, & ○ @ = $, and % ○ % = $. It is assumed the associative property holds as illustrated by the example: (@○$)○% = @○% = & and @○($○%) = @○% = &.

11. $59 \div 4 = 14$, remainder 3; $59 \equiv 3 \pmod 4$

12. $96 \div 11 = 8$, remainder 8; $96 \equiv 8 \pmod{11}$

13. 4; $4+5=9 \equiv 3 \pmod 6$

14. 2; $2-3=(5+2)-3=4 \equiv 4 \pmod 5$

15. 5; $3-5=(3+9)-5=7 \equiv 7 \pmod 9$

16. No solution or { }

$3 \cdot 0 = 0 \equiv 0 \pmod 6$

$3 \cdot 1 = 3 \equiv 3 \pmod 6$

$3 \cdot 2 = 6 \equiv 0 \pmod 6$

$3 \cdot 3 = 9 \equiv 3 \pmod 6$

$3 \cdot 4 = 12 \equiv 0 \pmod 6$

$3 \cdot 5 = 15 \equiv 3 \pmod 6$

17. a)

×	0	1	2	3	4
0	0	0	0	0	0
1	0	1	2	3	4
2	0	2	4	1	3
3	0	3	1	4	2
4	0	4	3	2	1

b) The system is closed. The identity is 1. However, 0 does not have an inverse, so the system is *not* a commutative group.

18. $A+B=\begin{bmatrix}2 & 1\\3 & -6\end{bmatrix}+\begin{bmatrix}-2 & 2\\5 & 3\end{bmatrix}=\begin{bmatrix}2+(-2) & 1+2\\3+5 & -6+3\end{bmatrix}=\begin{bmatrix}0 & 3\\8 & -3\end{bmatrix}$

19. $2A-3B=2\begin{bmatrix}2 & 1\\3 & -6\end{bmatrix}-3\begin{bmatrix}-2 & 2\\5 & 3\end{bmatrix}=\begin{bmatrix}4 & 2\\6 & -12\end{bmatrix}-\begin{bmatrix}-6 & 6\\15 & 9\end{bmatrix}=\begin{bmatrix}4-(-6) & 2-6\\6-15 & -12-9\end{bmatrix}=\begin{bmatrix}10 & -4\\-9 & -21\end{bmatrix}$

20. $A \times B=\begin{bmatrix}2 & 1\\3 & -6\end{bmatrix}\begin{bmatrix}-2 & 2\\5 & 3\end{bmatrix}=\begin{bmatrix}(2)(-2)+1(5) & (2)(2)+(1)(3)\\(3)(-2)+(-6)(5) & (3)(2)+(-6)(3)\end{bmatrix}=\begin{bmatrix}1 & 7\\-36 & -12\end{bmatrix}$

Chapter Ten: Consumer Mathematics

Section 10.1: Percent

1. 100

3. 100

5. Previous

7. $\frac{2}{5} = 0.40 = (0.4)(100)\% = 40.0\%$

9. $\frac{7}{20} = 0.35 = (0.35)(100)\% = 35.0\%$

11. $0.007654 = (0.007654)(100)\% = 0.8\%$

13. $3.78 = (3.78)(100)\% = 378\%$

15. $9\% = \frac{79}{100} = 0.09$

17. $7.24\% = \frac{7.24}{100} = 0.0724$

19. $\frac{1}{4}\% = 0.25\% = \frac{0.25}{100} = 0.0025$

21. $135.9\% = \frac{135.9}{100} = 1.359$

23. $\frac{5}{20} = \frac{25}{100} = 25\%$

25. $8(0.4125) = 3.3;\ 8.0 - 3.3 = 4.7$ g

27. $(1743)(0.27) = 471$ comedy DVDs

29. $(1743)(0.11) = 192$ horror DVDs

31. $(25)(0.8584) = 21.46$ ml of oxygen

33. $(500)(0.0194) = 9.7$ ml of chlorine

35. $\frac{921}{1960} \approx 0.470 = 47.0\%$

37. $\frac{274}{1960} \approx 0.140 = 14.0\%$

39. $\frac{146.8 - 148}{148} \approx -0.0081$, or a 0.81% decrease

41. a) $\frac{10,361.0 - 10,924.1}{10,924.1} \approx -0.052$,
    or a 5.2% decrease

    b) $\frac{11,129.4 - 10,361.0}{10,361.0} \approx 0.074$,
    or a 7.4% increase

    c) $\frac{11,377.5 - 11,129.4}{11,129.4} \approx 0.022$,
    or a 2.2% increase

    d) $\frac{11,072.2 - 11,377.5}{11,377.5} \approx -0.027$,
    or a 2.7% decrease

    e) $\frac{11,892.3 - 11,072.2}{11,072.2} \approx 0.074$,
    or a 7.4% increase

43. $(0.15)(75,000) = \$11,250.00$

45. $(0.08)(32) = 2.56$

47. $24/96 = 0.25;\ (0.25)(100\%) = 25\%$

49. $12/150 = 0.08;\ (0.08)(100\%) = 8\%$

51. $0.05x = 15;\ x = 15/0.05 = 300$

53. $42 = 0.28x;\ x = 42/0.28 = 150$

55. a) $0.06(63.50) = \$3.81$

    b) $0.15(63.50) = \$9.53$

    c) $63.50 + 3.81 + 9.53 = \$76.84$

57. Let $x =$ the pretax cost of the jacket
    $0.08x =$ the sales tax

    $x + 0.08x = 54.27$

    $1.08x = 54.27$

    $x = \frac{54.27}{1.08} = \$50.25$

    The pretax cost of the jacket was \$50.25.

59. Let $x =$ the number of A's on the second test

    $1.50x = 18$

    $x = \frac{18}{1.5} = 12$

    Twelve students got an A on the second test.

61. The increase in Mr. Brown's salary is $0.07(\$39,500) = \$2765$. Mr. Brown's new salary will be $39,500 + 2765 = \$42,265$.

63. $\frac{407-430}{430} \approx -0.053$, or a 5.3% decrease

65. $\frac{439-539.62}{539.62} \approx -0.186$, or an 18.6% decrease

67. $\frac{49-35}{35} \approx 0.40$, or a 40% markup

69. \$1000 increased by 10% is $\$1000 + 0.10(\$1000) = \$1000 + \$100 = \$1100$.
\$1100 decreased by 10% is $\$1100 - 0.10(\$1100) = \$1100 - \$110 = \$990$.
Therefore if he sells the car at the reduced price he will lose \$10.

71. a) No, the 25% discount is greater. See part (b).
   b) $\$189.99 - 0.10(\$189.99) = \$189.99 - \$19.00 = \$170.99$
   $\$170.99 - 0.15(\$170.99) = \$170.99 - \$25.65 = \$145.34$
   c) $\$189.99 - 0.25(\$189.99) = \$189.99 - \$47.50 = \$142.49$
   d) Yes

73. a) \$100 increased by 25%: $\$100 + 0.25(\$100) = \$125$
   \$200 decreased by 25%: $\$200 - 0.25(\$200) = \$150$
   \$200 decreased by 25% is $\$150 - \$125 = \$25$ greater.
   b) \$100 increased by 50%: $\$100 + 0.50(\$100) = \$150$
   \$200 decreased by 50%: $\$200 - 0.50(\$200) = \$100$
   \$100 increased by 50% is $\$150 - \$100 = \$50$ greater.
   c) \$100 increased by 100%: $\$100 + 1.00(\$100) = \$200$
   \$200 decreased by 100%: $\$200 - 1.00(\$200) = \$0$
   \$100 increased by 100% is $\$200 - \$0 = \$200$ greater.

Section 10.2: Personal Loans and Simple Interest

1. Principal
3. Interest
5. Rate
7. United States
9. $i = prt = (\$375)(0.0225)(4) = \$33.75$
11. $i = prt = (\$1100)(0.0875)\left(\frac{90}{360}\right) = \$24.06$
13. $i = prt = (\$587)(0.00045)(60) = \$15.85$
15. $i = prt = (\$550.31)(0.089)\left(\frac{67}{360}\right) = \$9.12$
17. $i = prt = (\$1372.11)(0.01375)(6) = \$113.20$
19. $$\begin{aligned} i &= prt \\ 82.80 &= (2000)r(3) \\ 6000r &= 82.80 \\ r &= \frac{82.80}{6000} = 0.0138 = 1.38\% \end{aligned}$$
21. $$\begin{aligned} i &= prt \\ 175 &= p(0.06)\left(\frac{5}{12}\right) \\ 0.025p &= 175 \\ p &= \frac{175}{0.025} = \$7000 \end{aligned}$$
23. $$\begin{aligned} i &= prt \\ 124.49 &= (957.62)(0.065)t \\ 62.2453t &= 124.49 \\ t &= \frac{124.49}{62.2453} \approx 2 \text{ years} \end{aligned}$$

25. Interest: $i = prt = (\$45,000)(0.015)(2) = \$1350$; Total repayment: $p + i = \$45,000 + \$1350 = \$46,350$

27. a) Interest: $i = prt = (\$3500)(0.075)\left(\frac{6}{12}\right)$
$= \$131.25$

b) Total repayment: $p + i = \$3500 + \$131.25$
$= \$3631.25$

29. a) Interest: $i = prt = (\$3650)(0.076)\left(\frac{18}{12}\right)$
$= \$416.10$

b) Amount received: $p - i = \$3650 - \$416.10$
$= \$3233.90$

c) Interest rate:
$i = prt$
$416.10 = (3233.90)r\left(\frac{18}{12}\right)$
$4850.85r = 416.10$
$r = \frac{416.10}{4850.85} \approx 0.0858$
$r = 8.58\%$

31. $i = \$280.50 - \$270.00 = \$10.50$
$10.50 = (270)(r)\left(\frac{7}{360}\right)$
$5.25r = 10.50$
$r = \frac{10.50}{5.25} = 2$, or 200%

33. [02/25 to 05/05]: $125 - 56 = 69$ days

35. [02/02 to 10/31]: $304 - 33 = 271$ days
Because of the leap year, $271 + 1 = 272$ days

37. [08/24 to 05/15]:
$(365 - 236) + 135 = 129 + 135 = 264$ days

39. [05/31] for 150 days: $151 + 150 = 301$, which is October 28.

41. [11/25] for 120 days: $329 + 120 = 449$;
$449 - 365 = 84$

Because of the leap year, $84 - 1 = 83$, which is March 24.

43. [06/01 to 07/01]: $182 - 152 = 30$ days
$(\$3200)(0.03)\left(\frac{30}{360}\right) = \$8.00$
$\$1200.00 - \$8.00 = \$1192.00$
$\$3200.00 - \$1192.00 = \$2008.00$
[07/01 to 09/29]: $272 - 182 = 90$ days
$(\$2008.00)(0.03)\left(\frac{90}{360}\right) = \$15.06$
$\$2008.00 + \$15.06 = \$2023.06$

45. [02/01 to 05/01]: $121 - 32 = 89$ days
$(\$2400)(0.055)\left(\frac{89}{360}\right) = \$32.63$
$\$1000.00 - \$32.63 = \$967.37$
$\$2400 - \$967.37 = \$1432.63$
[05/01 to 08/31]: $243 - 121 = 122$ days
$(\$1432.63)(0.055)\left(\frac{122}{360}\right) = \$26.70$
$\$1432.63 + \$26.70 = \$1459.33$

47. [07/15 to 12/27]: $361 - 196 = 165$ days
$(\$9000)(0.06)(165/360) = \$247.50$
$\$4000.00 - \$247.50 = \$3752.50$
$\$9000.00 - \$3752.50 = \$5247.50$
[12/27 to 02/01]: $(365 - 361) + 32 = 36$ days
$(\$5247.50)(0.06)\left(\frac{36}{360}\right) = \$31.49$
$\$5247.50 + \$31.49 = \$5278.99$

49. [08/01 to 09/01]: $244 - 213 = 31$ days
$(\$1800)(0.15)\left(\frac{31}{360}\right) = \$23.25$
$\$500.00 - \$23.25 = \$476.75$
$\$1800.00 - \$476.75 = \$1323.25$
[09/01 to 10/01]: $274 - 244 = 30$ days
$(\$1323.25)(0.15)\left(\frac{30}{360}\right) = \$16.54$
$\$500.00 - \$16.54 = \$483.46$
$\$1323.25 - \$483.46 = \$839.79$
[10/01 to 11/01]: $305 - 274 = 31$ days
$(\$839.79)(0.15)\left(\frac{31}{360}\right) = \$10.85$
$\$839.79 + \$10.85 = \$850.64$

51. [03/01 to 08/01]: $213-60=153$ days

$$(\$11,600)(0.06)\left(\frac{153}{360}\right)=\$295.80$$

$\$2000.00-\$295.80=\$1704.20$

$\$11,600.00-\$1704.20=\$9895.80$

[08/01 to 11/15]: $319-213=106$ days

$$(9895.80)(0.06)\left(\frac{106}{360}\right)=\$174.83$$

$\$4000.00-\$174.83=\$3825.17$

$\$9895.8-\$3825.17=\$6070.63$

[11/15 to 12/01]: $335-319=16$ days

$$(6070.63)(0.06)\left(\frac{16}{360}\right)=\$16.19$$

$\$6070.63+\$16.19=\$6086.82$

53. [03/01 to 05/01]: $121-60=61$ days

$$(\$6500)(0.105)\left(\frac{61}{360}\right)=\$115.65$$

$\$1750.00-\$115.65=\$1634.35$

$\$6500.00-\$1634.35=\$4865.65$

[05/01 to 07/01]: $182-121=61$ days

$$(\$4865.65)(0.105)\left(\frac{61}{360}\right)=\$86.57$$

$\$2350.00-\$86.57=\$2263.43$

$\$4865.65-\$2263.43=\$2602.22$

Since the loan was for 180 days, the maturity date occurs in $180-61-61=58$ days.

$$(\$2602.22)(0.105)\left(\frac{58}{360}\right)=\$44.02$$

$\$2602.22+\$44.02=\$2646.24$

55. a) November 4 is day 318; $318+364=682$; $682-365=317$; Day 317 is Nov. 13, 2019.

b) $i=(\$1000)(0.0015)\left(\frac{364}{360}\right)=\$1.52$

Amount paid: $\$1000-\$1.52=\$998.48$

c) Interest: \$1.52

d) $r=\frac{i}{pt}=\frac{1.52}{998.48\left(\frac{364}{360}\right)}\approx 0.001506$

$=0.1506\%$

57. Amount received: $\$743.21-\$39.95=\$703.26$

a) $i=prt$

$39.95=(703.26)(r)\left(\frac{5}{360}\right)$

$39.95=9.7675r$

$r=\frac{39.95}{9.7675}\approx 4.0901$

$r=409.01\%$

b) $i=prt$

$39.95=(703.26)(r)\left(\frac{10}{360}\right)$

$39.95=19.535r$

$r=\frac{39.95}{19.535}\approx 2.0450$

$r=204.50\%$

c) $i=prt$

$39.95=(703.26)(r)\left(\frac{20}{360}\right)$

$39.95=39.07r$

$r=\frac{39.95}{39.07}\approx 1.0225$

$r=102.25\%$

59. a) $(\$25,000)(0.03)\left(\frac{180}{360}\right)=\$375$

b) $(\$25,000)(0.03)\left(\frac{180}{365}\right)=\$369.86$

c) $\$375-\$369.86=\$5.14$

d) $\frac{5.14}{369.86}\approx 0.0139=1.39\%$

Section 10.3: Compound Interest

1. Profit

3. Variable

5. Compound

7. $n=2,\ \ r=0.024,\ \ t=4,\ \ p=\$850$

a) $A=850\left(1+\frac{0.024}{2}\right)^{(2\cdot 4)}\approx \$935.11$

b) $i=\$935.11-\$850=\$85.11$

9. $n=4,\ \ r=0.03,\ \ t=6,\ \ p=\$3000$

a) $A=3000\left(1+\frac{0.03}{4}\right)^{(4\cdot 6)}=\$3589.24$

b) $i=\$3589.24-\$3000=\$589.24$

11. $n=12,\ \ r=0.045,\ \ t=2,\ \ p=\$10,000$

a) $A=10,000\left(1+\frac{0.045}{12}\right)^{(12\cdot 2)}=\$10,939.90$

b) $i=\$10,939.90-\$10,000=\$939.90$

13. $n=360,\ \ r=0.04,\ \ t=2,\ \ p=\$8000$

a) $A=8000\left(1+\frac{0.04}{360}\right)^{(360\cdot 2)}\approx \$8666.26$

b) $i=\$8666.26-\$8000=\$666.26$

15. $n=12,\ \ r=0.036,\ \ t=10,\ \ A=\$25,000$

$$p=\frac{25,000}{\left(1+\frac{0.036}{12}\right)^{(12\cdot 10)}}\approx \$17451.31$$

17. $n=4,\ \ r=0.04,\ \ t=4,\ \ A=\$100,000$

$$p=\frac{100,000}{\left(1+\frac{0.04}{4}\right)^{(4\cdot 4)}}\approx \$85,282.13$$

19. $n=2,\ \ r=0.0257,\ \ t=5,\ \ p=\$8500$

$$A=8500\left(1+\frac{0.0257}{2}\right)^{(2\cdot 5)}\approx \$9657.62$$

21. $n=12,\ \ r=0.028,\ \ t=\frac{30}{12}=2.5,\ \ p=\$1500$

$$A=1500\left(1+\frac{0.028}{12}\right)^{(12\cdot 2.5)}\approx \$1608.63$$

23. $n=2,\ \ r=0.0175,\ \ t=10,\ \ p=\$10,000$

$$A=10,000\left(1+\frac{0.0175}{2}\right)^{(2\cdot 5)}\approx \$10,910.27$$

25. $n=4,\ \ r=0.0175,\ \ t=2,\ \ p=\$3000$

$$A=3000\left(1+\frac{0.0175}{4}\right)^{(4\cdot 2)}\approx \$3106.62$$

27. $n=360,\ \ r=0.02,\ \ t=2,$
$p=\$800+\$150+\$300+\$1000=\$2250$

$$A=2250\left(1+\frac{0.02}{360}\right)^{(360\cdot 2)}=\$2341.82$$

29. $i=A-p=1\left(1+\frac{0.035}{2}\right)^{(2\cdot 1)}-1\approx 0.0353,$ or 3.53%

31. a) $i=A-p=1\left(1+\frac{0.019}{4}\right)^{(4\cdot 1)}-1\approx 0.0191,$ or 1.91%

b) $i=A-p=1\left(1+\frac{0.018}{360}\right)^{(360\cdot 1)}-1\approx 0.0182,$ or 1.82%

c) Prospero Bank

33. a) $i=A-p=1\left(1+\frac{0.0223}{4}\right)^{(4\cdot 1)}-1\approx 0.0225,$ or 2.25%

b) $i=A-p=1\left(1+\frac{0.0225}{2}\right)^{(2\cdot 1)}-1\approx 0.0226,$ or 2.26%

c) Key Bank

35. $i=A-p=1\left(1+\frac{0.024}{12}\right)^{(12\cdot 1)}-1\approx 0.0243,$ or 2.43%; Yes, the APY should be 2.43%, not 2.6%.

37. $n = 4,\ r = 0.04,\ t = 15,\ A = \$55,000$

$$p = \frac{55,000}{\left(1+\frac{0.04}{4}\right)^{(4\cdot 15)}} \approx \$30,274.73$$

39. $n = 4,\ r = 0.035,\ t = 18,\ A = \$25,000$

$$p = \frac{25,000}{\left(1+\frac{0.035}{4}\right)^{(4\cdot 18)}} \approx \$13,351.33$$

41. a) $\$925,000 - \$370,000 = \$555,000$

b) $p = \dfrac{555,000}{\left(1+\frac{0.075}{12}\right)^{(12\cdot 30)}} \approx \$58,907.61$

c) The surcharge would be $\dfrac{\$58,907.61}{598} \approx \$98.51$ per homeowner.

43. a) $A = 1000\left(1+\dfrac{0.02}{2}\right)^{(2\cdot 2)} \approx \$1040.60$

$i = 1040.60 - 1000 = \$40.60$

b) $A = 1000\left(1+\dfrac{0.04}{2}\right)^{(2\cdot 2)} \approx \$1082.43$

$i = \$1082.43 - \$1000 = \$82.43$

c) $A = 1000\left(1+\dfrac{0.08}{2}\right)^{(2\cdot 2)} \approx \$1169.86$

$i = \$1169.86 - \$1000 = \$169.86$

d) No, there is not a predictable pattern.

45. Simple interest: $i = prt = \$100,000(0.05)(4) = \$20,000$

Compound interest: $A = 100,000\left(1+\dfrac{0.05}{360}\right)^{(360\cdot 4)} \approx \$122,138.58$

Interest earned: $122,138.58 - 100,000 = \$22,138.58$

Select investing at the compounded daily interest because the compound interest is greater by $\$22,138.58 - \$20,000 = \$2138.58$.

47. $p = \$2000,\ A = \$3586.58,\ n = 12,\ t = 5$

$$3586.58 = 2000\left(1+\frac{r}{12}\right)^{(12\cdot 5)}$$

$$\frac{3586.58}{2000} = \left(1+\frac{r}{12}\right)^{60}$$

$$1+\frac{r}{12} = \sqrt[60]{\frac{3586.58}{2000}}$$

$$\frac{r}{12} = \sqrt[60]{\frac{3586.58}{2000}} - 1$$

$$r = 12\left(\sqrt[60]{\frac{3586.58}{2000}} - 1\right) \approx 0.1174,$$

or 11.74%

49. Compound interest: $A = 2000\left(1+\dfrac{0.08}{2}\right)^{(2\cdot 3)}$

$\approx \$2530.64$

$i = \$2530.64 - \$2000 = \$530.64$

Simple interest: $i = prt$

$530.64 = 2000(r)(3)$

$530.64 = 6000r$

$r = \dfrac{530.64}{6000} \approx 0.0884$

The simple interest rate would be 8.84%.

Section 10.4: Installment Buying

1. Installment

3. Annual

5. Installment

7. From Table 10.2 the finance charge per $100 at 6% for 24 payments is 3.28.

Total finance charge: $(6.37)\left(\dfrac{825}{100}\right)=\$52.55$

Total amount due: $\$825+\$52.55=\$877.55$

Monthly payment: $\dfrac{\$877.55}{24}=\$36.56$

Using formula: $\dfrac{825\left(\dfrac{0.06}{12}\right)}{1-\left(1+\dfrac{0.06}{12}\right)^{(-12\cdot 2)}}=\$36.56$

9. From Table 10.2 the finance charge per $100 at 6.5% for 48 payments is 13.83.

Total finance charge: $(13.83)\left(\dfrac{15,000}{100}\right)=\$2074.50$

Total amount due: $\$15,000+\$2074.50=\$17,074.50$

Monthly payment: $\dfrac{\$17,074.50}{48}=355.72$

Using formula: $\dfrac{15,000\left(\dfrac{0.065}{12}\right)}{1-\left(1+\dfrac{0.065}{12}\right)^{(-12\cdot 4)}}=\$355.72$

11. Amount of loan: $\$6500-(0.10)(\$6500)=\$5850$

a) From Table 10.2 the finance charge per $100 at 4.5% for 48 payments is 9.46.

Total finance charge: $(9.46)\left(\dfrac{5850}{100}\right)=\$553.41$

Total amount due: $\$5850+\$553.41=\$6403.41$

b) Monthly payment: $\dfrac{\$6403.41}{48}=\$133.40$

Using formula: $\dfrac{5850\left(\dfrac{0.045}{12}\right)}{1-\left(1+\dfrac{0.045}{12}\right)^{(-12\cdot 4)}}=\$133.40$

13. a) Amount borrowed: $\$7500-(0.20)(\$7500)=\$6000$

Total installment price: $(36)(\$189.40)=\$6818.40$

Finance charge: $\$6818.40-\$6000=\$818.40$

b) $\left(\dfrac{\text{finance charge}}{\text{amount financed}}\right)(100)=\left(\dfrac{818.40}{6000}\right)(100)=\$13.64$

From Table 10.2 for 36 payments, the value of 13.64 corresponds with an APR of 8.5%.

15. a) Amount borrowed: $\$60,714-\$21,500=\$39,214$

Total installment price: $(60)(\$767.29)=\$46,037.40$

Finance charge: $\$46,037.40-\$39,214=\$6823.40$

b) $\left(\dfrac{\text{finance charge}}{\text{amount financed}}\right)(100)=\left(\dfrac{6823.40}{39214}\right)(100)\approx\$17.40$

From Table 10.2, for 60 payments, the value of $17.40 corresponds with an APR of 6.5%.

17. Total installment price: $(\$232)(60)=\$13,920$

Finance charge: $\$13,920-\$12,000=\$1920$

a) $\left(\dfrac{1920}{12,000}\right)(100)=\$16.00$; In Table 10.2, $16.00 yields an APR of 6.0%.

17. (continued)

b) From Table 10.2, the monthly payment per \$100 for the remaining 36 months at 6.0% APR is \$9.52.

$$u = \frac{n \cdot P \cdot V}{100 + V} = \frac{(36)(\$232)(9.52)}{100 + 9.52} \approx \$726.00$$

c) Remaining payments: $(\$232)(36) = \$8352$

Remaining balance: $\$8352 - \$726 = \$7626$

Total amount due: $\$7626 + \$232 = \$7858$

19. a) Amount financed: $\$32,000 - \$10,000 = \$22,000$

From Table 10.2, the finance charge per \$100 financed at 8.0% APR for 36 payments is \$12.81.

Total finance charge: $(12.81)\left(\frac{22,000}{100}\right) = \$2818.20$

b) Total amount due: $\$22,000 + \$2818.20 = \$24,818.20$

Monthly payment: $\frac{\$24,818.20}{36} = \$689.39$

c) From Table 10.2, the monthly payment per \$100 for the remaining 12 months at 8.0% APR is \$4.39.

$$u = \frac{n \cdot P \cdot V}{100 + V} = \frac{(12)(\$689.39)(4.39)}{100 + 4.39} = \$347.90$$

d) Remaining payments: $(\$689.39)(12) = \$8272.68$

Remaining balance: $\$8272.68 - \$347.90 = \$7924.78$

Total amount due: $\$7924.78 + \$689.39 = \$8614.17$

21. a) Minimum payment: $(0.01)(\$3000) = \$30.00$

b) Principal on which interest is charged during April: $\$3000 - \$500 = \$2500$

Interest for April: $\$2500(0.0004247)(30) = \$31.85$

Interest on outstanding balance: $(0.01)(\$2500) = \$25.0$

Minimum monthly payment: $\$25 + \$31.85 = \$56.85$, which rounds up to \$57.

23. a) Total charges: $\$677 + \$452 + \$139 + \$141 = \$1409$

Minimum payment: $(0.015)(\$1409) = \$21.14$, which rounds up to \$22.

b) Principal on which interest is charged during October: $\$1409 - \$300 = \$1109$

Interest for December: $\$1109(0.0005163)(31) = 17.75$

Interest on outstanding balance: $(0.015)(\$1109) = \$16.64$

Minimum monthly payment: $\$17.75 + \$16.64 = \$34.39$, which rounds up to \$35.

25. a) Finance charge: $(0.013)(\$2302.65) = \$29.93$

b) Balance on October 5: $\$2302.65 + \$106.72 + \$29.93 - \$550 = \$1889.30$

27. a) Finance charge: $(0.0125)(\$124.78) = \$1.56$

b) Balance on October 5: $\$124.78 + \$25.64 + \$67.23 + \$13.90 + \$1.56 - \$100 = \$133.11$

29. a)

Date	Balance Due	Days	(Balance)(Days)
May 12	378.50	1	$(\$378.50)(1) = \$378.50$
May 13	$378.50 + \$129.79 = \$508.29$	2	$(\$508.29)(2) = \$1016.58$
May 15	$508.29 - \$50 = \$458.29$	17	$(\$458.29)(17) = \$7790.93$
June 01	$458.29 + \$135.85 = \$594.14$	7	$(\$594.14)(7) = \$4158.98$
June 08	$594.14 + \$37.63 = \$631.77$	4	$(\$631.77)(4) = \$2527.08$
June 12		31	\$15,872.07

Average daily balance: $\dfrac{\$15,872.07}{31} = \$512.00$

b) Finance charge: $(0.013)(\$512.00) = \$6.66$

c) Balance due on June 12: $\$631.77 + \$6.66 = \$638.43$

31. a)

Date	Balance Due	Days	(Balance)(Days)
Feb 03	\$124.78	5	$(\$124.78)(5) = \$623.90$
Feb 08	$\$124.78 + \$25.64 = \$150.42$	4	$(\$150.42)(4) = \$601.68$
Feb 12	$\$150.42 - \$100.00 = \$50.42$	2	$(\$50.42)(2) = \$100.84$
Feb 14	$\$50.42 + \$67.23 = \$117.65$	11	$(\$117.65)(1) = \$117.65$
Feb 25	$\$117.65 + \$13.90 = \$131.55$	6	$(\$131.55)(6) = \$789.30$
Mar 03		28	\$3409.87

Average daily balance: $\dfrac{\$3409.87}{28} = \$121.78$

b) Finance charge: $(0.0125)(\$121.78) = \$1.52$

c) Balance due on March 3: $\$131.55 + \$1.52 = \$133.07$

d) The interest charged using the average daily balance method is \$0.04 less than the interest charged using the previous balance method.

33. a) $i = (\$875)(0.0004273)(32) = \$11.96$

b) $A = \$875 + \$11.96 = \$886.96$

35. a) State National Bank: $i = (\$1000)(0.05)(0.5) = \$25.00$

b) Consumers Credit Union: $(\$86.30)(12) = \$1035.60$; $i = \$1035.60 - \$1000.00 = \$35.60$

c) $\left(\dfrac{25}{1000}\right)(100) = 2.50$; In Table 10.2, \$2.49 is the closest value to \$2.50, which corresponds to an APR of 8.5 %.

d) $\left(\dfrac{35.60}{1000}\right)(100) = 3.56$; In Table 10.2, \$3.56 corresponds to an APR of 6.5 %.

37. Let $p =$ amount Ken borrowed

$p + 2500 =$ purchase price

Total price: $2500 + (379.50)(36) = \$16,162$

Interest (Total price – purchase price): $i = 16,162 - (p + 2500) = 16,162 - p - 2500 = 13,662 - p$

$$\begin{aligned} i &= prt \\ 13,662 - p &= (p)(0.06)(3) \\ 13,662 - p &= 0.18p \\ 13,662 &= 0.18p + p \\ 1.18p &= 13,662 \\ p &\approx 11,577.97 \end{aligned}$$

The purchase price is $\$11,577.97 + \$2500 = \$14,077.97$.

Section 10.5: Buying a House with a Mortgage

1. Mortgage
3. Points
5. Adjusted
7. From Table 10.4 the monthly principal and interest payment per \$1000 at 5% for 15 years is \$7.90794.

   Monthly principal and interest payment: $\left(\dfrac{\$95,000}{\$1000}\right)(\$7.90794) = \$751.25$

   Using formula: $\dfrac{95,000\left(\dfrac{0.05}{12}\right)}{1-\left(1+\dfrac{0.05}{12}\right)^{(-12\cdot 15)}} = \$751.25$

9. From Table 10.4 the monthly principal and interest payment per \$1000 at 5.5% for 25 years is \$6.14087.

   Monthly principal and interest payment: $\left(\dfrac{\$236,000}{\$1000}\right)(\$6.14087) = \$1449.25$

   Using formula: $\dfrac{236,000\left(\dfrac{0.055}{12}\right)}{1-\left(1+\dfrac{0.055}{12}\right)^{(-12\cdot 25)}} = \$1449.25$

11. a) Down payment: $(0.15)(\$275,000) = \$41,250$

    b) Amount of mortgage: $\$275,000 - \$41,250 = \$233,750$

    c) Table 10.4 yields \$4.77415 per \$1000 of mortgage.

    Monthly payment: $\left(\dfrac{\$233,750}{\$1000}\right)(\$4.77415) = \$1115.96$; Using formula: $\dfrac{233,750\left(\dfrac{0.04}{12}\right)}{1-\left(1+\dfrac{0.04}{12}\right)^{(-12\cdot 30)}} = \$1115.96$

13. a) Down payment: $(0.20)(\$2,337,500) = \$467,500$

    b) Amount of mortgage: $\$2,337,500 - \$467,500 = \$1,870,000$

    c) Table 10.4 yields \$6.32649 per \$1000 of mortgage.

    Monthly payment: $\left(\dfrac{\$1,870,000}{\$1000}\right)(\$6.32649) = \$11,830.54$

13. (continued)

Using formula: $\dfrac{1{,}870{,}000\left(\dfrac{0.045}{12}\right)}{1-\left(1+\dfrac{0.045}{12}\right)^{(-12\cdot 20)}} = \$11{,}830.54$

15. a) Down payment: $(0.20)(\$195{,}000) = \$39{,}000$

 b) Amount of mortgage: $\$195{,}000 - \$39{,}000 = \$156{,}000$

 c) Cost of points: $(\$156{,}000)(0.02) = \$3120$

17. Monthly income: $3200

 Adjusted monthly income: $\$3200 - \$335 = \$2865$

 a) $(0.28)(\$2865) = \$802.20$

 b) Table 10.4 yields $7.90794 per $1000 of mortgage.

 $\left(\dfrac{\$150{,}000}{\$1000}\right)(\$7.90794) = \$1186.19$; Using formula: $\dfrac{150{,}000\left(\dfrac{0.05}{12}\right)}{1-\left(1+\dfrac{0.05}{12}\right)^{(-12\cdot 15)}} = \$1186.19$

 Total monthly payment: $\$1186.19 + \$225.00 = \$1411.19$

 c) They do not qualify for the mortgage since $\$1411.19 > \$802.20$.

19. a) Total payments: $(\$555.33)(15)(12) = \$99{,}959.40$

 Amount paid: $\$99{,}959.40 + (\$75{,}000 - \$63{,}750) = \$111{,}209.40$

 b) $\$111{,}209.40 - \$75{,}000 = \$36{,}209.40$

 c) $i = prt = (\$63{,}750)(0.065)(1/12) = \$345.31$

 $\$555.33 - \$345.31 = \$210.02$

21. a) Down payment: $(0.20)(\$550{,}000) = \$110{,}000$

 b) Amount of mortgage: $\$550{,}000 - \$110{,}000 = \$440{,}000$;

 Cost of three points: $(0.03)(\$440{,}000) = \$13{,}200$

 c) Adjusted monthly income: $\$15{,}375 - \$995 = \$14{,}380$

 28% of adjusted income: $(0.28)(\$14{,}380) = \$4026.40$

 d) At a rate of 5.5% for 30 years, Table 10.4 yields $5.67789.

 Mortgage payment: $\left(\dfrac{\$440{,}000}{\$1000}\right)(\$5.67789) = \$2498.27$

 Using formula: $\dfrac{440{,}000\left(\dfrac{0.055}{12}\right)}{1-\left(1+\dfrac{0.055}{12}\right)^{(-12\cdot 30)}} = \$2498.27$

 e) Insurance and taxes: $\dfrac{\$5634 + \$2325}{12} = \$663.25$ per month

 Total monthly payment: $\$2498.27 + \$663.25 = \$3161.52$

21. (continued)

f) Yes, since \$4,026.40 is greater than \$3161.52, the Nejems qualify.

g) Interest on first payment: $i = prt = (\$440{,}000)(0.055)\left(\frac{1}{12}\right) = \$2016.67$

Amount applied to principal: $\$2498.27 - \$2016.67 = \$481.60$

23. Bank A

Down payment: $(0.10)(\$105{,}000) = \$10{,}500$

Amount of mortgage: $\$105{,}000 - \$10{,}500 = \$94{,}500$

At a rate of 4% for 30 years, Table 10.4 yields \$4.77415.

Monthly mortgage payment: $\left(\frac{\$94{,}500}{\$1000}\right)(\$4.77415) = \$451.16$

Using formula: $$\frac{94{,}500\left(\frac{0.04}{12}\right)}{1-\left(1+\frac{0.04}{12}\right)^{(-12\cdot 30)}} = \$451.16$$

Cost of three points: $(0.03)(\$94{,}500) = \$2835$

Total cost of the house: $\$10{,}500 + \$2835 + (\$451.16)(12)(30) = \$175{,}752.60$

Bank B

Down payment: $(0.20)(\$105{,}000) = \$21{,}000$

Amount of mortgage: $\$105{,}000 - \$21{,}000 = \$84{,}000$

At a rate of 5.5% for 25 years, Table 10.4 yields \$6.14087.

Monthly mortgage payment: $\left(\frac{\$84{,}000}{\$1000}\right)(\$6.14087) = \$515.83$

Using formula: $$\frac{84{,}000\left(\frac{0.055}{12}\right)}{1-\left(1+\frac{0.055}{12}\right)^{(-12\cdot 25)}} = \$515.83$$

Total cost of the house: $\$21{,}000 + (\$515.83)(12)(25) = \$175{,}749.00$

The Riveras should select Bank B.

25. Amount of mortgage: $\$235{,}000 - \$47{,}000 = \$188{,}000$

a) One-year Treasury Bill Rate + Add-on Rate = Initial ARM Rate = 1.5% + 2.5% = 4.0%

b) Monthly payment: $$\frac{188{,}000\left(\frac{0.04}{12}\right)}{1-\left(1+\frac{0.04}{12}\right)^{(-12\cdot 20)}} = \$1139.24$$

c) New ARM rate: 3.0% + 2.5% = 5.5%

27. Down payment: $(0.20)(\$450{,}000) = \$90{,}000$

Amount of mortgage: $\$450{,}000 - \$90{,}000 = \$360{,}000$

Cost of one point: $(0.01)(\$360{,}000) = \$3600$

a) Monthly payment: $\dfrac{360{,}000\left(\dfrac{0.10}{12}\right)}{1-\left(1+\dfrac{0.10}{12}\right)^{(-12\cdot 10)}} = \$4757.43$

Total payments: $\$90{,}000 + \$3600 + (10)(12)(\$4757.43) = \$664{,}491.60$

b) Monthly payment: $\dfrac{360{,}000\left(\dfrac{0.10}{12}\right)}{1-\left(1+\dfrac{0.10}{12}\right)^{(-12\cdot 20)}} = \$3474.08$

Total payments: $\$90{,}000 + \$3600 + (20)(12)(\$3474.08) = \$927{,}379.20$

c) Monthly payment: $\dfrac{360{,}000\left(\dfrac{0.10}{12}\right)}{1-\left(1+\dfrac{0.10}{12}\right)^{(-12\cdot 30)}} = \$3159.26$

Total payments: $\$90{,}000 + \$3600 + (30)(12)(\$3159.26) = \$1{,}230{,}933.60$

29. a) The variable rate mortgage would be the cheapest, since it will have a lower interest rate for the first years of the loan.

b) The total payments for the variable rate mortgage, are given below.

$12(\$568.86 + \$629.29 + \$692.02 + \$756.77 + \$823.27 + \$891.26) = 12(\$4361.47) = \$52{,}337.64.$

The payments over 6 years for the fixed rate mortgage are $(6)(12)\left(\dfrac{\$90{,}000}{\$1000}\right)(\$8.40854) = \$54{,}487.34.$

Using Table 10.4 for a 30-year mortgage at 9.5%. The variable rate saves $\$54{,}487.34 - \$52{,}337.64 = \$2149.70$. Using the payment formula for the fixed rate mortgage will yield total payments of $\$54{,}487.44$ and a savings of $2149.80.

Section 10.6: Ordinary Annuities, Sinking Funds, and Retirement Investments

1. Annuity

3. Sinking

5. Immediate

7. 401k

9. $A = \dfrac{5200\left[\left(1+\dfrac{0.035}{1}\right)^{(1)(30)} - 1\right]}{\dfrac{0.035}{1}}$

$= \$268{,}437.92$

11. $A = \dfrac{400\left[\left(1+\dfrac{0.08}{4}\right)^{(4)(35)} - 1\right]}{\dfrac{0.08}{4}} = \$299{,}929.32$

13. $p = \dfrac{40{,}000\left(\dfrac{0.055}{2}\right)}{\left(1+\dfrac{0.055}{2}\right)^{(2)(25)} - 1} = \$381.64$

15. $p = \dfrac{250{,}000\left(\dfrac{0.06}{12}\right)}{\left(1+\dfrac{0.06}{12}\right)^{(12)(35)} - 1} = \$175.48$

17. $A = \dfrac{150\left[\left(1+\dfrac{0.033}{12}\right)^{(12)(40)} - 1\right]}{\dfrac{0.033}{12}}$
$= \$149,271.58$

19. $A = \dfrac{1500\left[\left(1+\dfrac{0.06}{2}\right)^{(2)(10)} - 1\right]}{\dfrac{0.06}{2}} = \$40,305.56$

21. $p = \dfrac{24,000\left(\dfrac{0.075}{12}\right)}{\left(1+\dfrac{0.075}{12}\right)^{(12)(6)} - 1} = \$264.97$

23. $p = \dfrac{1,000,000\left(\dfrac{0.08}{4}\right)}{\left(1+\dfrac{0.08}{4}\right)^{(4)(20)} - 1} = \$5160.71$

25. a) $A = \dfrac{100\left[\left(1+\dfrac{0.12}{12}\right)^{(12)(10)} - 1\right]}{\dfrac{0.12}{12}}$
$= \$23,003.87$

b) $A = 23,003.87\left(1+\dfrac{0.12}{12}\right)^{(12)(30)}$
$= \$826,980.88$

c) $A = \dfrac{100\left[\left(1+\dfrac{0.12}{12}\right)^{(12)(30)} - 1\right]}{\dfrac{0.12}{12}}$
$= \$349,496.41$

d) $(\$100)(12)(10) = \$12,000$

e) $(\$100)(12)(30) = \$36,000$

f) Alberto

Review Exercises

1. $\dfrac{7}{20} = 0.35 = (0.35)(100)\% = 35\%$

2. $\dfrac{7}{12} \approx 0.583 = (0.583)(100)\% = 58.3\%$

3. $\dfrac{5}{8} = 0.625 = (0.625)(100)\% = 62.5\%$

4. $0.041 = 0.041(100)\% = 4.1\%$

5. $0.0098 = 0.0098(100)\% = 0.98\% \approx 1.0\%$

6. $3.141 = 3.1418(100)\% = 314.1\%$

7. $8\% = \dfrac{8}{100} = 0.08$

8. $22.9\% = \dfrac{22.9}{100} = 0.229$

9. $123\% = \dfrac{123}{100} = 1.23$

10. $\dfrac{1}{4}\% = 0.25\% = \dfrac{0.25}{100} = 0.0025$

11. $\dfrac{5}{6}\% = 0.8\overline{3}\% = \dfrac{0.8\overline{3}}{100} = 0.008\overline{3}$

12. $0.00045\% = \dfrac{0.00045}{100} = 0.0000045$

13. $\dfrac{42,745 - 41,500}{41,500} = 0.03 = 3.0\%$

14. $\dfrac{2800 - 3200}{3200} = -0.125 = -12.5\%$,
or a 12.5% decrease

15. Let $x =$ the percentage
$x(80) = 16$
$x = \dfrac{16}{80} = 0.20$
Sixteen is 20.0% of 80.

16. Let $x =$ the number
$0.55(x) = 44$
$x = \dfrac{44}{0.55} = 80$
Forty-four is 55% of 80.

17. $0.17(540) = 91.8$

18. $0.15(52.19) = \$7.83$

19. Let $x =$ the original number in the class
$0.20(x) = 8$
$x = \dfrac{8}{0.20} = 40$
The original number was 40 people.

20. $\dfrac{126}{1050} = 0.12 = 12.0\%$

21. $p = \$4400,\ \ r = 3.2\%,\ \ t = 4$

$i = \$4400(0.032)(4) = \$563.20$

22. $p = \$2700,\ \ t = 100 \text{ days},\ \ i = \$37.50$

$i = prt$

$37.50 = 2700(r)\left(\dfrac{100}{360}\right)$

$37.50 = 750r$

$p = \dfrac{37.50}{750} = 0.05 = 5.0\%$

23. $r = 5.5\%,\ \ t = 3,\ \ i = \$280.50$

$i = prt$

$280.50 = p(0.055)(3)$

$280.50 = 0.165p$

$p = \dfrac{280.50}{0.165} = \$1700$

24. $p = \$3600,\ \ r = 3\tfrac{1}{4}\%,\ \ i = \$555.75$

$i = prt$

$555.75 = 3600(0.0325)(t)$

$555.75 = 117t$

$t = \dfrac{555.75}{117} = 4.75 \text{ years}$

25. Interest: $i = (\$7500)(0.025)\left(\tfrac{24}{12}\right) = \$375$

Total amount due at maturity:
$\$7500 + \$375 = \$7875$

26. a) Interest: $i = (\$6000)(0.115)\left(\tfrac{24}{12}\right) = \$1380$

b) Amount received: $\$6000 - \$1380 = \$4620$

c) $i = prt$

$1380 = 4620(p)\left(\tfrac{24}{12}\right)$

$1380 = 9240p$

$p \approx 0.149$, or 14.9%

27. Partial payment on June 1 (61 days)

$i = (\$8400)(0.045)\left(\tfrac{61}{360}\right) = \$64.05$

$\$2900.00 - \$64.05 = \$2835.95$

$\$8400.00 - \$2835.95 = \$5564.045$

$i = (\$5564.05)(0.045)\left(\tfrac{191}{360}\right) = \$82.77$

$\$5564.05 + \$82.77 = \$5646.82$

28. Partial payment on Aug 31 (108 days)

$i = (\$6700)(0.061)\left(\tfrac{108}{360}\right) = \$122.61$

$\$3250.00 - \$122.61 = \$3{,}127.39$

$\$6700.00 - \$3127.39 = \$3572.61$

$i = (\$3572.61)(0.061)\left(\tfrac{102}{360}\right) = \$61.75$

$\$3572.61 + \$61.75 = \$3634.36$

29. a) $A = 5000\left(1 + \dfrac{0.06}{1}\right)^{(1)(5)} = \$6691.13$

Interest: $\$6691.13 - \$5000 = \$1691.13$

b) $A = 5000\left(1 + \dfrac{0.06}{2}\right)^{(2)(5)} = \$6719.58$

Interest: $\$6719.58 - \$5000 = \$1719.58$

c) $A = 5000\left(1 + \dfrac{0.06}{4}\right)^{(4)(5)} = \$6734.28$

Interest: $\$6734.28 - \$5000 = \$1734.28$

d) $A = 5000\left(1 + \dfrac{0.06}{12}\right)^{(12)(5)} = \$6744.25$

Interest: $\$6744.25 - \$5000 = \$1744.25$

e) $A = 5000\left(1 + \dfrac{0.06}{360}\right)^{(360)(5)} = \$6749.13$

Interest: $\$6749.13 - \$5000 = \$1749.13$

30. $A = p\left(1 + \dfrac{r}{n}\right)^{(n \cdot t)} = 2500\left(1 + \dfrac{0.0175}{12}\right)^{(12 \cdot 2)}$

$= \$2588.98$

31. $1\left(1 + \dfrac{0.056}{360}\right)^{(360 \cdot 1)} \approx 1.0576$; Effective Annual Yield: $1.0576 - 1 = 0.0576$, or 5.76%

32. $p = \dfrac{5500}{\left(1 + \dfrac{0.031}{4}\right)^{(4 \cdot 5)}} = \$4713.10$

33. a) Down payment: $(0.30)(\$26,000) = \$7800$

b) Amount Financed: $\$26,000 - \$7800 = \$18,200$

c) Finance charge: $\left(\frac{18,200}{100}\right)(13.23) = \$2407.86$

d) Monthly payment: $\frac{\$18,200 + \$2407.86}{60} = \$343.46$

34. a) Down payment: $(0.20)(\$13,220) = \$2644$

b) Amount Financed: $\$13,220 - \$2644 = \$10,576$

c) From Table 10.2, the finance charge per \$100 at 6.0% for 36 payments is \$9.52.

Finance charge: $\left(\frac{10,576}{100}\right)(9.52) = \$1006.84$

d) Monthly payment: $\frac{\$10,576 + \$1006.84}{36} = \$321.74$

35. a) Down payment: $(0.40)(\$140,000) = \$56,000$

b) Amount Financed: $\$140,000 - \$56,000 = \$84,000$

c) Total installment price: $(60)(\$1624) = \$97,440$

Total finance charge: $\$97,440 - \$84,000 = \$13,440$

d) $\left(\frac{\text{finance charge}}{\text{amount financed}}\right)(100) = \left(\frac{13,440}{84,000}\right)(100) = \$16.00$

From Table 10.2, for 60 payments, the value of \$16.00 corresponds with an APR of 6.0%.

36. Total installment price: $(\$176.14)(48) = \$8454.72$

Finance charge: $\$8454.72 - \$7500 = \$954.72$

a) $\left(\frac{954.72}{7500}\right)(100) \approx \$12.73$; In Table 10.2, \$12.73 yields an APR of 6.0%.

b) From Table 10.2, the monthly payment per \$100 for the remaining 24 months at 6.0% APR is \$6.37.

$$u = \frac{n \cdot P \cdot V}{100 + V} = \frac{(24)(\$176.14)(6.37)}{100 + 6.37} = \$253.16$$

c) Remaining payments: $(\$176.14)(24) = \$4227.36$

Remaining balance: $\$4227.36 - \$253.16 = \$3974.2$

Total amount due: $\$3974.20 + \$176.14 = \$4150.34$

37. a) Amount financed: $\$3420 - \$860 = \$2560$

Total installment price: $(\$111.73)(24) = \$2681.52$

Finance charge: $\$2681.52 - \$2560 = \$121.52$

$\left(\frac{\text{finance charge}}{\text{amount financed}}\right)(100) = \left(\frac{121.52}{2560}\right)(100) \approx \$4.75$

From Table 10.2, for 24 payments, the value of \$4.75 corresponds with an APR of 4.5%.

b) From Table 10.2, the monthly payment per \$100 for the remaining 12 months at 4.5% APR is \$2.45.

$$u = \frac{n \cdot P \cdot V}{100 + V} = \frac{(12)(\$111.73)(2.45)}{100 + 2.45} = \$32.06$$

37. (continued)

c) Remaining payments: $(\$111.73)(12) = \$1340.76$

Remaining balance: $\$1340.76 - \$32.06 = \$1308.7$

Total amount due: $\$1308.7 + \$111.73 = \$1420.43$

38. a) Finance charge: $(0.013)(\$485.75) = \$6.31$

b) Balance on July 01: $\$485.75 - \$375.00 + \$370.00 + \$175.80 + \$184.75 + \$6.31 = \$847.61$

c)

Date	Balance Due	Days	(Balance)(Days)
June 01	$\$485.75$	3	$(\$485.75)(3) = \$1457.25$
June 04	$\$485.75 - \$375.00 = \$110.75$	4	$(\$110.75)(4) = \$443.00$
June 08	$\$110.75 + \$370.00 = \$480.75$	13	$(\$480.75)(13) = \$6249.75$
June 21	$\$480.75 + \$175.80 = \$656.55$	7	$(\$656.55)(7) = \$4595.85$
June 28	$\$656.55 + \$184.75 = \$841.30$	3	$(\$841.30)(3) = \$2523.99$
July 01		30	$\$15,269.75$

Average daily balance: $\dfrac{\$15,269.75}{30} = \$508.99$

d) Finance charge: $(0.013)(\$508.99) = \$6.62$

e) Balance due on July 1: $\$841.30 + \$6.62 = \$847.92$

39. a) Finance charge: $(0.014)(\$185.75) = \$2.60$

b) Balance on September 5: $\$185.75 + \$85.72 - \$75.00 + \$72.85 + \$275.00 + \$2.60 = \$546.92$

c)

Date	Balance Due	Days	(Balance)(Days)
August 05	$\$185.72$	3	$(\$185.72)(3) = \$557.16$
August 08	$\$185.75 + \$85.72 = \$271.47$	2	$(\$271.47)(2) = \$542.94$
August 10	$\$271.47 - \$75.00 = \$196.47$	5	$(\$196.47)(5) = \$982.35$
August 15	$\$196.47 + \$72.85 = \$269.32$	6	$(\$269.32)(6) = \$1615.92$
August 21	$\$269.32 + \$\$275.00 = \$544.32$	15	$(\$544.32)(15) = \$8164.80$
September 5		31	$\$11,863.17$

Average daily balance: $\dfrac{\$11,863.17}{31} = \$382.68$

d) Finance charge: $(0.014)(\$382.68) = \$5.36$

e) Balance due on September 5: $\$544.32 + \$5.36 = \$549.68$

40. a) Down payment: $(0.25)(\$135,700) = \$33,925$

b) Gross monthly income: $\dfrac{\$64,000}{12} = \$5333.33$

Adjusted monthly income: $\$5333.33 - \$218 - \$120 - \$190 = \$4805.33$

28% of adjusted monthly income: $(0.28)(\$4805.33) = \$1345.49$

40. (continued)

c) From Table 10.4 the monthly principal and interest payment per $1000 at 3% for 30 years is $5.06685.
Mortgage amount: $\$135,700-\$33,925=\$101,775$

Monthly principal and interest payment: $\left(\frac{101,775}{1000}\right)(\$5.06685)=\$515.68$

Using formula: $\dfrac{\$101,775\left(\frac{0.045}{12}\right)}{1-\left(1+\frac{0.045}{12}\right)^{(-12\cdot 30)}}=\$515.68$

d) Total monthly payment: $\$515.68+\frac{\$3450}{12}+\frac{\$350}{12}=\$832.35$

e) Yes, $832.35 is less than $1345.49.

41. a) Down payment: $(0.15)(\$89,900)=\$13,485$

b) Amount of mortgage: $\$89,900-\$13,485=\$76,415$

Monthly payment: $\dfrac{\$76,415\left(\frac{0.115}{12}\right)}{1-\left(1+\frac{0.115}{12}\right)^{(-12\cdot 30)}}=\$756.73$

c) Interest: $(\$76,415)(0.115)\left(\frac{1}{12}\right)=\$732.31$

Amount applied to principal: $\$756.73-\$732.31=\$24.42$

d) Total cost of house: $\$13,485+(\$756.73)(12)(30)=\$285,907.80$

e) Total interest paid: $\$285,907.80-\$89,900=\$196,007.80$

42. a) $\text{InitialARMrate}=\text{One-year Treasury Bill Rate}+\text{Add-on Rate}=1.0\%+3.5\%=4.5\%$

b) Amount of mortgage: $\$375,000-\$140,000=\$235,000$

Monthly principal and interest payment: $\left(\frac{235,000}{1000}\right)(\$7.64993)=\$1797.73$

Using formula: $\dfrac{235,000\left(\frac{0.045}{12}\right)}{1-\left(1+\frac{0.045}{12}\right)^{(-12\cdot 15)}}=1797.73$

c) New ARM rate: $2.5\%+3.5\%=6.0\%$

43. $A=\dfrac{250\left[\left(1+\frac{0.09}{12}\right)^{(12)(10)}-1\right]}{\frac{0.09}{12}}=\$48,378.57$

44. $p=\dfrac{80,000\left(\frac{0.036}{4}\right)}{\left(1+\frac{0.036}{4}\right)^{(4)(5)}-1}=\$3668.72$

Chapter Test

1. a) $i = (\$1600)(0.027)\left(\dfrac{27}{12}\right) = \$97.20$

   b) $268.80 = (4200)(0.032)(t)$

   $351 = 117t$

   $t = \dfrac{351}{117} = 2$ years

2. $i = prt = (\$1900)(0.0315)\left(\dfrac{18}{12}\right) = \$89.78$

3. The total amount paid to the bank is $\$1900 + \$89.78 = \$1989.78$.

4. Partial payment on Sept. 15 (45 days)

   $i = (\$5400)(0.125)\left(\frac{45}{360}\right) = \$84.375$

   $\$3000.00 - \$84.375 - \$2,915.62$

   $\$5400.00 - \$2915.625 = \$2484.375$

   $i = (\$2484.375)(0.125)\left(\frac{45}{360}\right) = \$38.82$

   $\$2484.38 + \$38.82 = \$2523.20$

5. $\$84.38 + \$38.82 = \$123.20$

6. a) $n = 4,\ \ r = 0.03,\ \ t = 2,\ \ p = \$7500$

   $A = 7500\left(1 + \dfrac{0.03}{4}\right)^{(4\cdot 2)} \approx \$7961.99$

   Interest: $\$7961.99 - \$7500 = \$461.99$

   b) $n = 12,\ \ r = 0.065,\ \ t = 3,\ \ p = \$2500$

   $A = 2500\left(1 + \dfrac{0.065}{12}\right)^{(12\cdot 3)} \approx \$3036.68$

   Interest: $\$3036.68 - \$2500 = \$536.68$

7. $(0.15)(\$2350) = \$352.50;\ \ \$2350 - \$352.5 = \$1997.50$

8. Total payment: $(\$90.79)(24) = \$2178.96$; Finance charge: $\$2178.96 - \$1997.50 = \$181.46$

9. $\left(\dfrac{181.46}{1997.50}\right)(100) = \$9.08$; In Table 10.2, \$7.09 is closest to \$9.09 which yields an APR of 8.5%.

10. Total installment price: $(\$223.10)(36) = \$8031.60$

    Finance charge: $\$8031.60 - \$7500.00 = \$531.60$

    a) $\left(\dfrac{531.60}{7500}\right)(100) = \$7.09$; In Table 10.2, \$7.09 yields an APR of 4.5%.

    b) From Table 10.2, the monthly payment per \$100 for the remaining 12 months at 4.5% APR is \$2.45.

    $$u = \frac{n \cdot P \cdot V}{100 + V} = \frac{(12)(\$223.10)(2.45)}{100 + 2.45} = \$64.02$$

    c) Remaining payments: $(\$223.10)(12) = 2677.20$

    Remaining balance: $\$2677.20 - \$64.02 = \$2613.18$

    Total amount due: $\$2613.18 + \$223.10 = \$2836.28$

11. a) Finance charge on March 23: $(\$878.25)(0.014) = \$12.30$

b) Account balance on April 23: $\$878.25 + \$95.89 + \$68.76 - \$450.00 + \$90.52 + 450.85 + 12.30 = \$1146.57$

c)

Date	Balance	Days	(Balance)(Days)
March 23	$878.25	3	$(\$878.25)(3) = \$2634.75$
March 26	$\$878.25 + \$95.89 = \$974.14$	4	$(\$974.14)(4) = \$3896.56$
March 30	$\$974.14 + \$68.76 = \$1042.90$	4	$(\$1042.90)(4) = \$4171.60$
April 03	$\$1042.90 - \$450.00 = \$592.90$	12	$(\$592.90)(12) = \$7114.80$
April 15	$\$592.90 + \$90.52 = \$683.42$	7	$(\$683.42)(7) = \$4783.94$
April 22	$\$683.42 + \$450.85 = \$1134.27$	1	$(\$1134.27)(1) = \$1134.27$
April 23		31	$23,735.92

Average daily balance: $\dfrac{\$23,735.92}{31} = \$765.67$

d) Finance charge on April 23: $(\$765.67)(0.014) = \$10.72$

e) Account balance on April 23: $\$1134.27 + \$10.72 = \$1144.99$

12. Down payment: $(0.15)(\$215,000) = \$32,250$

13. Loan amount: $\$215,000 - \$32,250 = \$182,750$; Points: $(0.02)(\$182,750) = \$3655$

14. Gross monthly income: $\dfrac{\$122,740}{12} = \$10,228.33$

Adjusted monthly income: $\$10,228.33 - \$220 - \$175 - \$210 = \$9623.33$

Maximum monthly payment: $0.28(\$9623.33) = \$2694.53$

15. Monthly payment: $\dfrac{\$182,750\left(\dfrac{0.055}{12}\right)}{1-\left(1+\dfrac{0.055}{12}\right)^{(-12\cdot 30)}} \approx \$1037.63$

16. Total monthly payment: $\$1037.63 + \dfrac{\$3200}{12} + \dfrac{\$450}{12} = \$1037.63 + \$266.67 + \$37.50 = \$1341.80$

17. Yes, since $2694.53 is greater than $1341.80, the Leungs qualify for the mortgage.

18. a) Total cost of the house: $\$32,250 + \$3655 + (\$1037.63)(12)(30) = \$409,451.80$

b) Interest and points: $\$409,451.80 - \$215,000 = \$194,451.80$

19. $A = \dfrac{400\left(\left(1+\dfrac{0.03}{12}\right)^{(12\cdot 5)} - 1\right)}{\left(\dfrac{0.03}{12}\right)} \approx \$25,858.69$

20. $p = \dfrac{15,000\left(\dfrac{0.06}{12}\right)}{\left(1+\dfrac{0.06}{12}\right)^{(12\cdot 4)} - 1} \approx \$277.28$

Chapter Eleven: Probability

Section 11.1: Empirical and Theoretical Probabilities

1. Experiment

3. Event

5. Theoretical

7. a) 0
   b) 1
   c) 0, 1

9. 1

11. a) $P(s) = \frac{14}{30} = \frac{7}{15}$

    b) $P(k) = \frac{10}{30} = \frac{1}{3}$

    c) $P(r) = \frac{6}{30} = \frac{1}{5}$

13. a) $P(\text{dog}) = \frac{45}{105} = \frac{3}{7}$

    b) $P(\text{cat}) = \frac{40}{105} = \frac{8}{21}$

    c) $P(\text{rabbit}) = \frac{5}{105} = \frac{1}{21}$

15. a) $P(\text{Red Redemption 2}) = \frac{19,711,000}{63,707,000} \approx 0.3094$

    b) $P(\text{FIFA 19}) = \frac{12,119,000}{63,707,000} \approx 0.1902$

    c) $P(\text{Spider-Man}) = \frac{8,758,000}{63,707,000} \approx 0.1375$

17. a) $P(\text{General Motors}) = \frac{2,954,000}{17,300,000} \approx 0.1708$

    b) $P(\text{not General Motors}) = 1 - 0.8292 = 0.8292$

    c) $P(\text{Honda}) = \frac{1,605,000}{17,300,000} \approx 0.0928$

    d) $P(\text{not Honda}) = 1 - 0.0928 = 0.9072$

19. a) $P(\text{Favorable with type } A) = \frac{24}{80} = 0.53$

    b) $P(\text{Favorable with type } B) = \frac{11}{22} = 0.5$

    c) $P(\text{Favorable with type } AB) = \frac{8}{8} = 1$

    d) $P(\text{Favorable with type } O) = \frac{0}{90} = 0$

21. a) $P(\text{affecting circular}) = \frac{0}{150} = 0$

    b) $P(\text{affecting elliptical}) = \frac{50}{250} = 0.2$

    c) $P(\text{affecting irregular}) = \frac{100}{100} = 1$

23. a) $P(\text{correct}) = \frac{1}{5}$

    b) $P(\text{incorrect}) = 1 - \frac{1}{5} = \frac{4}{5}$

25. $P(\text{club}) = \frac{13}{52} = \frac{1}{4}$

27. $P(\text{not club}) = 1 - \frac{1}{4} = \frac{3}{4}$

29. $P(\text{face card}) = \frac{12}{52} = \frac{3}{13}$

31. $P(\text{heart and jack}) = P(\text{jack of hearts}) = \frac{1}{52}$

33. $P(\text{red and king}) = P(\text{king of hearts or king of diamonds}) = \frac{2}{52} = \frac{1}{26}$

35. a) $P(\text{red}) = \frac{1}{4}$

b) $P(\text{green}) = \frac{1}{2}$

c) $P(\text{yellow}) = \frac{1}{4}$

d) $P(\text{not yellow}) = \frac{3}{4}$

37. a) $P(\text{red}) = \frac{4}{8} = \frac{1}{2}$

b) $P(\text{green}) = \frac{3}{8}$

c) $P(\text{yellow}) = \frac{1}{8}$

d) $P(\text{not yellow}) = \frac{7}{8}$

39. $P(\text{Mint Chip bar}) = \frac{20}{100} = \frac{1}{5}$

41. $P(\text{Rocky Road bar}) = \frac{10}{100} = \frac{1}{10}$

43. $P(\text{Vanilla, Mint Chocolate Chip, or Krunch bar}) = \frac{40+20+30}{100} = \frac{90}{100} = \frac{9}{10}$

45. $P(\$2500) = \frac{1}{12}$

47. $P(\text{Lose a Turn or Bankrupt}) = \frac{2}{12} = \frac{1}{6}$

49. $P(\text{A number less than } \$3000) = \frac{9}{12} = \frac{3}{4}$

51. $P(S) = \frac{2}{9}$

53. $P(\text{consonant}) = \frac{5}{9}$

55. $P(W) = 0$

57. $P(\text{Coca-Cola}) = \frac{24}{66} = \frac{4}{11}$

59. $P(\text{not Coca-Cola}) = 1 - \frac{4}{11} = \frac{7}{11}$

61. $P(\text{diet soda}) = P(\text{Diet Coke or Diet Pepsi}) = \frac{12+6}{66} = \frac{18}{66} = \frac{3}{11}$

63. $P(\text{Frontera}) = \frac{10}{36} = \frac{5}{18}$

65. $P(\text{not Frontera}) = 1 - \frac{5}{18} = \frac{13}{18}$

67. $P(\text{mild or medium salsa}) = \frac{18+15}{36} = \frac{33}{36} = \frac{11}{12}$

69. $P(\text{La Victoria hot salsa}) = \frac{0}{36} = 0$

71. Not necessarily, but it does mean that if a coin was flipped many times, about one-half of the tosses would land heads up.

73. a) Roll a die many times and then determine the relative frequency of 5's to the total number of rolls.
b) Answers will vary.
c) Answers will vary.

75. $P(\text{red}) = \frac{2}{18} + \frac{1}{12} + \frac{1}{6} = \frac{4}{36} + \frac{3}{36} + \frac{6}{36} = \frac{13}{36}$

77. $P(\text{not red}) = 1 - \frac{13}{36} = \frac{23}{36}$

79. $P(\text{yellow}) = \frac{1}{6} + \frac{1}{12} + \frac{1}{12} = \frac{2}{12} + \frac{2}{12} = \frac{4}{12} = \frac{1}{3}$

81. $P(\text{yellow or green}) = \frac{1}{3} + \frac{11}{36} = \frac{12}{36} + \frac{11}{36} = \frac{23}{36}$

83. a) Answers will vary.
b) Answers will vary.
c) Answers will vary.

85. a) Answers will vary.
b) Answers will vary.
c) Answers will vary.

87. a) Answers will vary.
b) Answers will vary.
c) Answers will vary.

Section 11.2: Odds

1. against
3. $4:1$
5. $2:1$
7. $\frac{1}{1+3}=\frac{1}{4}$

9. a) $P(\text{grand prize})=\frac{1}{100}$

   b) $P(\text{not grand prize})=1-P(\text{grand prize})=1-\frac{1}{100}=\frac{99}{100}$

   c) The odds against Tito winning the grand prize are $\frac{P(\text{not grand prize})}{P(\text{grand prize})}=\frac{99/100}{1/100}=\frac{99}{1}$, or $99:1$.

   d) The odds in favor Tito winning the grand prize are $1:99$.

11. a) $P(\text{liberal arts})=\frac{18}{30}=\frac{3}{5}$

    b) $P(\text{not liberal arts})=1-\frac{3}{5}=\frac{2}{5}$

    c) The odds against liberal arts are $\frac{P(\text{not liberal arts})}{P(\text{liberal arts})}=\frac{2/5}{3/5}=\frac{2}{3}$, or $2:3$.

    d) The odds in favor of liberal arts are $3:2$.

13. Since there is only one 2, the odds against a 2 are $5:1$.

15. The odds against rolling less than 5 are $\frac{P(5\text{ or greater})}{P(\text{less than }5)}=\frac{2/6}{4/6}=\frac{2}{6}\cdot\frac{6}{4}=\frac{2}{4}=\frac{1}{2}$ or $1:2$.

17. The odds against a jack are $\frac{P(\text{failure to pick a jack})}{P(\text{pick a jack})}=\frac{48/52}{4/52}=\frac{48}{52}\cdot\frac{52}{4}=\frac{48}{4}=\frac{12}{1}$ or $12:1$. Therefore, the odds in favor of picking a jack are $1:12$.

19. The odds against a face card are $\frac{P(\text{failure to pick a face card})}{P(\text{pick a face card})}=\frac{40/52}{12/52}=\frac{40}{12}=\frac{10}{3}$ or $10:1$. Therefore, odds in favor of picking a face card are $3:10$.

21. The odds against red are $\frac{P(\text{not red})}{P(\text{red})}=\frac{1/2}{1/2}=\frac{2}{2}=\frac{1}{1}$ or $1:1$.

23. The odds against red are $\frac{P(\text{not red})}{P(\text{red})}=\frac{5/8}{3/8}=\frac{5}{3}$ or $5:3$.

25. The odds against a stripe are $\frac{P(\text{not a stripe})}{P(\text{stripe})}=\frac{8/15}{7/15}=\frac{8}{15}\cdot\frac{15}{7}=\frac{8}{7}$ or $8:7$.

27. The odds in favor of even are $\frac{P(\text{even})}{P(\text{not even})}=\frac{7/15}{8/15}=\frac{7}{8}$ or $7:8$.

29. The odds against a ball with 9 or greater are $\frac{P(\text{less than }9)}{P(9\text{ or greater})}=\frac{8/15}{7/15}=\frac{8}{7}$ or $8:7$.

31. $P(B)=\frac{15}{75}=\frac{1}{5}$

33. The odds in favor of B are $\frac{P(B)}{P(\text{not } B)} = \frac{1/5}{4/5} = \frac{1}{4}$ or 1 : 4.

35. The odds against $G50$ are $\frac{P(\text{not } G50)}{P(G50)} = \frac{74/75}{1/75} = \frac{74}{1}$ or 74 : 1.

37. $P(\text{A+}) = 0.36$

39. The odds against A+ are $\frac{P(\text{not A+})}{P(\text{A+})} = \frac{64/100}{36/100} = \frac{16}{9}$ or 16 : 9.

41. The odds in favor of A–, B–, AB–, or O– are $\frac{P(\text{A}-, \text{B}-, \text{AB}-, \text{O}-)}{P(\text{not A}-, \text{B}-, \text{AB}-, \text{O}-)} = \frac{0.06+0.02+0.01+0.07}{0.36+0.08+0.03+0.37}$

$= \frac{16/100}{84/100} = \frac{4}{21}$ or 4 : 21.

43. a) $P(\text{sells out}) = \frac{9}{9+2} = \frac{9}{11}$

b) $P(\text{not sold out}) = \frac{2}{9+2} = \frac{2}{11}$

45. a) $P(\text{Wendy wins}) = \frac{7}{7+4} = \frac{7}{11}$

b) $P(\text{Wendy loses}) = \frac{4}{7+4} = \frac{4}{11}$

47. a) $P(\text{win Super Bowl}) = \frac{1}{100+1} = \frac{1}{101}$

b) $P(\text{not win Super Bowl}) = 1 - \frac{1}{101} = \frac{100}{101}$

49. a) $P(\text{less than four minutes}) = \frac{197}{197+2} = \frac{197}{199}$

b) $P(\text{four minutes or greater}) = 1 - \frac{197}{199} = \frac{2}{199}$

51. a) The odds against the person having diabetes are $\frac{P(\text{no diabetes})}{P(\text{diabetes})} = \frac{91/100}{9/100} = \frac{91}{9}$ or 91 : 9.

b) The odds in favor of the person having diabetes are 9 : 91.

53. The odds against the mechanic fixing a car correctly are $\frac{P(\text{fixes incorrectly})}{P(\text{fixes correctly})} = \frac{0.1}{0.9} = \frac{1/10}{9/10} = \frac{1}{9}$ or 1 : 9.

55. a) $P(\text{congenital disability}) = \frac{1}{33}$

b) The odds against congenital disability are $\frac{P(\text{does not have C.D.})}{P(\text{has C.D.})} = \frac{32/33}{1/33} = \frac{32}{1}$ or 32 : 1.

57. $P(\#2 \text{ wins}) = \frac{1}{1+15} = \frac{1}{16}$

$P(\#3 \text{ wins}) = \frac{1}{1+1} = \frac{1}{2}$

$P(\#4 \text{ wins}) = \frac{1}{1+3} = \frac{1}{4}$

$P(\#5 \text{ wins}) = \frac{3}{3+13} = \frac{3}{16}$

59. If multiple births are approximately 3.5% of births, then single births are 96.5% of births, and the odds against a multiple birth are 96.5 : 3.5 or 193 : 7.

Section 11.3: Expected Value (Expectation)

1. Expected

3. Positive

5. $E = P_1A_1 + P_2A_2 = 0.20(20) + 0.80(60) = 52$ hot dogs

7. $E = P_1A_1 + P_2A_2 + P_3A_3 = 0.70(15{,}000) + 0.10(0) + 0.20(-8500) = 10{,}500 + 0 - 1700 = \$8800$

9. a) $E = P(\text{sunny})(2.5) + P(\text{cloudy})(1.5) = 0.70(2.5) + 0.30(1.5) = 1.75 + 0.45 = 2.2$ inches

   b) $(2.2 \text{ inches/day})(31 \text{ days}) = 68.2$ inches of growth during July is expected.

11. a) $E = P_1A_1 + P_2A_2 + P_3A_3 = 0.7(1) + 0.2(2) + 0.1(5) = \$1.60$

    b) $45 - 1.60 = \$43.40$

13. a) Dexter's expectation is $E = P(\text{black})(-5) + P(\text{white})(10) = \frac{7}{10}(-5) + \frac{3}{10}(10) = -\frac{1}{2} = -\$0.50.$

    b) Thanh's expectation is the negative of Chi's, or \$0.50.

15. a) Alyssa's expectation is $E = P(1, 2, \text{ or } 3)(3) + P(4 \text{ or } 5)(2) + P(6)(-14) = \frac{3}{6}(3) + \frac{2}{6}(2) + \frac{1}{6}(-14)$

    $= \frac{3}{2} + \frac{2}{3} + -\frac{7}{3} = -\frac{1}{6} \approx -\$0.17.$

    b) Gabriel's expectation is the negative of Alyssa's, or about \$0.17.

17. a) $E = P(\text{correct})(5) + P(\text{incorrect})(-1) = \frac{1}{5}(5) + \frac{4}{5}(-1) = 1 - \frac{4}{5} = \frac{1}{5}$

    Yes, because you have a positive expectation of $\frac{1}{5}$.

    b) $E = P(\text{correct})(5) + P(\text{incorrect})(-1) = \frac{1}{4}(5) + \frac{3}{4}(-1) = \frac{5}{4} - \frac{3}{4} = \frac{1}{2}$

    Yes, because you have a positive expectation of $\frac{1}{2}$.

19. $P(5)(5) + P(10)(10) = \frac{1}{2}(5) + \frac{1}{2}(10) = \frac{15}{2} = \$7.50$

21. $P(-15)(-15) + P(-2)(-2) + P(5)(5) = \frac{1}{4}(-15) + \frac{1}{4}(-2) + \frac{1}{2}(5) = -\frac{7}{4} = -\$1.75$

23. $P(\$500)(500) + P(\$1000)(1000) = \frac{1}{2}(500) + \frac{1}{2}(1000) = \$750$

25. $P(\$100)(100) + P(\$200)(200) + P(\$300)(300) + P(\$400)(400) + P(\$500)(500) + P(\$1000)(1000)$

    $= \frac{1}{6}(100) + \frac{1}{6}(200) + \frac{1}{6}(300) + \frac{1}{6}(400) + \frac{1}{6}(500) + \frac{1}{6}(1000) \approx \$416.67$

27. $P(\$10)(10 - 15) + P(\$25)(25 - 15) = \frac{1}{2}(-5) + \frac{1}{2}(10) = \$2.50$

29. $P(\$0)(0 - 15) + P(\$2)(2 - 15) + P(\$5)(5 - 15) + P(\$25)(20 - 15) = \frac{1}{4}(-15) + \frac{1}{4}(-13) + \frac{1}{4}(-10) + \frac{1}{4}(5)$

    $= -\$8.25$

31. $E = 0.20(300) + 0.70(400) + 0.10(-500) = 290,$ or a gain of 290 employees

33. $E = P(\text{approved})(75) + P(\text{not approved})(20) = 0.65(75) + 0.35(20) = 55.75 \approx 56$ employees

35. $E = P(\text{profit})(200,000) + P(\text{break even})(0) + P(\text{loss})(-30,000)$

    $= 0.85(200,000) + 0.50(0) + 0.10)(-30,000) = \$167,000$

37. $E = P(\text{clear day})(110) + P(\text{rainy day})(160) + P(\text{snowy day})(210)$
$= \frac{200}{365}(110) + \frac{100}{365}(160) + \frac{65}{365}(210) \approx 141.51$ service calls

39. $E = P(\text{sells house})(20,000 - 2400) + P(\text{does not sell house})(-2400)$
$= 0.80(17600) + 0.20)(-2400) = \$13,600$

41. a) $P(\$1) = \frac{1}{2} + \frac{1}{16} = \frac{9}{16}$; $P(\$10) = \frac{1}{4}$; $P(\$20) = \frac{1}{8}$; $P(\$100) = \frac{1}{16}$

b) $E = P_1A_1 + P_2A_2 + P_3A_3 + P_4A_4 = \frac{9}{16}(1) + \frac{1}{4}(10) + \frac{1}{8}(20) + \frac{1}{16}(100) = \$11.81$

43. Fair Price = expected value + cost to play = $-\$4.50 + \$10.00 = \$5.50$

45. a) $E = P(\text{wins})(500 - 3) + P(\text{does not win})(-3) = \frac{1}{500}(497) + \frac{499}{500}(-3) = -\$2.00$

b) Fair Price = expected value + cost to play = $-\$2.00 + \$3.00 = \$1.00$

47. a) $E = P(\text{win } \$1000)(1000 - 3) + P(\text{win } \$500)(500 - 3) + P(\text{does not win})(-2)$
$= \frac{1}{2000}(997) + \frac{2}{2000}(497) + \frac{1997}{2000}(-3) = -\$2.00$

b) Fair Price = expected value + cost to play = $-\$2.00 + \$3.00 = \$1.00$

49. a) $P(1)(1-2) + P(10)(10-2) = \frac{1}{2}(-1) + \frac{1}{2}(8) = \$3.50$

b) Fair Price = expected value + cost to play = $3.50 + 2.00 = \$5.50$

51. a) $P(1)(1-2) + P(5)(5-2) + P(10)(5-2) = \frac{2}{4}(-1) + \frac{1}{4}(3) + \frac{1}{4}(8) = \$2.25$

b) Fair Price = expected value + cost to play = $2.25 + 2.00 = \$4.25$

53. $E = P(\text{insured lives})(\text{cost}) + P(\text{insured dies})(\text{cost} - \$40,000)$
$= 0.97(\text{cost}) + 0.03(\text{cost} - 40,000) = 0.97(\text{cost}) + 0.03(\text{cost}) - 1200 = 1.00(\text{cost}) - 1200$
Thus, in order for the company to make a profit, the cost must exceed \$1200

55. $E = P(\text{win})(\text{amount won}) + P(\text{lose})(\text{amount lost}) = \frac{1}{38}(35) + \frac{37}{38}(-1) = \frac{35}{38} - \frac{37}{38} = -\frac{2}{38} \approx -\$0.053$ or $-5.3¢$

57. a) $E = \frac{1}{12}(100) + \frac{1}{12}(200) + \frac{1}{12}(300) + \frac{1}{12}(400) + \frac{1}{12}(500)$
$+ \frac{1}{12}(600) + \frac{1}{12}(700) + \frac{1}{12}(800) + \frac{1}{12}(900) + \frac{1}{12}(1000) = \$458.33$

b) $E = 458.33 + \frac{1}{12}(-1800) = \$308.33$

Section 11.4: Tree Diagrams

1. Sample

3. $(3)(8) = 24$

5. a) $(26)(26) = 676$
b) $(26)(25) = 650$

7. a) $(24)(24) = 576$
b) $(24)(23) = 552$

9\. a) $(2)(2) = 4$ points

b) Sample Space

HH
HT
TH
TT

c) $P(\text{two tails}) = \frac{1}{4}$

d) $P(\text{exactly one tail}) = \frac{2}{4} = \frac{1}{2}$

e) $P(\text{no tails}) = \frac{1}{4}$

f) $P(\text{at least one tail}) = \frac{3}{4}$

11\. a) $(3)(3) = 9$ sample points

b) Sample Space

SS
SQ
SA
QS
QQ
QA
AS
AQ
AA

c) $P(\text{two apples}) = \frac{1}{9}$

d) $P(\text{no apples}) = \frac{4}{9}$

e) $P(\text{at least one apple}) = \frac{5}{9}$

13\. a) $(4)(3) = 12$ sample points

b) Sample Space

YR
YB
YG
RY
RB
RG
BY
BR
BG
GY
GR
GB

c) $P(\text{exactly one red}) = \frac{6}{12} = \frac{1}{2}$

d) $P(\text{at least one is not red}) = \frac{12}{12} = 1$

e) $P(\text{no green}) = \frac{6}{12} = \frac{1}{2}$

15\. a) $(3)(3) = 9$ sample points

b) Sample Space

SA
SW
SO
JA
JW
JO
CA
CW
CO

c) $P(\text{Jade}) = \frac{1}{3}$

d) $P(\text{Jade Opal}) = \frac{1}{9}$

e) $P(\text{not Jade}) = \frac{2}{3}$

17\. a) $(6)(6) = 36$ sample points

b)

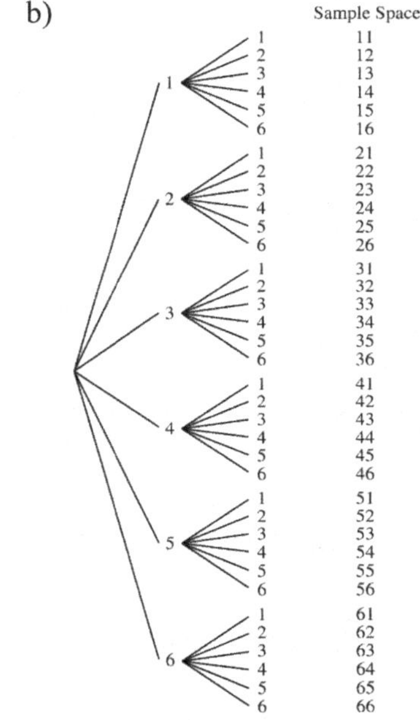

c) $P(\text{double}) = \frac{6}{36} = \frac{1}{6}$

d) $P(\text{sum of } 8) = \frac{5}{36}$

e) $P(\text{sum of } 2) = \frac{1}{36}$

f) No; $P(\text{sum of } 2) < P(\text{sum of } 8)$

19. a) $(2)(2)(3) = 12$ sample points

b) Sample Space

VSP
VSB
VSC
VLP
VLB
VLC
GSP
GSB
GSC
GLP
GLB
GLC

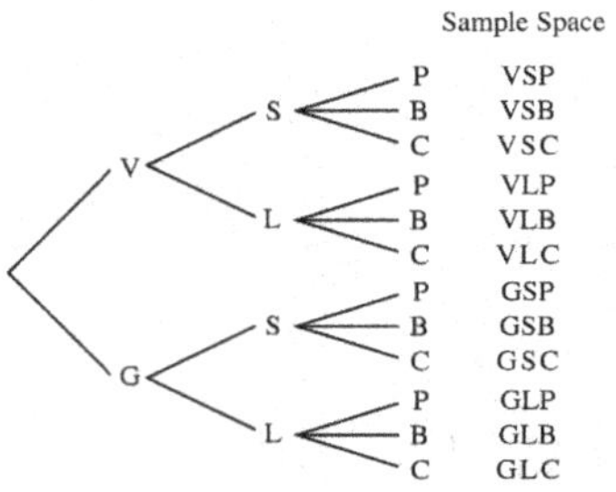

c) $P(\text{guacamole}) = \dfrac{6}{12} = \dfrac{1}{2}$

d) $P(\text{lemonade and brownies}) = \dfrac{2}{12} = \dfrac{1}{6}$

e) $P(\text{not pie}) = \dfrac{8}{12} = \dfrac{2}{3}$

21. a) $(4)(2)(2) = 16$ sample points

b) Sample Space

BCA
BCS
BLA
BLS
GCA
GCS
GLA
GLS
PCA
PCS
PLA
PLS
OCA
OCS
OLA
OLS

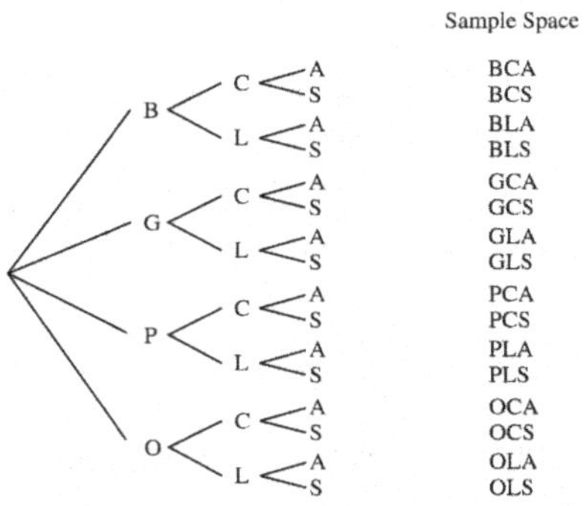

c) $P(\text{physics}) = \dfrac{4}{16} = \dfrac{1}{4}$

d) $P(\text{geology and literature}) = \dfrac{2}{16} = \dfrac{1}{8}$

e) $P(\text{not oceanography}) = \dfrac{12}{16} = \dfrac{3}{4}$

23. a) $(3)(3)(3) = 27$ sample points

b) Sample Space

DFC
DFS
DFP
DHC
DHS
DHP
DJC
DJS
DJP
MFC
MFS
MFP
MHC
MHS
MHP
MJC
MJS
MJP
BFC
BFS
BFP
BHC
BHS
BHP
BJC
BJS
BJP

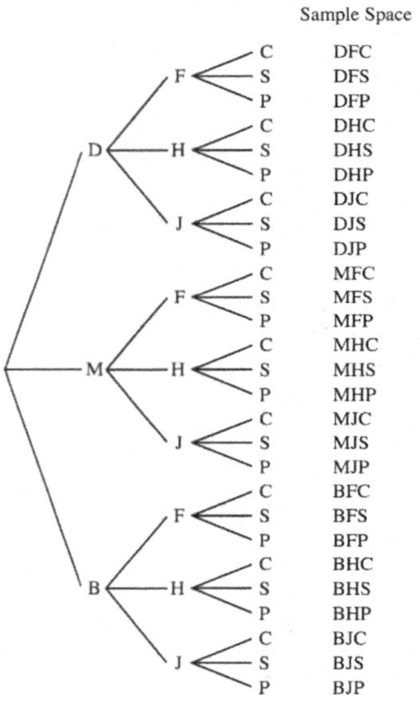

c) $P(\text{maple}) = \dfrac{9}{27} = \dfrac{1}{3}$

d) $P(\text{dogwood and holly shrub}) = \dfrac{3}{27} = \dfrac{1}{9}$

e) $P(\text{lilac bush not Sensation}) = \dfrac{18}{27} = \dfrac{2}{3}$

25. a) $(3)(3)(2) = 18$ options

b) Sample Space

WRN
WRK
WBN
WBK
WGN
WGK
PRN
PRK
PBN
PBK
PGN
PGK
YRN
YRK
YBN
YBK
YGN
YGK

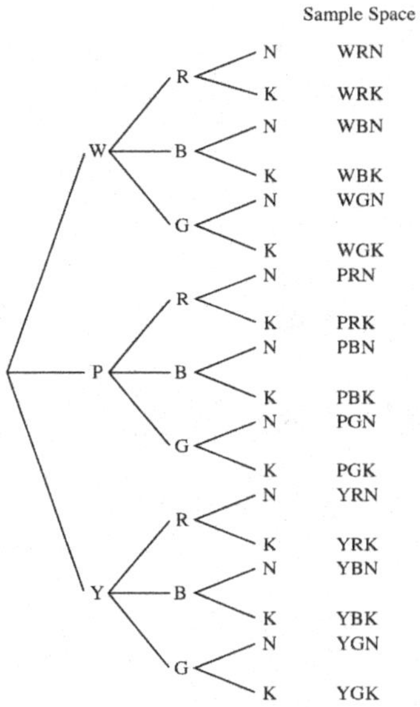

c) $P(\text{white shirt, red tie, navy slacks}) = \dfrac{1}{18}$

d) $P(\text{gray tie}) = \dfrac{1}{3}$

e) $P(\text{not white shirt}) = \dfrac{2}{3}$

27. a) $P(\text{white}) = \frac{1}{3}$

b) The odds against white are $2:1$.

c) $P(\text{red}) = \frac{2}{3}$

d) No; $P(\text{white}) \neq P(\text{red})$

e) Answers will vary.

29. 3 ties; 1 red, 1 blue, and 1 brown

31. a) $5! = 5 \cdot 4 \cdot 3 \cdot 2 \cdot 1 = 120$

b) STORE

Section 11.5: *OR* and *AND* Problems

1. Compound

3. And

5. Independent

7. Independent

9. $P(A) + P(B) - P(A \text{ and } B)$

11. $P(\text{odd or } > 5) = \frac{5}{10} + \frac{5}{10} - \frac{2}{10} = \frac{4}{5}$

13. $P(A \text{ or } B) = P(A) + P(B) - P(A \text{ and } B)$
$= 0.3 + 0.5 - 0.2 = 0.6$

15. $P(A \text{ or } B) = P(A) + P(B) - P(A \text{ and } B)$
$0.9 = 0.8 + 0.2 - P(A \text{ and } B)$
$0.9 = 1.0 - P(A \text{ and } B)$
$-0.1 = -P(A \text{ and } B)$
$P(A \text{ and } B) = 0.1$

17. $P(A \text{ or } B) = P(A) + P(B) - P(A \text{ and } B)$
$0.7 = 0.6 + P(B) - 0.3$
$0.7 = P(B) + 0.3$
$P(B) = 0.4$

19. $P(5 \text{ or } 6) = \frac{1}{6} + \frac{1}{6} = \frac{2}{6} = \frac{1}{3}$

21. $P(\text{odd number or greater than 4}) = P(1,\ 3,\ 5) + P(5,\ 6) - P(5) = \frac{3}{6} + \frac{2}{6} - \frac{1}{6} = \frac{2}{3}$

23. Since these events are mutually exclusive, $P(2 \text{ or } 3) = P(2) + P(3) = \frac{4}{52} + \frac{4}{52} = \frac{2}{13}$.

25. Since it is possible to obtain a card that is a 4 card and a diamond, these events are not mutually exclusive, so $P(4 \text{ or diamond}) = P(4) + P(\text{diamond}) - P(4 \text{ and diamond}) = \frac{4}{52} + \frac{13}{52} - \frac{1}{52} = \frac{4}{13}$.

27. Since it is possible to obtain a card that is a face card and a black card, these events are not mutually exclusive, so $P(\text{face card or an ace}) = P(\text{face card}) + P(\text{black card}) - P(\text{face card and black card}) = \frac{12}{52} + \frac{26}{52} - \frac{6}{52} = \frac{8}{13}$.

29. a) $P(\text{bird and bird}) = \frac{5}{20} \cdot \frac{5}{20} = \frac{1}{4} \cdot \frac{1}{4} = \frac{1}{16}$

b) $P(\text{bird and bird}) = \frac{5}{20} \cdot \frac{4}{19} = \frac{1}{4} \cdot \frac{4}{19} = \frac{1}{19}$

31. a) $P(\text{bird then monkey}) = \frac{5}{20} \cdot \frac{5}{20} = \frac{1}{4} \cdot \frac{1}{4} = \frac{1}{16}$

b) $P(\text{bird then monkey}) = \frac{5}{20} \cdot \frac{5}{19} = \frac{1}{4} \cdot \frac{5}{19}$
$= \frac{5}{76}$

33. a) $P(\text{yellow bird then lion}) = \frac{2}{20} \cdot \frac{5}{20} = \frac{1}{10} \cdot \frac{1}{4}$
$= \frac{1}{40}$

b) $P(\text{yellow bird then lion}) = \frac{2}{20} \cdot \frac{5}{19} = \frac{1}{10} \cdot \frac{5}{19}$
$= \frac{1}{38}$

35. a) $P(\text{odd then odd}) = \frac{12}{20} \cdot \frac{12}{20} = \frac{3}{5} \cdot \frac{3}{5} = \frac{9}{25}$

b) $P(\text{odd then odd}) = \frac{12}{20} \cdot \frac{11}{19} = \frac{3}{5} \cdot \frac{11}{19} = \frac{33}{95}$

37. $P(\text{lion or even}) = \frac{5}{20} + \frac{8}{20} - \frac{2}{20} = \frac{11}{20}$

39. $P(\text{lion or a 3}) = \frac{5}{20} + \frac{4}{20} - \frac{1}{20} = \frac{8}{20} = \frac{2}{5}$

41. $P(\text{red and green}) = \frac{1}{4} \cdot \frac{1}{2} = \frac{1}{8}$

43. $P(\text{2 yellows}) = P(\text{yellow and yellow})$
$= \frac{3}{8} \cdot \frac{3}{8} = \frac{9}{64}$

45. $P(\text{2 reds}) = \frac{1}{4} \cdot \frac{3}{8} = \frac{3}{32}$

47. $P(\text{both not red}) = \frac{3}{4} \cdot \frac{5}{8} = \frac{15}{32}$

49. $P(\text{yellow or blue}) = \frac{3}{7}$

51. $P(\text{none are red}) = \frac{4}{7} \cdot \frac{4}{7} = \frac{16}{49}$

53. $P(\text{at least one is red}) = 1 - P(\text{none red})$
$= 1 - \frac{4}{7} \cdot \frac{4}{7} = 1 - \frac{16}{49} = \frac{33}{49}$

55. $P(\text{all red}) = \frac{3}{7} \cdot \frac{2}{6} \cdot \frac{1}{5} = \frac{1}{35}$

57. $P(\text{at least one is red}) = 1 - P(\text{none red})$
$= 1 - \frac{4}{7} \cdot \frac{3}{6} \cdot \frac{2}{5} = 1 - \frac{4}{35} = \frac{31}{35}$

59. $P(\text{3 girls}) = P(\text{1st girl})P(\text{2nd girl})P(\text{3rd girl})$
$= \frac{1}{2} \cdot \frac{1}{2} \cdot \frac{1}{2} = \frac{1}{8}$

61. $P(\text{at least one girl}) = 1 - P(\text{all boys}) = 1 - \frac{1}{8} = \frac{7}{8}$

63. a) $P(\text{5 boys}) = P(\text{boy})P(\text{boy})P(\text{boy})P(\text{boy})P(\text{boy}) = \frac{1}{2} \cdot \frac{1}{2} \cdot \frac{1}{2} \cdot \frac{1}{2} \cdot \frac{1}{2} = \frac{1}{32}$

b) $P(\text{next child is a boy}) = \frac{1}{2}$

65. a) $P(\text{Diamond then Wilson}) = \frac{18}{36} \cdot \frac{4}{36} = \frac{1}{18}$

b) $P(\text{Diamond then Wilson}) = \frac{18}{36} \cdot \frac{4}{35} = \frac{2}{35}$

67. a) $P(\text{no Diamonds}) = \frac{18}{36} \cdot \frac{18}{36} = \frac{1}{4}$

b) $P(\text{no Diamonds}) = \frac{18}{36} \cdot \frac{17}{35} = \frac{17}{70}$

69. $P(\text{neither had traditional insurance}) = \frac{48}{75} \cdot \frac{47}{74} = \frac{376}{925}$

71. $P(\text{at least one traditional insurance}) = 1 - P(\text{neither had traditional insurance}) = 1 - \frac{48}{75} \cdot \frac{47}{74} = 1 - \frac{376}{925} = \frac{549}{925}$

73. $P(\text{all recommended}) = \frac{23}{30} \cdot \frac{22}{29} \cdot \frac{21}{28} = \frac{253}{580}$

75. $P(\text{no then no then not sure}) = \frac{3}{30} \cdot \frac{2}{29} \cdot \frac{4}{28}$
$= \frac{1}{1015}$

77. $P(\text{all favorable}) = \frac{16}{24} \cdot \frac{15}{23} \cdot \frac{14}{22} = \frac{70}{253}$

79. $P(\text{none favorable}) = \frac{8}{24} \cdot \frac{7}{23} \cdot \frac{6}{22} = \frac{7}{253}$

81. $P(\text{outer blue and inner blue}) = \frac{2}{12} \cdot \frac{2}{8} = \frac{1}{24}$

83. $P(\text{not outer red and not inner red}) = \frac{8}{12} \cdot \frac{5}{8} = \frac{5}{12}$

85. $P(\text{M or E}) = P(\text{M}) + P(\text{E}) - P(\text{M and E}) = 0.7 + 0.6 - 0.55 = 1.3 - 0.55 = 0.75$

87. No; The probability of the second outcome depends on the first outcome.

89. $P(\text{both miss}) = (0.6)(0.6) = 0.36$

91. $P(\text{both hit}) = (0.4)(0.9) = 0.36$

93. No; The probability of the second outcome depends on the first outcome.

95. $P(\text{both afflicted}) = (0.001)(0.04) = 0.00004$

97. $P(\text{1st not afflicted and 2nd afflicted}) = (0.999)(0.001) = 0.000999$

99. $P(\text{2 of same color}) = P(\text{2 red}) + P(\text{2 blue}) + P(\text{2 yellow}) = \frac{5}{10}\cdot\frac{4}{9} + \frac{3}{10}\cdot\frac{2}{9} + \frac{2}{10}\cdot\frac{1}{9} = \frac{2}{9} + \frac{1}{15} + \frac{1}{45} = \frac{14}{45}$

101. $P(\text{at least one diamond}) = 1 - P(\text{no diamonds}) = 1 - \frac{39}{52}\cdot\frac{38}{51} = \frac{15}{34} = 0.44$; The game favors the dealer since the probability of at least one diamond is less than 0.5.

103. $P(\text{two 2's}) = \frac{2}{6}\cdot\frac{2}{6} = \frac{1}{9}$

105. $P(\text{even or} < 3) = \frac{2}{6} + \frac{3}{6} - \frac{2}{6} = \frac{3}{6} = \frac{1}{2}$

107. Answers will vary.

Section 11.6: Conditional Probability

1. Conditional

3. $P(E_2 \mid E_1) = \frac{n(E_1 \text{ and } E_2)}{n(E_1)} = \frac{3}{7}$

5. a) $P(\text{club}) = \frac{13}{52} = \frac{1}{4}$

   b) $P(\text{club} \mid \text{black}) = \frac{13}{26} = \frac{1}{2}$

7. a) $P(\text{jack or king}) = \frac{8}{52} = \frac{2}{13}$

   b) $P(\text{jack or king} \mid \text{face card}) = \frac{8}{12} = \frac{2}{3}$

9. a) $P(5) = \frac{1}{6}$

   b) $P(5 \mid \text{orange}) = \frac{1}{3}$

11. a) $P(\text{even}) = \frac{3}{6} = \frac{1}{2}$

    b) $P(\text{even} \mid \text{orange}) = \frac{1}{3}$

13. a) $P(\text{red}) = \frac{2}{6} = \frac{1}{3}$

    b) $P(\text{red} \mid \text{yellow}) = \frac{0}{2} = 0$

15. $P(\text{circle} \mid \text{odd}) = \frac{3}{4}$

17. $P(\text{odd} \mid \text{circle}) = \frac{2}{2} = 1$

19. $P(\text{circle or square} \mid < 4) = \frac{2}{3}$

21. $P(\text{even} \mid \text{purple}) = \frac{3}{5}$

23. $P(\text{purple} \mid \text{even}) = \frac{3}{6} = \frac{1}{2}$

25. $P(> 4 \mid \text{purple}) = \frac{3}{5}$

27. $P(\text{gold} \mid > 5) = \frac{1}{7}$

29. $P(\$1 \text{ and } \$1) = \frac{1}{16}$

31. $P(\text{both } \$5 \mid \text{at least one } \$5) = \frac{1}{7}$

33. $P(3) = \frac{1}{18}$

35. $P(3 \mid \text{1st die is 3}) = 0$

37. $P(> 7 \mid \text{2nd die is 5}) = \frac{4}{6} = \frac{2}{3}$

39. a) $P(\text{prefers Famous Coffee}) = \frac{115}{225} = \frac{23}{45}$

    b) $P(\text{Famous} \mid \text{woman}) = \frac{75}{125} = \frac{3}{5}$

41. a) $P(\text{woman}) = \frac{125}{225} = \frac{5}{9}$

    b) $P(\text{woman} \mid \text{Famous}) = \frac{75}{115} = \frac{15}{23}$

43. a) $P(\text{car}) = \dfrac{1462}{2461} \approx 0.5941$

b) $P(\text{car} \mid \text{E-Z}) = \dfrac{527}{843} \approx 0.6251$

45. a) $P(\text{E-Z}) = \dfrac{843}{2461} \approx 0.3425$

b) $P(\text{E-Z} \mid \text{car}) = \dfrac{527}{1462} \approx 0.3605$

47. $P(\text{agg}) = \dfrac{350}{650} = \dfrac{7}{13}$

49. $P(\text{no sale} \mid \text{pass}) = \dfrac{80}{300} = \dfrac{4}{15}$

51. $P(\text{sale} \mid \text{pass}) = \dfrac{220}{300} = \dfrac{11}{15}$

53. $P(\text{Malibu} \mid \text{coupe}) = \dfrac{34}{55}$

55. $P(\text{sedan} \mid \text{Impala}) = \dfrac{59}{80}$

57. $P(\text{Malibu} \mid \text{sedan}) = \dfrac{72}{131}$

59. $P(\text{good}) = \dfrac{300}{330} = \dfrac{10}{11}$

61. $P(\text{defective} \mid 20 \text{ watts}) = \dfrac{15}{95} = \dfrac{3}{19}$

63. $P(\text{good} \mid 50 \text{ or } 100 \text{ watts}) = \dfrac{220}{235} = \dfrac{44}{47}$

65. $P(\text{Apple} \mid \text{man}) = \dfrac{127}{242} \approx 0.5248$

67. $P(\text{woman} \mid \text{LG}) = \dfrac{26}{42} \approx 0.6190$

69. $P(\text{Samsung} \mid \text{woman}) = \dfrac{68}{395} \approx 0.1722$

71. $P(\text{large company stock}) = \dfrac{93}{200}$

73. $P(\text{blend} \mid \text{medium company stock}) = \dfrac{15}{52}$

75. a) $n(A) = 60 + 80 = 140$

b) $n(B) = 40 + 80 = 120$

c) $P(A) = \dfrac{140}{200} = \dfrac{7}{10}$

d) $P(B) = \dfrac{120}{200} = \dfrac{3}{5}$

e) $P(A \mid B) = \dfrac{n(A \text{ and } B)}{n(B)} = \dfrac{80}{120} = \dfrac{2}{3}$

f) $P(B \mid A) = \dfrac{n(A \text{ and } B)}{n(A)} = \dfrac{80}{140} = \dfrac{4}{7}$

g) From part (e), $P(A \mid B) = \dfrac{2}{3}$, but

$P(A) \cdot P(B) = \dfrac{7}{10} \cdot \dfrac{3}{5} = \dfrac{21}{50}$.

77. a) $P(A \mid B) = \dfrac{P(A \text{ and } B)}{P(B)} = \dfrac{0.15}{0.5} = 0.3$

b) $P(B \mid A) = \dfrac{P(A \text{ and } B)}{P(A)} = \dfrac{0.15}{0.3} = 0.5$

c) Yes, $P(A \mid B) = P(A)$ and $P(B \mid A) = P(B)$

79. $P(+ \mid \text{orange circle}) = \dfrac{1}{2}$

81. $P(\text{green} + \mid +) = \dfrac{1}{3}$

83. $P(\text{yellow circle with green} + \mid +) = \dfrac{1}{3}$

Section 11.7: The Fundamental Counting Principle and Permutations

1. Permutation

3. $n!$

5. $\dfrac{n!}{(n-r)!}$

7. ${}_5P_3$

9. $6! = 720$

11. $0! = 1$

13. ${}_3P_2 = \dfrac{3!}{(3-2)!} = \dfrac{3!}{1!} = 3 \cdot 2 \cdot 1 = 6$

15. ${}_8P_0 = \dfrac{8!}{(8-0)!} = \dfrac{8!}{8!} = 1$

17. $_4P_4 = \frac{4!}{(4-4)!} = \frac{4!}{0!} = 4 \cdot 3 \cdot 2 \cdot 1 = 24$

19. $_8P_4 = \frac{8!}{(8-4)!} = \frac{8!}{4!} = 8 \cdot 7 \cdot 6 \cdot 5 = 1680$

21. $(10)(10)(10)(10) = 10,000$ codes

23. $(3)(4)(2) = 24$ outfits

25. $(5)(4)(7)(2) = 280$ systems

27. a) $(10)(10)(10)(10)(10)(10) = 1,000,000$ codes

 b) $\frac{1}{1,000,000} = 0.000001$

29. $(10)(10)(10)(10)(10)(10)(10)(10)(10)$
 $= 1,000,000,000$ numbers

31. a) $5! = 120$
 b) $5! = 120$
 c) $4! = 24$
 d) $3! = 6$

33. a) $8! = 40,320$ different ways

 b) $\frac{1}{40,320}$

35. a) $6! = 720$ different ways

 b) $\frac{(1)(4)(3)(2)(1)(1)}{6!} = \frac{1}{30}$

37. a) $(10)(10)(10)(26)(26) = 676,000$
 b) $(10)(9)(8)(26)(25) = 468,000$
 c) $(5)(4)(8)(26)(25) = 104,000$
 d) $(9)(9)(8)(26)(25) = 421,200$

39. a) $9! = 362,880$
 b) $(5)(7)(6)(5)(4)(3)(2)(1)(4) = 100,800$
 (The first and last songs are chosen first)
 c) $(4)(3)(7)(6)(5)(4)(3)(2)(1) = 60,480$

41. $_{10}P_3 = \frac{10!}{7!} = 10 \cdot 9 \cdot 8 = 720$

43. $_9P_5 = \frac{9!}{4!} = 9 \cdot 8 \cdot 7 \cdot 6 \cdot 5 = 15,120$

45. $_{16}P_4 = \frac{16!}{11!} = 16 \cdot 15 \cdot 14 \cdot 13 \cdot 12 = 524,160$

47. $9! = 362,880$

49. $\frac{10!}{2!3!3!} = 50,400$

51. $\frac{9!}{2!2!2!2!} = 22,680$

53. $\frac{7!}{2!2!2!} = 630$

55. a) $5^5 = 3125$ different keys

 b) $\frac{400,000}{3125} = 128$ cars

 c) $\frac{1}{3125} = 0.00032$

57. $_7P_5 = \frac{7!}{2!} = 2520$ different letter permutations; Time $= 2520 \times 5$ sec $= 12,600$ sec, or $\frac{12,600}{3600} = 3.5$ hr

59 No, for example, $_5P_2 = \frac{5!}{3!} = 20$, but $_5P_3 = \frac{5!}{2!} = 60$.

61. $(25)(24) = 600$ tickets

63. a) $\frac{7!}{2!} = 2520$
 b) STUDENT

Section 11.8: Combinations

1. Combination

3. Permutations

5. a) Permutation
 b) Combination

7. $_5C_3 = \dfrac{5!}{(5-3)!3!} = \dfrac{5!}{2!3!} = 10$

9. a) $_6C_4 = \dfrac{6!}{(6-4)!4!} = \dfrac{6!}{2!4!} = 15$

b) $_6P_4 = \dfrac{6!}{(6-4)!} = \dfrac{6!}{2!} = 360$

11. a) $_8C_0 = \dfrac{8!}{(8-0)!0!} = \dfrac{8!}{8!0!} = 1$

b) $_8P_0 = \dfrac{8!}{(8-0)!} = \dfrac{8!}{8!} = 1$

13. a) $_{10}C_3 = \dfrac{10!}{(10-3)!3!} = \dfrac{10!}{7!3!} = 120$

b) $_{10}P_3 = \dfrac{10!}{(10-3)!} = \dfrac{10!}{7!} = 720$

15. $\dfrac{_3C_2}{_{13}C_2} = \dfrac{\frac{3!}{2!1!}}{\frac{13!}{11!2!}} = \dfrac{3!}{2!1!} \cdot \dfrac{11!2!}{13!} = \dfrac{1}{26}$

17. $\dfrac{_8C_5}{_{14}C_5} = \dfrac{\frac{8!}{3!5!}}{\frac{14!}{9!5!}} = \dfrac{8!}{3!5!} \cdot \dfrac{9!5!}{14!} = \dfrac{4}{143}$

19. $\dfrac{_4C_3}{_{10}C_3} = \dfrac{\frac{4!}{1!3!}}{\frac{10!}{7!3!}} = \dfrac{4!}{1!3!} \cdot \dfrac{7!3!}{10!} = \dfrac{1}{30}$

21. $_{10}C_3 = \dfrac{10!}{(10-3)!3!} = \dfrac{10!}{7!3!} = 120$ ways

23. $_6C_4 = \dfrac{6!}{(6-4)!4!} = \dfrac{6!}{2!4!} = 15$ ways

25. $_8C_3 = \dfrac{8!}{(8-3)!3!} = \dfrac{8!}{5!3!} = 56$ ways

27. $_{49}C_6 = \dfrac{49!}{43!6!} = 13{,}983{,}816$ sets

29. $_{12}C_6 = \dfrac{12!}{8!4!} = 495$ ways

31. $(_4C_3)(_4C_2) = \left(\dfrac{4!}{1!3!}\right)\left(\dfrac{4!}{2!2!}\right) = (4)(6) = 24$

33. $(_9C_5)(_{15}C_5) = \left(\dfrac{9!}{4!5!}\right)\left(\dfrac{15!}{10!5!}\right) = (126)(3003)$
$= 378{,}378$ ways

35. $(_9C_3)(_8C_2) = \left(\dfrac{9!}{6!3!}\right)\left(\dfrac{8!}{6!2!}\right) = (84)(28) = 2352$

37. $(_7C_2)(_{11}C_4) = \left(\dfrac{7!}{5!2!}\right)\left(\dfrac{11!}{7!4!}\right) = (21)(330)$
$= 6930$

39. $(_{10}C_5)(_7C_3) = \left(\dfrac{10!}{5!5!}\right)\left(\dfrac{7!}{4!3!}\right) = (252)(35)$
$= 8820$

41. $(_6C_3)(_5C_3) = \left(\dfrac{6!}{3!3!}\right)\left(\dfrac{5!}{2!3!}\right) = (70)(10) = 700$

43. $(_5C_2)(_4C_3)(_6C_2) = \left(\dfrac{5!}{3!2!}\right)\left(\dfrac{4!}{1!3!}\right)\left(\dfrac{6!}{4!2!}\right)$
$= (10)(4)(15) = 600$

45. a) $_{10}C_8 = \dfrac{10!}{2!8!} = 45$

b) $_{10}C_8 + {_{10}C_9} + {_{10}C_{10}} = \dfrac{10!}{2!8!} + \dfrac{10!}{1!9!} + \dfrac{10!}{0!10!}$
$= 45 + 10 + 1 = 56$

47.

1
1 1
1 2 1
1 3 3 1
1 4 6 4 1
1 5 10 10 5 1

49. a) $4! = 24$
b) $4! = 24$

51. $(15)(14)(_{13}C_3) = (15)(14)\left(\dfrac{13!}{10!3!}\right) = 60{,}060$

53. a) The order of the numbers is important. For example: if the combination is 12–4–23, the lock will not open if 4–12–23 is used. Since repetition is permitted, it is not a true permutation problem.

b) $(40)(40)(40) = 64{,}000$

c) $(40)(39)(38) = 59{,}280$

Section 11.9: Solving Probability Problems by Using Combinations

1. $\dfrac{{}_{12}C_4}{{}_{20}C_4}$

3. $\dfrac{{}_4C_3}{{}_{52}C_3}$

5. $\dfrac{{}_8C_5}{{}_{15}C_5}$

7. $\dfrac{{}_{23}C_5}{{}_{120}C_5}$

9. $\dfrac{{}_3C_2}{{}_{13}C_2} = \dfrac{3}{78} = \dfrac{1}{26}$

11. $\dfrac{{}_8C_5}{{}_{14}C_5} = \dfrac{56}{2002} = \dfrac{4}{143}$

13. $\dfrac{{}_4C_3}{{}_{10}C_3} = \dfrac{4}{120} = \dfrac{1}{30}$

15. $\dfrac{({}_8C_1)({}_5C_2)}{{}_{13}C_3} = \dfrac{(8)(10)}{286} = \dfrac{40}{143}$

17. $P(\text{no cars}) = \dfrac{({}_2C_0)({}_3C_2)}{{}_5C_2} = \dfrac{(1)(3)}{10} = \dfrac{3}{10}$

19. $P(\text{at least one car}) = 1 - P(\text{no cars}) = 1 - \dfrac{3}{10}$

$= \dfrac{7}{10}$ (See Exercise 17)

21. $\dfrac{({}_{26}C_5)({}_{24}C_5)({}_3C_0)}{{}_{53}C_{10}} = \dfrac{(65,780)(42,504)(1)}{19,499,099,620}$

≈ 0.1434

23. $\dfrac{({}_{26}C_4)({}_{24}C_5)({}_3C_2)}{{}_{53}C_{10}} = \dfrac{(14,950)(10626)(3)}{19,499,099,620}$

≈ 0.0244

25. $\dfrac{({}_3C_2)({}_4C_0)({}_5C_2)}{{}_{12}C_4} = \dfrac{(3)(1)(10)}{495} = \dfrac{2}{33}$

27. $\dfrac{({}_3C_2)({}_4C_1)({}_5C_1)}{{}_{12}C_4} = \dfrac{(3)(4)(5)}{495} = \dfrac{4}{33}$

29. $\dfrac{({}_{12}C_1)({}_2C_1)({}_6C_4)({}_5C_6)}{{}_{25}C_9} = \dfrac{(12)(2)(15)(10)}{2,042,975}$

$= \dfrac{144}{81,719}$

31. a) $P(\text{royal spade flush}) = \dfrac{{}_{47}C_2}{{}_{52}C_7}$

$= \dfrac{1081}{133,784,560}$

$= \dfrac{1}{123,760}$

b) $P(\text{royal flush}) = 4 \cdot \dfrac{1}{123,760} = \dfrac{1}{30,940}$

33. $\dfrac{({}_{13}C_1)({}_{13}C_1)({}_{13}C_1)({}_{13}C_2)}{{}_{52}C_5} = \dfrac{(13)(13)(13)(78)}{2,598,960}$

$= \dfrac{2197}{33,320}$

35. $1 - \dfrac{(752,538,150)(1)}{10,40,465,790} = \dfrac{44}{159}$

37. a) $\dfrac{({}_4C_2)({}_4C_2)({}_{44}C_1)}{{}_{52}C_5} = \dfrac{(6)(6)(44)}{2,598,960} = \dfrac{33}{54,145}$

b) $\dfrac{1}{{}_{52}C_5} = \dfrac{1}{2,598,960}$

39. a) $\dfrac{(1)(1)(1)(5!)}{{}_{15}P_8} = \dfrac{120}{259,459,200} = \dfrac{1}{2,162,160}$

b) $\dfrac{({}_3P_8)(5!)}{{}_{15}P_8} = \dfrac{(336)(120)}{259,459,200} = \dfrac{1}{6435}$

41. Since there are more people than hairs, 2 or more people must have the same number of hairs, so the probability is 1.

Section 11.10: Binomial Probability Formula

1. Trials

3. Success

5. $P(3) = ({}_6C_3)(0.2)^3(0.8)^{6-3} = \dfrac{6!}{3!3!}(0.2)^3(0.8)^3$

≈ 0.08192

7. $P(2) = ({}_5C_2)\left(\dfrac{1}{3}\right)^2\left(\dfrac{2}{3}\right)^{5-2} = \dfrac{5!}{2!3!}\left(\dfrac{1}{3}\right)^2\left(\dfrac{2}{3}\right)^3$

≈ 0.32922

9. $P(0) = ({}_6C_0)(0.5)^0(0.5)^{6-0} = \dfrac{6!}{0!6!}(0.5)^0(0.5)^6$
$= 0.015625$

11. a) $P(x) = ({}_nC_x)(0.14)^n(0.86)^{n-x}$

b) $P(2) = ({}_{12}C_2)(0.14)^2(0.86)^{12-2}$
$= ({}_{12}C_2)(0.14)^2(0.86)^{10}$

c) $P(2) = \dfrac{12!}{2!10!}(0.14)^2(0.86)^{10} \approx 0.28628$

13. a) $P(1) = ({}_{10}C_1)(0.5)^1(0.5)^{10-1}$
$= \dfrac{10!}{1!9!}(0.5)^1(0.5)^9$
≈ 0.00977

b) $P(5) = ({}_{10}C_5)(0.5)^5(0.5)^{10-5}$
$= \dfrac{10!}{5!5!}(0.5)^5(0.5)^5$
≈ 0.24609

c) $P(9) = ({}_{10}C_9)(0.5)^9(0.5)^{10-9}$
$= \dfrac{10!}{9!1!}(0.5)^9(0.5)^1$
≈ 0.00977

15. a) $P(0) = ({}_4C_0)(0.2)^4(0.8)^{4-0}$
$= \dfrac{4!}{0!4!}(0.2)^0(0.8)^4$
$= 0.4096$

b) $P(\text{at least one red}) = 1 - P(0)$
$= 1 - 0.4096$
$= 0.5904$

17. $P(6) = ({}_{10}C_6)(0.53)^6(0.47)^{10-6}$
$= \dfrac{10!}{6!4!}(0.53)^6(0.47)^4$
≈ 0.22713

19. $P(5) = ({}_8C_5)(0.6)^5(0.4)^{8-5}$
$= \dfrac{8!}{5!3!}(0.6)^5(0.4)^3$
≈ 0.27869

21. $P(4) = ({}_6C_4)(0.92)^4(0.08)^{6-4}$
$= \dfrac{6!}{4!2!}(0.92)^4(0.08)^2$
≈ 0.06877

23. $P(100) = ({}_{100}C_{100})(0.99)^{100}(0.01)^{100-100}$
$= \dfrac{100!}{100!0!}(0.99)^{100}(0.01)^0$
≈ 0.36603

25. a) $P(10) = ({}_{20}C_{10})(0.27)^{10}(0.73)^{20-10}$
$= \dfrac{20!}{10!10!}(0.27)^{10}(0.73)^{10}$
≈ 0.01635

b) $P(0) = ({}_{20}C_0)(0.27)^0(0.73)^{20-0}$
$= \dfrac{20!}{0!20!}(0.27)^0(0.73)^{20}$
≈ 0.00185

c) $P(\text{at least one online}) = 1 - P(0)$
$= 1 - 0.00185$
$= 0.99815$

27. a) $P(3) = ({}_6C_3)\left(\frac{12}{52}\right)^3\left(\frac{40}{52}\right)^{6-3}$
$= \dfrac{6!}{3!3!}\left(\frac{3}{13}\right)^3\left(\frac{10}{13}\right)^3$
≈ 0.11188

b) $P(3) = ({}_6C_2)\left(\frac{13}{52}\right)^2\left(\frac{39}{52}\right)^{6-2}$
$= \dfrac{6!}{2!4!}\left(\frac{1}{4}\right)^2\left(\frac{3}{4}\right)^4$
≈ 0.29663

Review Exercises

1. Relative frequency over the long run can accurately be predicted, not individual events or totals.
2. Roll the die many times then compute the relative frequency of each outcome and compare with the expected probability of 1/6.
3. $P(\text{smartphone}) = \dfrac{40}{45} = \dfrac{8}{9}$
4. Answers will vary.
5. $P(\text{watches ABC}) = \dfrac{80}{250} = \dfrac{8}{25}$
6. $P(\text{even}) = \dfrac{5}{10} = \dfrac{1}{2}$
7. $P(\text{odd or} > 3) = \dfrac{5}{10} + \dfrac{6}{10} - \dfrac{3}{10} = \dfrac{8}{10} = \dfrac{4}{5}$

8. $P(>2 \text{ or } <6)=\frac{7}{10}+\frac{6}{10}-\frac{3}{10}=\frac{10}{10}=1$

9. $P(\text{even and} >4)=\frac{2}{10}=\frac{1}{5}$

10. $P(\text{cheddar})=\frac{21}{60}=\frac{7}{20}$

11. $P(\text{Gouda})=\frac{13}{60}$

12. $P(\text{Cheddar or Monterey Jack})=\frac{21}{60}+\frac{17}{60}=\frac{38}{60}$
$=\frac{19}{30}$

13. $P(\text{not Mozzarella})=1-P(\text{Mozzarella})=1-\frac{9}{60}$
$=\frac{17}{20}$

14. a) $69:31$
 b) $31:69$

15. $5:3$

16. $P(\text{wins Triple Crown})=\frac{3}{3+82}=\frac{3}{85}$

17. The odds in favor of the pizza restaurant succeeding are $\frac{P(\text{succeeding})}{P(\text{failing})}=\frac{3/4}{1/4}=\frac{3}{1}$ or $3:1$.

18. a) $E=P(\text{win\$200})(\$198)+P(\text{win\$100})(\$98)+P(\text{lose})(-\$2)$
$=(0.003)(198)+(0.002)(98)-(0.995)(2)$
$=0.594+0.196-1.990=-\$1.20$

 b) The expectation of a person who purchases three tickets would be $3(-1.20)=-\$3.60$.

19. a) $E_{\text{Cameron}}=P(\text{picture card})(\$9)+P(\text{not picture card})(-\$3)=\frac{12}{52}(9)-\frac{40}{52}(3)=-\frac{3}{13}\approx-\$0.23$

 b) $E_{\text{Lindsey}}=P(\text{not picture card})(\$3)+P(\text{picture card})(-\$9)=\frac{40}{52}(3)-\frac{12}{52}(9)=\frac{3}{13}\approx\$0.23$

 c) Cameron can expect to lose $(100)\left(\frac{3}{13}\right)\approx\$23.08$.

20. $E=P(\text{sunny})(1000)+P(\text{cloudy})(500)+P(\text{rain})(100)=0.4(1000)+0.5(500)+0.1(100)$
$=400+250+10=660$ people

21. Fair price = expected value + cost to play $=-\$2.50+\$6.50=\$4.00$

22. Fair price = expected value + cost to play $=-\$1.50+\$5.00=\$3.50$

23. a)

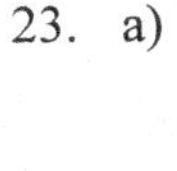

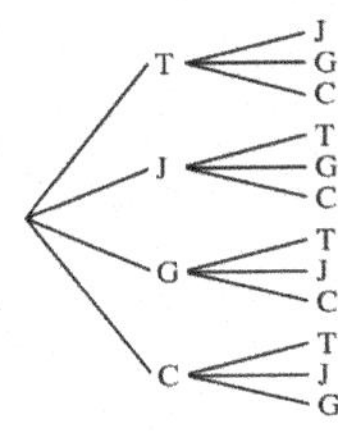

 b) Sample space: {TJ, TG, TC, JT, JG, JC, GT, GJ, GC, CT, CJ, CG}

 c) $P(\text{Gina president and Jake V.P})=\frac{1}{12}$

24. a)

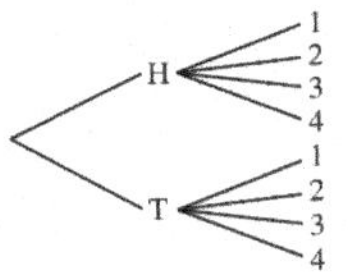

 b) Sample space: {H1, H2, H3, H4, T1, T2, T3, T4}

 c) $P(\text{heads and odd})=\frac{2}{8}=\frac{1}{4}$

 d) $P(\text{heads or odd})=\frac{6}{8}=\frac{3}{4}$

25. $P(\text{outer even and inner even})=P(\text{outer even})\cdot P(\text{inner even})=\frac{4}{8}\cdot\frac{4}{8}=\frac{1}{4}$

26. $P(\text{outer} > 5 \text{ and inner} > 5) = P(\text{outer} > 5) \cdot P(\text{inner} > 5) = \frac{3}{8} \cdot \frac{3}{8} = \frac{9}{64}$

27. $P(\text{outer odd and inner} < 6) = P(\text{outer odd}) \cdot P(\text{inner} < 6) = \frac{4}{8} \cdot \frac{5}{8} = \frac{5}{16}$

28. $P(\text{outer even or less than } 6) = P(\text{outer even}) + P(\text{outer} < 6) - P(\text{even and} < 6) = \frac{4}{8} + \frac{5}{8} - \frac{2}{8} = \frac{7}{8}$

29. $P(\text{inner even and not green}) = P(\text{inner even}) + P(\text{inner not green}) - P(\text{inner even and not green})$
$= \frac{1}{2} + \frac{6}{8} - \frac{2}{8} = 1$

30. $P(\text{outer gold and inner not gold}) = \frac{2}{8} \cdot \frac{6}{8} = \frac{3}{16}$

31. $P(\text{all 3 are cola}) = \frac{5}{12} \cdot \frac{4}{11} \cdot \frac{3}{10} = \frac{1}{22}$

32. $P(\text{none are root beer}) = \frac{8}{12} \cdot \frac{7}{11} \cdot \frac{6}{10} = \frac{14}{55}$

33. $P(\text{at least one is root beer}) = 1 - P(\text{none are root beer}) = 1 - \frac{14}{55}$

34. $P(\text{cola, then root beer, then ginger ale}) = \frac{5}{12} \cdot \frac{4}{11} \cdot \frac{3}{10} = \frac{1}{22}$

35. $P(\text{not green}) = \frac{1}{4} + \frac{1}{4} + \frac{1}{8} = \frac{5}{8}$

36. The odds in favor of green are $3:5$.
The odds against green are $5:3$.

37. $E = P(\text{green})(\$10) + P(\text{red})(\$5) + P(\text{yellow})(-\$20) = \frac{3}{8}(10) + \frac{1}{2}(5) - \frac{1}{8}(20) = \frac{15}{4} + \frac{10}{4} - \frac{10}{4} = \frac{15}{4}$ or \$3.75

38. $P(\text{at least one red}) = 1 - P(\text{none are red})$
$= 1 - \frac{1}{2} \cdot \frac{1}{2} \cdot \frac{1}{2} = 1 - \frac{1}{8} = \frac{7}{8}$

39. $P(\text{rated poor}) = \frac{30}{200} = \frac{3}{20}$

40. $P(\text{good} \mid \text{lunch}) = \frac{75}{90} = \frac{5}{6}$

41. $P(\text{poor} \mid \text{breakfast}) = \frac{15}{110} = \frac{3}{22}$

42. $P(\text{breakfast} \mid \text{poor}) = \frac{15}{30} = \frac{1}{2}$

43. $P(\text{right handed}) = \frac{230}{400} = \frac{23}{40}$

44. $P(\text{left brained} \mid \text{left handed}) = \frac{30}{170} = \frac{3}{17}$

45. $P(\text{right handed} \mid \text{no predominance}) = \frac{60}{80} = \frac{3}{4}$

46. $P(\text{right brained} \mid \text{left handed}) = \frac{120}{170} = \frac{12}{17}$

47. a) $4! = (4)(3)(2)(1) = 24$

b) $E = \left(\frac{1}{4}\right)(10K) + \left(\frac{1}{4}\right)(5K) + \left(\frac{1}{4}\right)(2K) + \left(\frac{1}{4}\right)(1K) = \left(\frac{1}{4}\right)(18K) = \$4500$

48. $({}_5C_2)({}_3C_2)({}_1C_1) = \frac{5!}{2!3!} \cdot \frac{3!}{2!1!} \cdot \frac{1!}{1!0!} = 30$

49. ${}_{10}P_3 = \frac{10!}{(10-3)!} = \frac{10!}{7!} = 720$ ways

50. ${}_9P_3 = \frac{9!}{(9-3)!} = \frac{9!}{6!} = 504$ ways

51. $({}_6C_3) = \frac{6!}{3!3!} = 20$ ways

52. $10! = 3,628,800$ arrangements

53. a) $P(\text{match 5 numbers}) = \dfrac{1}{{}_{70}C_5} = \dfrac{1}{\frac{70!}{5!65!}} = \dfrac{5!65!}{70!} = \dfrac{1}{12{,}103{,}014}$

b) $P(\text{jackpot}) = P(\text{match 5 and match Mega Ball}) = P(\text{match 5}) \cdot P(\text{match Mega Ball})$

$= \dfrac{1}{12{,}103{,}014} \cdot \dfrac{1}{25} = \dfrac{1}{302{,}575{,}350}$

54. $({}_{10}C_2)({}_{12}C_5) = \dfrac{10!}{8!2!} \cdot \dfrac{12!}{7!5!} = \dfrac{10!12!}{8!2!7!5!} = 35{,}640$ possible committees

55. $({}_{5}C_3)({}_{14}C_3) = \dfrac{8!}{5!3!} \cdot \dfrac{5!}{2!3!} = \dfrac{8!5!}{5!3!2!3!} = 560$ selections

56. $P(\text{two aces}) = \dfrac{{}_4C_2}{{}_{52}C_2} = \dfrac{\frac{4!}{2!2!}}{\frac{52!}{50!2!}} = \dfrac{4!50!2!}{2!2!52!} = \dfrac{1}{221}$

57. $P(\text{all three are red}) = \dfrac{5}{10} \cdot \dfrac{4}{9} \cdot \dfrac{3}{8} = \dfrac{1}{12}$

58. $P(\text{1st and 2nd are red and 3rd is blue}) = \dfrac{5}{10} \cdot \dfrac{4}{9} \cdot \dfrac{2}{8} = \dfrac{1}{18}$

59. $P(\text{1st red, 2nd white, 3rd blue}) = \dfrac{5}{10} \cdot \dfrac{3}{9} \cdot \dfrac{2}{8} = \dfrac{1}{24}$

60. $P(\text{at least one red}) = 1 - P(\text{none are red}) = 1 - \dfrac{5}{10} \cdot \dfrac{4}{9} \cdot \dfrac{3}{8} = 1 - \dfrac{1}{12} = \dfrac{11}{12}$

61. $P(3\ \textit{Motor Trend}) = \dfrac{{}_5C_3}{{}_{14}C_3} = \dfrac{\frac{5!}{3!2!}}{\frac{14!}{3!11!}} = \dfrac{5!3!11!}{14!3!2} = \dfrac{5}{182}$

62. $P(2\ \textit{Parenting}\ \&\ 1\ \textit{SI}) = \dfrac{({}_6C_2)({}_{3n}C_1)}{{}_{14}C_3} = \dfrac{\frac{6!}{2!4!} \cdot \frac{3!}{1!2!}}{\frac{14!}{3!11!}} = \dfrac{6!3!3!11!}{2!4!1!2!14!} = \dfrac{45}{364}$

63. $P(\text{no } \textit{Parenting}) = \dfrac{{}_8C_3}{{}_{14}C_3} = \dfrac{\frac{8!}{3!5!}}{\frac{14!}{3!11!}} = \dfrac{8!3!11!}{3!5!14!} = \dfrac{2}{13}$

64. $P(\text{at least one } \textit{Parenting}) = 1 - P(\text{no } \textit{Parenting}) = 1 - \dfrac{2}{13} = \dfrac{11}{13}$

65. a) $P(x) = ({}_nC_x)(0.6)^n (0.4)^{n-x}$

b) $P(50) = ({}_{100}C_{50})(0.6)^{50} (0.4)^{100-50} = ({}_{100}C_{50})(0.6)^{50} (0.4)^{50}$

c) $P(50) = \dfrac{100!}{50!50!}(0.6)^{50} (0.4)^{50} \approx 0.01034$

66. $P(50) = ({}_5C_3)(0.2)^3 (0.8)^{5-3} = \dfrac{5!}{3!2!}\left(\frac{1}{5}\right)^3 \left(\frac{4}{5}\right)^2 \approx 0.0512$

67. a) $P(0) = ({}_5C_0)(0.6)^0(0.4)^{5-0} = \frac{5!}{0!5!}(0.6)^0(0.4)^5 \approx 0.01024$

b) $P(\text{at least 1}) = 1 - P(0) = 1 - 0.01024 = 0.98976$

Chapter Test

1. $P(\text{glazed}) = \frac{11}{25}$

2. $P(>5) = \frac{4}{9}$

3. $P(\text{odd}) = \frac{5}{9}$

4. $P(\text{even or } < 4) = \frac{6}{9} = \frac{2}{3}$

5. $P(\text{both } < 3) = \frac{2}{9}\cdot\frac{1}{8} = \frac{2}{72} = \frac{1}{36}$

6. $P(\text{1st odd, 2nd even}) = \frac{5}{9}\cdot\frac{4}{8} = \frac{20}{72} = \frac{5}{18}$

7. $P(\text{neither} > 6) = \frac{6}{9}\cdot\frac{5}{8} = \frac{5}{12}$

8. $P(\text{red or face}) = P(\text{red}) + P(\text{face}) - P(\text{red and face}) = \frac{26}{52} + \frac{12}{52} - \frac{6}{52} = \frac{32}{52} = \frac{8}{13}$

9. $(6)(3) = 18$

10.

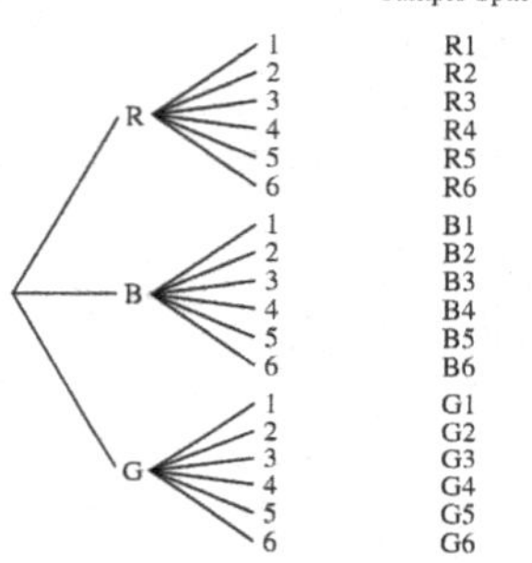

11. $P(\text{green and 2}) = \frac{1}{18}$

12. $P(\text{red or 1}) = \frac{8}{18} = \frac{4}{9}$

13. $P(\text{not red or even}) = \frac{15}{18} = \frac{5}{6}$

14. Number of codes $= (26)(9)(10)(26)(26)$
$= 1{,}581{,}840$

15. a) $5:4$

b) $4:5$

16. $E = P(\text{club})(\$8) + P(\text{heart})(\$4) + P(\text{spade or diamond})(-\$6) = \left(\frac{1}{4}\right)(8) + \left(\frac{1}{4}\right)(4) + \left(\frac{2}{4}\right)(-6) = \frac{8}{4} + \frac{4}{4} - \frac{12}{4} = \$0$

17. a) $P(\text{SUV}) = \frac{242}{456} = \frac{121}{228}$

b) $P(\text{George Washington}) = \frac{226}{456} = \frac{113}{228}$

c) $P(\text{SUV} \mid \text{George Washington}) = \frac{106}{226} = \frac{53}{113}$

d) $P(\text{Golden Gate} \mid \text{car}) = \frac{94}{214} = \frac{47}{107}$

18. ${}_6P_3 = \frac{6!}{(6-3)!} = \frac{6!}{3!} = 120$

19. a) $P(\text{neither is good}) = \frac{6}{20}\cdot\frac{5}{19} = \frac{3}{38}$

b) $P(\text{at least one good})$
$= 1 - P(\text{neither is good})$
$= 1 - \frac{3}{38} = \frac{35}{38}$

20. $P(2) = ({}_5C_2)(0.6)^2(0.4)^3 = 0.2304$

Chapter Twelve: Statistics

Section 12.1: Sampling Techniques and Misuses of Statistics

1. Statistics
3. Descriptive
5. Sample
7. Random
9. Stratified
11. Stratified sample
13. Cluster sample
15. Systematic sample
17. Convenience sample
19. Random sample
21. The patients may have improved on their own without taking honey.
23. Half the students in a population are expected to be below average.
25. A recommended toothpaste may not be better than all other types of toothpaste.
27. Most driving is done close to home. Thus, one might expect more accidents close to home.
29. We don't know how many of each professor's students were surveyed. Perhaps more of Professor Fogal's students than Professor Bond's students were surveyed. Also, because more students prefer a teacher does not mean that he or she is a better teacher. For example, a particular teacher may be an easier grader and that may be why that teacher is preferred.
31. Just because they are more expensive does not mean they will last longer.
33. There may be deep sections in the pond, so it may not be safe to go wading.
35. a)

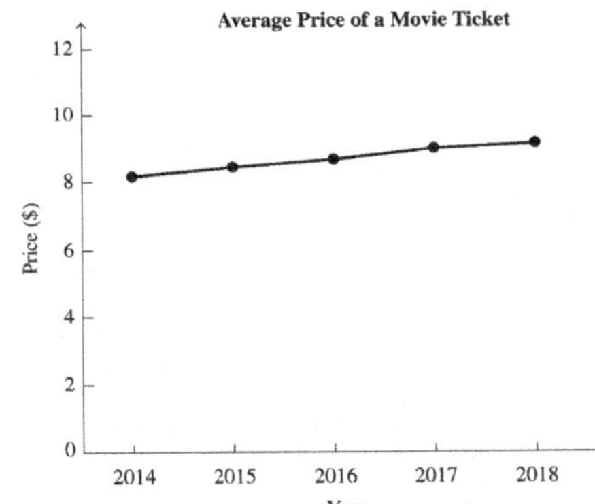

b)

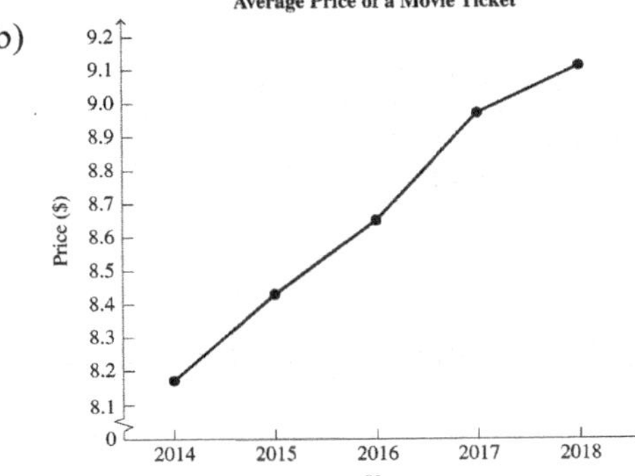

37. a)

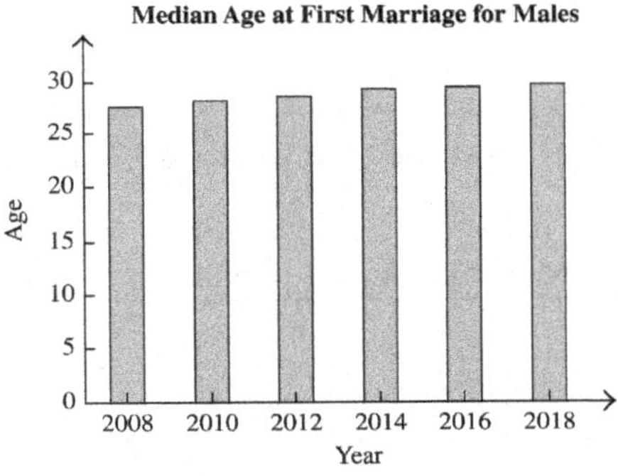

b)

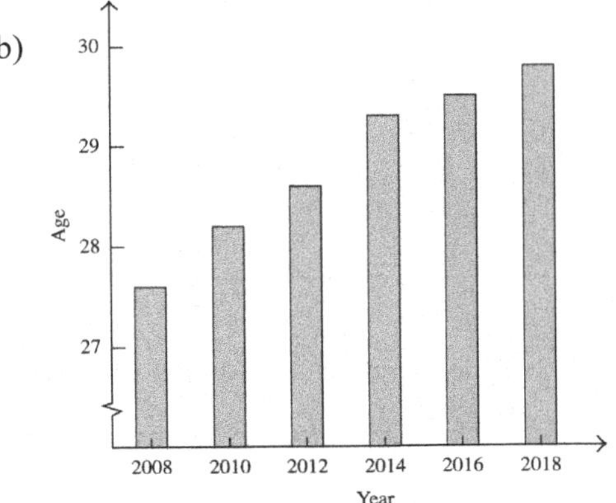

39. a)

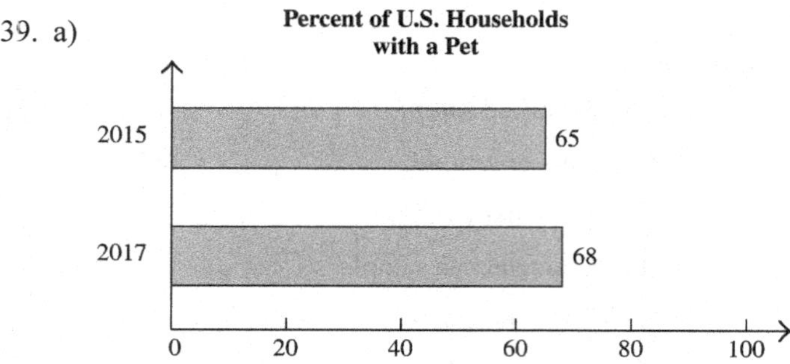

b) Answers will vary.

41. Yes, the sum of its parts is 121%. The sum of the parts of a circle graph should be 100%. When the total percent of responses is more than 100%, a circle graph is not an appropriate graph to display the data. A bar graph is more appropriate in this situation.

43. Biased because the subscribers of *Consumer Reports* are not necessarily representative of the entire population.

Section 12.2: Frequency Distributions and Statistical Graphs

1. Frequency

3. Mark

5. Histogram

7. a) Stem
   b) Leaf

9. a) $4+7+1+0+3+5=20$
   b) $16-9=7$
   c) $\dfrac{16+22}{2}=\dfrac{38}{2}=19$
   d) The modal class is the class with the greatest frequency. Thus, the modal class is 16–22.
   e) Since the class widths are 7, the next class would be 51–57.

11.

Number of Visits	Number of Students
0	3
1	8
2	3
3	5
4	2
5	7
6	2
7	3
8	4
9	1
10	2

13.

Copies Sold (millions)	Number of Books
0–99	21
100–199	7
200–299	1
300–399	0
400–499	0
500–599	1

15.

Copies Sold (millions)	Number of Books
50–99	21
100–149	6
150–199	1
200–249	1
250–299	0
300–349	0
350–399	0
400–499	0
450–449	0
500–549	1

17.

Population (millions)	Number of Cities
12.5–17.4	9
17.5–22.4	8
22.5–27.4	2
27.5–32.4	0
32.5–37.4	0
37.5–42.4	1

19.

Population (millions)	Number of Cities
12.0–17.9	10
18.0–23.9	8
24.0–29.9	1
30.0–35.9	0
36.0–41.9	1

21.

Cost of Living	Number of States
80.0–94.9	21
95.0–109.9	17
110.0–124.9	4
125.0–139.9	7
140.0–154.9	0
155.0–169.9	0
170.0–184.9	0
185.0–199.9	1

27. a)

Ticket Sales (million $)	Number of Movies
877–996	7
997–1116	0
1117–1236	4
1237–1356	2
1357–1476	0
1477–1596	0
1597–1716	1
1717–1836	1

b) and c)

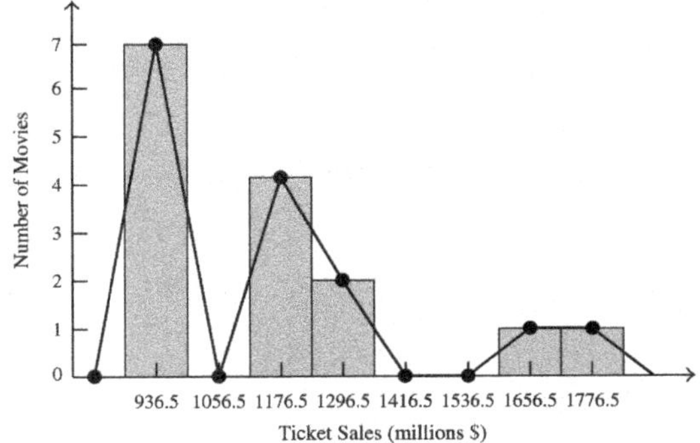

23.

Cost of Living	Number of States
85.7–100.6	30
100.7–115.6	8
115.7–130.6	6
130.7–145.6	5
145.7–160.6	0
160.7–175.6	0
175.7–190.6	1

25. 1 | 2 represents 12

```
0 | 4 6 7 8
1 | 2 2 3 5 6 7 8 9
2 | 1 2 3 5 7
3 | 3 4
4 | 0
```

29. a)

Age	Number of People
20–30	10
31–41	14
42–52	9
53–63	4
64–74	3

b) and c)

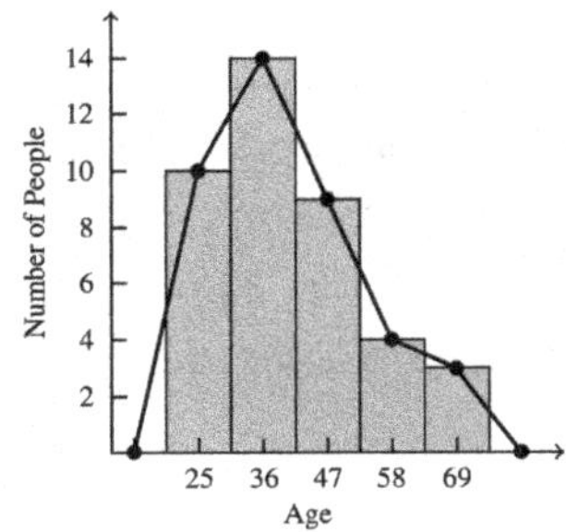

d) 2 | 3 represents 23

```
2 | 0 0 3 3 6 6 9
3 | 0 0 0 1 1 2 3 4 4 5 5 7 8 9
4 | 0 0 0 2 5 7 9 9 9
5 | 0 1 1 4 7
6 | 2 3 6 9
7 | 2
```

31. a) and b)

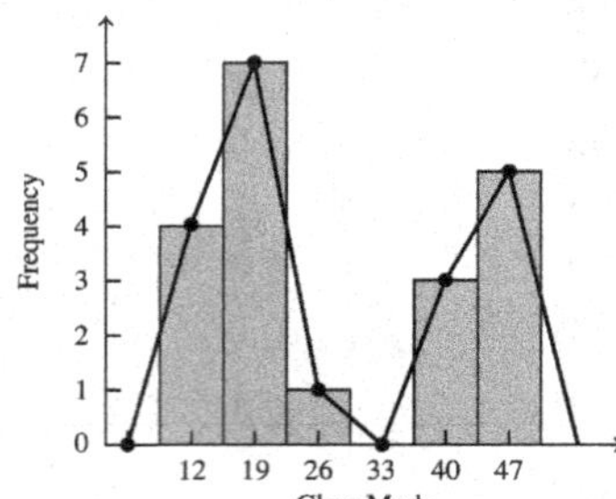

33. a) $2+4+8+6+4+3+1=28$
    b) 4
    c) 2
    d) $2(0)+1(4)+2(8)+6(3)+4(4)+3(5)+1(6)$
    $=75$
    e)

Number of Televisions	Number of Homes
0	2
1	4
2	8
3	6
4	4
5	3
6	1

35. a) 7 people
    b) Adding the number of people who sent 6, 5, 4, or 3 messages gives $4+7+3+2=16$ people
    c) The total number of people in the survey is $2+3+7+4+3+8+6+3=36$.
    d)

Number of Messages	Number of People
3	2
4	3
5	7
6	4
7	3
8	8
9	6
10	3

    e) Number of E-mail Messages Sent

37. Tuition: $0.41(25,440)=\$10,430.40$
    Room/board: $0.39(25,440)=\$9921.60$
    Books/supplies: $0.03(25,440)=\$763.20$
    Other: $0.17(25,440)=\$4324.80$

39. a) Answers will vary.
    b) Answers will vary.
    c) Answers will vary.
    d) Answers will vary.
    e) Answers will vary.

41. a) There are 6 F's.
    b) Answers will vary.

Section 12.3: Measures of Central Tendency and Position

1. Average

3. Mean

5. Mode

7. Quartiles

9. a) $\bar{x}$
   b) μ

11. $\bar{x}=\dfrac{11+12+12+14+16+16+16+27+29}{9}=\dfrac{153}{9}=17$

Median: 16 Midrange: $\dfrac{11+29}{2}=20$ Mode: 16

13. $\bar{x} = \frac{37+40+61+67+79+81+85}{7} = \frac{450}{7} \approx 64.3$

Median: 67 Midrange: $\frac{37+85}{2} = 61$ Mode: none

15. $\bar{x} = \frac{1+3+5+7+9+11+13+15}{8} = \frac{64}{8} = 8$

Median: $\frac{7+9}{2} = 8$ Midrange: $\frac{1+15}{2} = 8$ Mode: no mode

17. $\bar{x} = \frac{1+1+7+9+11+12+14+27+36}{9} = \frac{118}{9} \approx 13.1$

Median: 11 Midrange: $\frac{1+36}{2} = 18.5$ Mode: 1

19. $\bar{x} = \frac{6+8+11+12+13+13+15+17}{8} = \frac{95}{8} \approx 11.9$

Median: $\frac{12+13}{2} = 12.5$ Midrange: $\frac{6+17}{2} = 11.5$ Mode: 13

21. $\bar{x} = \frac{764}{10} = 76.4$

Median: $\frac{70+76}{2} = 73$ Midrange: $\frac{64+94}{2} = 79$ Mode: none

23. a) $\bar{x} = \frac{34}{7} \approx 4.9$
Median: 5
Mode: 5
Midrange: $\frac{1+11}{2} = 6$

b) $\bar{x} = \frac{37}{7} \approx 5.3$
Median: 5
Mode: 5
Midrange: $\frac{1+11}{2} = 6$

c) Only the mean was affected.

d) The mean and midrange will be affected.

25. a) $\bar{x} = \frac{421,000}{10} = \$42,100$

b) Median: $\frac{33,000+34,000}{2} = \$33,500$

c) Mode: $32,000

d) Midrange: $\frac{30,000+87,000}{2} = \$58,500$

e) The median, because it is lower.

f) The mean, because it is higher.

27. a) $\bar{x} = \frac{47.9}{10} \approx 4.8$ million

b) Median: $\frac{4+4.1}{2} = 4.1$ million

c) Mode: 3.5 million

d) Midrange: $\frac{3+11.4}{2} = 7.2$ million

29. a) $\bar{x} = \frac{62.19}{7} \approx \$8.88$

b) Median: $8.25

c) Mode: none

d) Midrange: $\frac{6.99+12.75}{2} = \$9.87$

31. Let $x =$ the sum of his scores

$\frac{x}{5} = 81 \Rightarrow x = 81(5) = 405$

33. a) Yes
 b) Cannot be found since we do not know the middle two numbers in the ranked list.
 c) Cannot be found without knowing all of the numbers.
 d) Yes

35. A total of $80 \times 5 = 400$ points are needed for a grade of B. Jorge earned $73 + 69 + 85 + 80 = 307$ points on his first four exams. Thus, he needs $400 - 307 = 93$ or higher on the fifth exam to get a B.

37. a) For a mean average of 60 on 7 exams, she must have a total of $60 \times 7 = 420$ points. Sheryl presently has $52 + 72 + 80 + 65 + 57 + 69 = 395$ points. Thus, to pass the course, her last exam must be $420 - 395 = 25$ or greater.
 b) A C average requires a total of $70 \times 7 = 490$ points. Sheryl has 395. Therefore, she would need $490 - 395 = 95$ or greater on her last exam.
 c) For a mean average of 60 on 6 exams, she must have a total of $60 \times 6 = 360$ points. If the lowest score on an exam she has already taken is dropped, she will have a total of $72 + 80 + 65 + 57 + 69 = 343$ points. Thus, to pass the course, her last exam must be $360 - 343 = 17$ or greater.
 d) For a mean average of 70 on 6 exams, she must have a total of $70 \times 6 = 420$ points. If the lowest score on an exam she has already taken is dropped, she will have a total of 343 points. Thus, to obtain a C, her last exam must be $420 - 343 = 77$ or greater.

39. Let x = sum of the values

$$\frac{x}{12} = \$85.20$$
$$x = 85.20(12) = \$1022.40$$
$$1022.40 - 47 + 74 = \$1049.40$$
$$\frac{1049.40}{12} = \$87.45 \text{ is the correct mean.}$$

41. List the values in ascending order.

 \$76 \$79 \$80 \$81 \$85 \$87 \$88 \$90 \$91 \$91 \$92 \$92 \$95 \$97 \$99 \$100 \$108 \$113 \$115

 a) $Q_2 = \$91$
 b) Q_1 = Median of the first 9 data values = \$85
 c) Q_3 = Median of the second 9 data values = \$99

43. a) No, the percentile only indicates relative position of the score and not the value of it.
 b) Yes, Kendra was in a better relative position.

45. a) \$600
 b) \$610
 c) 25%
 d) 25%
 e) 17%
 f) $100 \times \$620 = \$62{,}000$

47. Answers will vary. The National Center for Health uses the median for averages in this exercise.

49. One example is 1, 1, 2, 5, 6: Mean: $\frac{15}{5} = 3$, Median: 2, and Mode: 1.

51. One example is 81, 82, 83, 85, 86, 87.

53. The mode is the only measure which must be an actual piece of data since it is the most frequently occurring piece of data.

55. The data must be ranked.

57. He is taller than approximately 35% of all kindergarten children.

59. Second quartile, median

61. a) Ruth: $104/359 \approx 0.290$, $186/518 \approx 0.359$, $138/459 \approx 0.301$, $37/136 \approx 0.272$, $128/406 \approx 0.315$

Mantle: $163/543 \approx 0.300$, $173/474 \approx 0.365$, $158/519 \approx 0.304$, $145/527 \approx 0.275$, $121/377 \approx 0.321$

b) Mantle's is greater in every case.

c) Ruth: $\frac{593}{1878} \approx 0.316$

Mantle: $\frac{760}{2440} \approx 0.311$; Ruth's is greater.

d) Answers will vary.

e) Ruth: $\frac{1.537}{5} \approx 0.307$;

Mantle: $\frac{1.565}{5} = 0.313$; Mantle's is greater.

f) Answers will vary.

g) Answers will vary.

63. $\Sigma xw = 84(0.40) + 94(0.60) = 33.6 + 56.4 = 90$

$\Sigma w = 0.40 + 0.60 = 1.00$

weighted average $= \frac{\Sigma xw}{\Sigma w} = \frac{90}{1.00} = 90$

65. a) Answers will vary.

b) Answers will vary.

c) Answers will vary.

Section 12.4: Measures of Dispersion

1. Variability

3. Standard deviation

5. Sample

7. Range: $13 - 2 = 11$; $\bar{x} = \frac{\Sigma x}{n} = \frac{35}{5} = 7$

x	$x - \bar{x}$	$(x - \bar{x})^2$
2	–5	25
5	–2	4
7	0	0
8	1	1
13	6	36
35	0	66

$$s = \sqrt{\frac{\Sigma(x - \bar{x})^2}{n-1}} = \sqrt{\frac{66}{4}} = \sqrt{16.5} \approx 4.06$$

9. Range: $156 - 150 = 6$; $\bar{x} = \frac{\Sigma x}{n} = \frac{1071}{7} = 153$

x	$x - \bar{x}$	$(x - \bar{x})^2$
150	–3	9
151	–2	4
152	–1	1
153	0	0
154	1	1
155	2	4
156	3	9
1071	0	28

$$s = \sqrt{\frac{\Sigma(x - \bar{x})^2}{n-1}} = \sqrt{\frac{28}{6}} \approx \sqrt{4.667} \approx 2.16$$

11. Range: $15 - 4 = 11$; $\bar{x} = \frac{\Sigma x}{n} = \frac{60}{6} = 10$

x	$x - \bar{x}$	$(x - \bar{x})^2$
4	–6	36
8	–2	4
9	–1	1
11	1	1
13	3	9
15	5	25
60	0	76

$$s = \sqrt{\frac{\Sigma(x - \bar{x})^2}{n-1}} = \sqrt{\frac{76}{5}} = \sqrt{15.2} \approx 3.90$$

13. Range: $12 - 7 = 5$; $\bar{x} = \frac{\Sigma x}{n} = \frac{63}{7} = 9$

x	$x - \bar{x}$	$(x - \bar{x})^2$
7	–2	4
7	–2	4
9	0	0
9	0	0
9	0	0
10	1	1
12	3	9
63	0	18

$$s = \sqrt{\frac{\Sigma(x - \bar{x})^2}{n-1}} = \sqrt{\frac{18}{6}} = \sqrt{3} \approx 1.73$$

15. Range: $90-30=\$60$; $\overline{x}=\frac{\Sigma x}{n}=\frac{520}{8}=\$65$

x	$x-\overline{x}$	$(x-\overline{x})^2$
30	–35	1225
40	–25	625
50	–15	225
60	–5	25
75	10	100
85	20	400
90	25	625
90	25	625
520	0	3850

$s=\sqrt{\frac{\Sigma(x-\overline{x})^2}{n-1}}=\sqrt{\frac{3850}{7}}=\sqrt{550}\approx\$23.45$

17. Range: $250-60=\$190$;

$\overline{x}=\frac{\Sigma x}{n}=\frac{1044}{9}=\$116$

x	$x-\overline{x}$	$(x-\overline{x})^2$
60	–56	3136
60	–56	3136
60	–56	3136
80	–36	1296
100	–16	256
109	–7	49
115	–1	1
210	94	8836
250	134	17956
1044	0	37802

$s=\sqrt{\frac{\Sigma(x-\overline{x})^2}{n-1}}=\sqrt{\frac{37,802}{8}}=\sqrt{4725.25}$

$\approx\$68.74$

19. a) Range: $68-5=63$; $\overline{x}=\frac{\Sigma x}{n}=\frac{204}{6}=34$

x	$x-\overline{x}$	$(x-\overline{x})^2$
5	–29	841
14	–20	400
25	–9	81
32	–2	4
60	26	676
68	34	1156
204	0	3158

$s=\sqrt{\frac{\Sigma(x-\overline{x})^2}{n-1}}=\sqrt{\frac{3158}{5}}=\sqrt{631.6}$

≈ 25.13

b) Answers will vary.

c) Range: $78-15=63$; $\overline{x}=\frac{\Sigma x}{n}=\frac{264}{6}=44$

x	$x-\overline{x}$	$(x-\overline{x})^2$
15	–29	841
24	–20	400
35	–9	81
42	–2	4
70	26	676
78	34	1156
264	0	3158

$s=\sqrt{\frac{\Sigma(x-\overline{x})^2}{n-1}}=\sqrt{\frac{3158}{5}}=\sqrt{631.6}$

≈ 25.13

The range and standard deviation remain the same.

21. a) Answers will vary.
    b) Answers will vary.
    c) Answers will vary.
    d) If each number in a distribution is multiplied by n, both the mean and standard deviation of the new distribution will be n times that of the original distribution.
    e) The mean of the second set is $4\times 5=20$, and the standard deviation of the second set is $2\times 5=10$.

25. The first set of data will have the greater standard deviation because the scores have a greater spread about the mean.

27. The sum of the values in the $(\text{Data}-\text{Mean})^2$ column will always be greater than or equal to 0.

29. a) The standard deviation increases. There is a greater spread from the mean as they get older.
    b) The mean weight is about 100 pounds and the normal range is about 60 to 140 pounds.
    c) The mean height is about 62 inches and the normal range is about 53 to 68 inches.
    d) ≈ 140 lb
    e) $\approx \frac{40}{2} = 20$ lb
    f) $100\% - 95\% = 5\%$

31. a)

East

Number of Oil Changes Made	Number of Days
15–20	2
21–26	2
27–32	5
33 –38	4
39–44	7
45–50	1
51–56	1
57–62	2
63–68	1

West

Number of Oil Changes Made	Number of Days
15 –20	0
21–26	0
27–32	6
33–38	9
39–44	4
45–50	6
51–56	0
57–62	0
63–68	0

b)

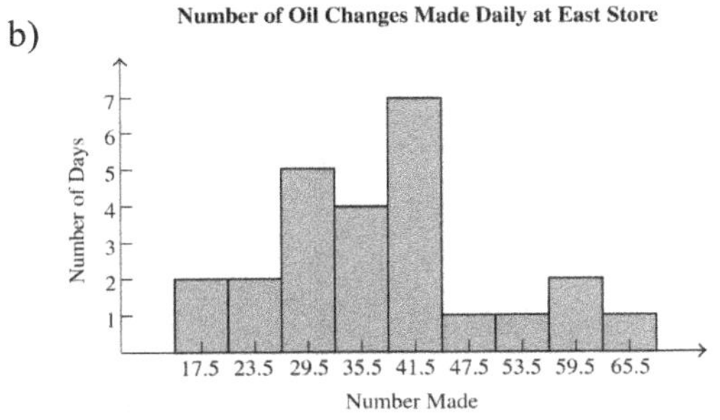

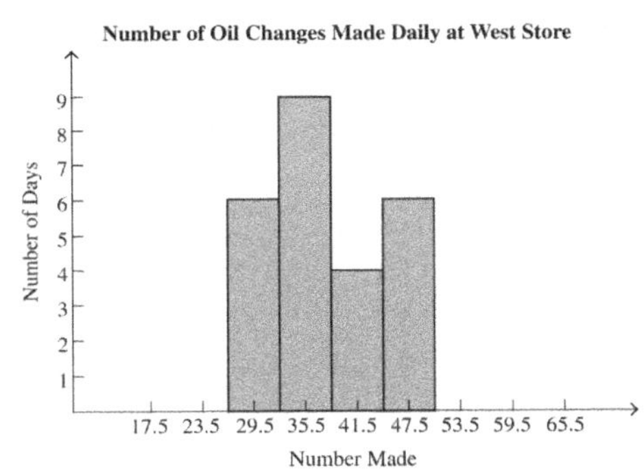

c) They appear to have about the same mean, since they are both centered around 38.

d) The distribution for East is more spread out. Therefore, East has a greater standard deviation.

31. (continued)

e) East Store: $\bar{x} = \dfrac{\Sigma x}{n} = \dfrac{950}{25} = 38$

x	$x - \bar{x}$	$(x - \bar{x})^2$
15	–23	529
19	–19	361
22	–16	256
25	–13	169
27	–11	121
27	–11	121
30	–8	64
31	–7	49
32	–6	36
33	–5	25
35	–3	9
37	–1	1
38	0	0
40	2	4
41	3	9
42	4	16
42	4	16
43	5	25
43	5	25
44	6	36
49	11	121
52	14	196
57	19	361
59	21	441
67	29	841
950	0	3832

$$s = \sqrt{\frac{\Sigma(x - \bar{x})^2}{n-1}} = \sqrt{\frac{3832}{24}} \approx 12.64$$

West Store: $\bar{x} = \dfrac{\Sigma x}{n} = \dfrac{950}{25} = 38$

x	$x - \bar{x}$	$(x - \bar{x})^2$
28	–10	100
29	–9	81
30	–8	64
30	–8	64
31	–7	49
31	–7	49
34	–4	16
36	–2	4
37	–1	1
37	–1	1
38	0	0
38	0	0
38	0	0
38	0	0
38	0	0
39	1	1
39	1	1
40	2	4
42	4	16
45	7	49
45	7	49
46	8	64
46	8	64
47	9	81
48	10	100
950	0	858

$$s = \sqrt{\frac{\Sigma(x - \bar{x})^2}{n-1}} = \sqrt{\frac{858}{24}} \approx 5.98$$

32. Answers will vary.

33. 6, 6, 6, 6, 6

Section 12.5: The Normal Curve

1. Rectangular

3. a) Right
   b) Left

5. Bell

7. 0

9. a) 68%
   b) 95%
   c) 99.7%

11. Answers will vary.

13. Answers will vary.

15. Normal

17. Skewed right

19. Bimodal

21. Rectangular

23. (area to the left of 0) = 0.5

25. (area between -1 and 2) = (area to the left of 2) − (area to the left of -1)
$= 0.9772 - 0.1587 = 0.8185$

27. (area to the left of 1.62) = 0.9474

29. (area to the left of -2.19) = 0.0143

31. (area to the right of -2.25) = 1 − (area to the left of -2.25) $= 1 - 0.0122 = 0.9878$

33. (area between -1.64 and -1.32) = (area to the left of -1.32) − (area to the left of -1.64)
$= 0.0934 - 0.0505 = 0.0429$

35. (area to the left of 0.83) = 0.7967 = 79.67%

37. (area to the right of -1.90) = 1 − (area to the left of -1.90) $= 1 - 0.0287 = 0.9713 = 97.13\%$

39. (area between -1.84 and 1.84) = (area to the left of 1.84) − (area to the left of -1.84)
$= 0.9671 - 0.0329 = 0.9342 = 93.42\%$

41. (area between 1.96 and 2.14) = (area to the left of 2.14) − (area to the left of 1.96)
$= 0.9838 - 0.9750 = 0.0088 = 0.88\%$

43. (area between 0.72 and 3.21) = (area to the left of 2.14) − (area to the left of 0.72)
$= 0.9838 - 0.7642 = 0.2196 = 21.96\%$

45. a) Luisa, Sarah, and Eleanor are taller than the mean because their *z*-scores are positive.
b) Jenny and Shenice are at the mean because their *z*-scores are zero.
c) Sadaf, Heather, and Kim-Liu are shorter than the mean because their *z*-scores are negative.

47. $z_{250} = \frac{250-250}{48} = 0$; (area to the right of 0) = 1 − (area to the left of 0) $= 1 - 0.5 = 0.5 = 50\%$

49. $z_{250} = \frac{250-250}{48} = 0$; (area to the left of 0) = 0.5 = 50%

51. $z_{202} = \frac{202-250}{48} = -1.00$; $z_{346} = \frac{346-250}{48} = 2.00$;
(area between -1.00 and 2.00) = (area to the left of 2.00) − (area to the left of -1.00)
$= 0.9772 - 0.1587 = 0.8185 = 81.85\%$

53. $z_{310} = \frac{310-250}{48} = 1.25$;
(area to the right of 1.25) = 1 − (area to the left of 1.25) $= 1 - 0.8944 = 0.1056 = 10.56\%$
$(500)(0.1056) \approx 53$ adults

55. $z_{550} = \frac{550-500}{100} = 0.50$; (area to the left of 0.50) = 0.6915 = 69.15%

57. $z_{550} = \frac{550-500}{100} = 0.50$; $z_{650} = \frac{650-500}{100} = 1.50$;
(area between 0.5 and 1.5) = (area to the left of 1.50) − (area to the left of 0.50)
$= 0.9332 - 0.6915 = 0.2417 = 24.17\%$

59. $z_{300} = \frac{300-500}{100} = -2.00$; (area to the left of -2.00) = 0.0228 = 2.28%

61. $z_{7.0} = \dfrac{7.0-7.6}{0.4} = -1.50;$
(area to the right of -1.50) = 1 − (area to the left of -1.50) = 1 − 0.0668 = 0.9332 = 93.32%

63. $z_{7.7} = \dfrac{7.7-7.6}{0.4} = 0.25;$ (area to the left of 0.25) = 0.5987 = 59.87%

65. $z_{62} = \dfrac{62-62}{5} = 0.0;$ (area to the left of 0) = 0.5 = 50%

67. $z_{56} = \dfrac{56-62}{5} = -1.20;$ (area to the left of -1.20) = 0.1151 = 11.51%

69. $200(0.1151) \approx 23$ cars (See Exercise 67)

71. $z_{70} = \dfrac{70-74.0}{12.5} = -0.32;$ $z_{78} = \dfrac{78-74.0}{12.5} = 0.32;$
(area between -0.32 and 0.32) = (area to the left of 0.32) − (area to the left of -0.32)
= 0.6255 − 0.3745 = 0.2510 = 25.1%

73. $z_{70} = \dfrac{70-74.0}{12.5} = -0.32;$ (area to the left of -0.32) = 0.3745 = 37.45% and $500(0.3745) \approx 187$ females

75. $z_{13,200} = \dfrac{13,200-12,000}{2000} = 0.60;$
(area to the right of 0.60) = 1 − (area to the left of 0.60) = 1 − 0.7257 = 0.2743 = 27.43%

77. $z_{17,000} = \dfrac{17,000-12,000}{2000} = 2.50;$
(area to the right of 2.50) = 1 − (area to the left of 2.50) = 1 − 0.9938 = 0.0062 = 0.62%

79. $z_{11,000} = \dfrac{11,000-12,000}{2000} = -0.50;$
(area to the right of -0.50) = 1 − (area to the left of -0.50) = 1 − 0.3085 = 0.6915 = 69.15%
$(120)(0.6915) \approx 83$ families

81. $z_5 = \dfrac{5-6.7}{0.81} \approx -2.10$ (area to the left of -2.10) = 0.0179 = 1.79%

83. The standard deviation is too large. There is too much variation.

85. a) *B*; It is the center point of the distribution.
b) *C*; It is above the mean.
c) *A*; It is below the mean.

87. The mean is the greatest value. The median is lower than the mean. The mode is the lowest value. The greatest frequency appears on the left side of the curve. Since the mode is the value with the greatest frequency, the mode would appear on the left side of the curve (where the lowest values are). Every value in the set of data is considered in determining the mean. The values on the far right of the curve would increase the value of the mean. Thus, the value of the mean would be farther to the right than the mode. The median would be between the mode and the mean.

89. Answers will vary.

91. a) Katie: $z_{28,408} = \dfrac{28,408 - 23,200}{2170} = \dfrac{5208}{2170} = 2.4$

Stella: $z_{29,510} = \dfrac{29,510 - 25,600}{2300} = \dfrac{3910}{2300} = 1.7$

b) Katie. Her z-score is higher than Stella's z-score. This means her sales are further above the mean than Stella's sales.

93. Using Table 12.8, the answer is -1.18.

95. The mean of 12.0 is centered between 9.6 and 14.4 and $\dfrac{0.77}{2} = 0.385$. Using Table 12.8, the area of $0.5 + 0.385 = 0.885$ is to the left of a z-score of 1.20, which corresponds to a score of 14.4.

$$z = \frac{x - \overline{x}}{s}$$

$$1.20 = \frac{14.4 - 12}{s}$$

$$1.20 = \frac{2.4}{s}$$

$$\frac{1.20s}{1.20} = \frac{2.4}{1.20}$$

$$s = 2$$

Section 12.6: Linear Correlation and Regression

1. Coefficient

3. a) 1
   b) -1
   c) 0

5. Positive

7. No correlation

9. Strong positive correlation

11. Yes; $|0.92| > 0.765$

13. No; $|-0.59| < 0.602$

15. Yes; $|-0.32| > 0.254$

17. No; $|0.75| < 0.917$

19. a)

b) $r = \dfrac{5(398) - (32)(58)}{\sqrt{5(226) - (32)^2}\sqrt{5(708) - (58)^2}} \approx 0.981$

x	y	x^2	y^2	xy
4	8	16	64	32
5	10	25	100	50
6	12	36	144	72
7	12	49	144	84
10	16	100	256	160
32	58	226	708	398

c) Yes; $|0.981| > 0.878$

d) Yes; $|0.981| > 0.959$

21. a)

b) $r = \dfrac{5(5720)-(181)(156)}{\sqrt{5(6965)-(181)^2}\sqrt{5(5114)-(156)^2}} \approx 0.228$

x	y	x^2	y^2	xy
23	30	529	900	690
35	38	1225	1444	1330
31	27	961	729	837
43	21	1849	441	903
49	40	2401	1600	1960
181	156	6965	5114	5720

c) No; $|0.228| < 0.878$

d) No; $|0.228| < 0.959$

23. a)

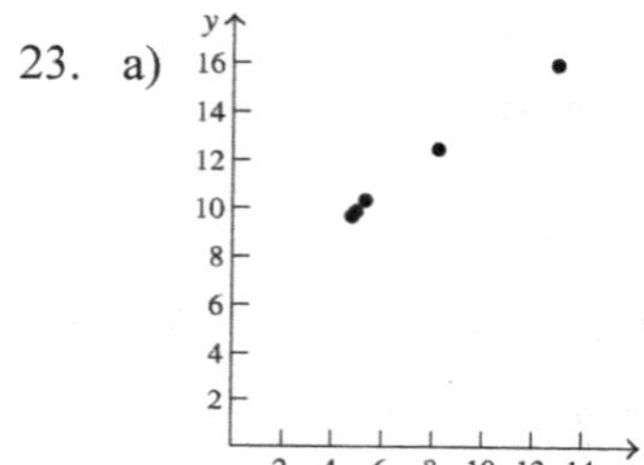

b) $r = \dfrac{5(458.47)-(36)(58.4)}{\sqrt{5(306.04)-(36)^2}\sqrt{5(712.98)-(58.4)^2}} \approx 0.999$

x	y	x^2	y^2	xy
5.3	10.3	28.09	106.09	54.59
4.7	9.6	22.09	92.16	45.12
8.4	12.5	70.56	156.25	105
12.7	16.2	161.29	262.44	205.74
4.9	9.8	24.01	96.04	48.02
36	58.4	306.04	712.98	458.47

c) Yes; $|0.999| > 0.878$

d) Yes; $|0.999| > 0.959$

25. a)

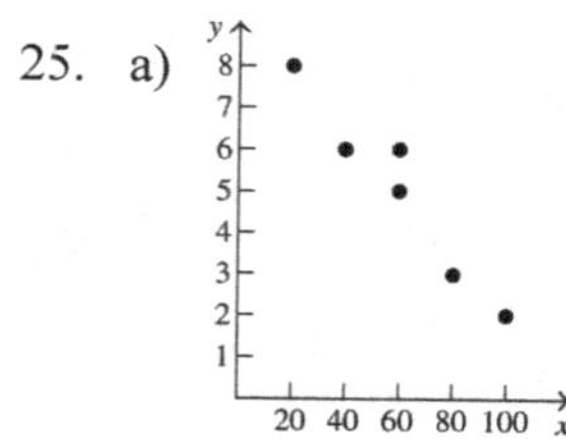

25. (continued)

b) $r = \dfrac{6(1500)-(360)(30)}{\sqrt{6(25{,}600)-(360)^2}\sqrt{6(174)-(30)^2}} \approx -0.968$

x	y	x^2	y^2	xy
100	2	10,000	4	200
80	3	6400	9	240
60	5	3600	25	300
60	6	3600	36	360
40	6	1600	36	240
20	8	400	64	160
360	30	25,600	174	1500

c) Yes; $|-0.968| > 0.811$

d) Yes; $|-0.968| > 0.917$

27. (See table in Exercise 19.)

$m = \dfrac{5(398)-32(58)}{5(226)-(32)^2} \approx 1.26$

$b = \dfrac{58-1.264(32)}{5} \approx 3.51$

$y = 1.26x + 3.51$

29. (See table in Exercise 21.)

$m = \dfrac{5(5720)-181(156)}{5(6965)-(181)^2} \approx 0.18$

$b = \dfrac{156-0.176(181)}{5} \approx 24.82$

$y = 0.18x + 24.82$

31. (See table in Exercise 23.)

$m = \dfrac{5(458.47)-36(58.4)}{5(306.04)-(36)^2} \approx 0.81$

$b = \dfrac{58.4-0.811(36)}{5} \approx 5.84$

$y = 0.81x + 5.84$

33. (See table in Exercise 25.)

$m = \dfrac{6(1500)-360(30)}{6(25{,}600)-(360)^2} \approx -0.08$

$b = \dfrac{30-(-0.075)(360)}{6} \approx 9.50$

$y = -0.08x + 9.50$

35. a) $r = \dfrac{6(1830)-(68)(125)}{\sqrt{6(1032)-(68)^2}\sqrt{6(3275)-(125)^2}} \approx 0.987$

x	y	x^2	y^2	xy
2	5	4	25	10
15	25	225	625	375
16	30	256	900	480
9	20	81	400	180
21	35	441	1225	735
5	10	25	100	50
68	125	1032	3275	1830

b) Yes; $|0.987| > 0.811$

c) $m = \dfrac{6(1830)-68(125)}{6(1032)-(68)^2} \approx 1.58;\ b = \dfrac{125-1.582(68)}{6} \approx 2.91;\ y = 1.58x + 2.91$

37. a) $r=\dfrac{7(10,230)-(210)(359)}{\sqrt{7(6806)-(210)^2}\sqrt{7(19,013)-(359)^2}}\approx -0.979$

x	y	x^2	y^2	xy
42	35	1764	1225	1470
38	45	1444	2025	1710
20	60	400	3600	1200
30	52	900	2704	1560
28	55	784	3025	1540
35	47	1225	2209	1645
17	65	289	4225	1105
210	359	6806	19,013	10,230

b) Yes; $|-0.979| > 0.875$

c) $m=\dfrac{7(10,230)-210(359)}{7(6806)-(210)^2}\approx -1.07;\;\; b=\dfrac{359-(-1.067)(210)}{7}\approx 83.30;\;\; y=-1.07x+83.30$

d) $y=-1.07(33)+83.30=47.99\approx 48$ cups

39. a) $r=\dfrac{6(1511.94)-(214.26)(36.75)}{\sqrt{6(8644.2954)-(214.26)^2}\sqrt{6(268.5625)-(36.75)^2}}\approx 0.961$

x	y	x^2	y^2	xy
35.2	5.5	1239.04	30.25	193.6
22.93	4.5	525.7849	20.25	103.185
16.45	2	270.6025	4	32.9
45.82	9	2099.472	81	412.38
54.16	10	2933.306	100	541.6
39.7	5.75	1576.09	33.0625	228.275
214.26	36.75	8644.295	268.5625	1511.94

b) Yes; $|0.961| > 0.811$

c) $m=\dfrac{6(1511.94)-214.26(36.75)}{6(8644.2954)-(214.26)^2}\approx 0.20;\;\; b=\dfrac{36.75-0.201(214.26)}{6}\approx -1.05;\;\; y=0.20x-1.05$

d) $y=0.20(25)-1.05=\$3.95$

41. a) $r=\dfrac{6(878.7)-(33.9)(144)}{\sqrt{6(208.21)-(33.9)^2}\sqrt{6(3714)-(144)^2}}\approx 0.993$

x	y	x^2	y^2	xy
6.5	27	42.25	729	175.5
7.0	30	49.00	900	210.0
6.0	25	36.00	625	150.0
2.0	10	4.00	100	20.0
6.4	28	40.96	784	179.2
6.0	24	36.00	576	144.0
33.9	144	208.21	3714	878.7

b) Yes; $|0.993| > 0.811$

41. (continued)

c) $m = \dfrac{6(878.7) - 33.9(144)}{6(208.21) - (33.9)^2} \approx 3.90;\ b = \dfrac{144 - 3.904(33.9)}{6} \approx 1.94;\ y = 3.90x + 1.94$

d) $y = 3.90(5) + 1.94 \approx 21$ kilocalories

43. a) Answers will vary. b) Answers will vary.

c)

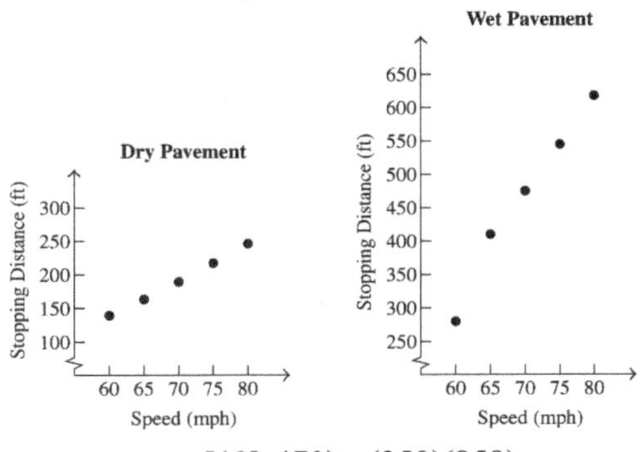

d) $r = \dfrac{5(68,470) - (350)(959)}{\sqrt{5(24,750) - (350)^2}\sqrt{5(191,129) - (959)^2}} \approx 0.999$

x	y	x^2	y^2	xy
60	140	3600	19,600	8400
65	164	4225	26,896	10,660
70	190	4900	36,100	13,300
75	218	5625	47,524	16,350
80	247	6400	61,009	19,760
350	959	24,750	191,129	68,470

e) $r = \dfrac{5(167,015) - (350)(2328)}{\sqrt{5(24,750) - (350)^2}\sqrt{5(1,151,074) - (2328)^2}} \approx 0.990$

x	y	x^2	y^2	xy
60	280	3600	78,400	16,800
65	410	4225	168,100	26,650
70	475	4900	225,625	33,250
75	545	5625	297,025	40,875
80	618	6400	381,924	49,440
350	2328	24,750	1,151,074	167,015

f) Answers will vary.

g) $m = \dfrac{5(68,470) - 350(959)}{5(24,750) - (350)^2} \approx 5.36;\ b = \dfrac{959 - 5.36(350)}{5} \approx -183.40;\ y = 5.36x - 183.40$

h) $m = \dfrac{5(167,015) - 350(2328)}{5(24,750) - (350)^2} \approx 16.22;\ b = \dfrac{2328 - 16.22(350)}{5} \approx -669.80;\ y = 16.22x - 669.80$

i) Dry: $y = 5.36(77) - 183.40 = 229.2$ ft

Wet: $y = 16.22(77) - 669.80 = 579.1$ ft

45. a) The correlation coefficient will not change because $\Sigma xy = \Sigma yx$, $(\Sigma x)(\Sigma y) = (\Sigma y)(\Sigma x)$, and the square roots in the denominator will be the same.
    b) Answers will vary.

47. a) Answers will vary.
    b) Answers will vary.
    c) Answers will vary.
    d) Answers will vary.
    e) Answers will vary.
    f) Answers will vary.
    g) Answers will vary.

49. a) $SS(xy) = \Sigma xy - \dfrac{(\Sigma xy)(\Sigma xy)}{n}$

$= 387 - \dfrac{17(106)}{6} = 86.67$

$SS(x) = \Sigma x^2 - \dfrac{(\Sigma x)^2}{n} = 75 - \dfrac{289}{6}$

$= 26.83$

$SS(y) = \Sigma y^2 - \dfrac{(\Sigma y)^2}{n} = 2202 - \dfrac{11236}{6}$

$= 329.33$

$r = \dfrac{SS(xy)}{\sqrt{SS(x)SS(y)}} = \dfrac{86.67}{\sqrt{(26.83)(329.33)}}$

≈ 0.922

    b) The result is the same.

Review Exercises

1. a) A population consists of all items or people of interest.
   b) A sample is a subset of the population.
2. a) A random sample is one where every item in the population has the same chance of being selected.
   b) A systematic sample is obtained by selecting a random starting point and then selecting every nth item in a population.
   c) A cluster sample consists of dividing the population into sections. Then randomly select sections to use and either select all the items in the selected sections or a random sample of items from the selected sections.
   d) A stratified sample consists of dividing the population into strata and then taking a random sample from each strata.
   e) A convenience sample uses data that are easily or readily obtained.
3. The candy bars may have lots of calories, or fat, or sodium. Therefore, it may not be healthy to eat them.
4. Sales may not necessarily be a good indicator of profit. Expenses must also be considered.
5. a)

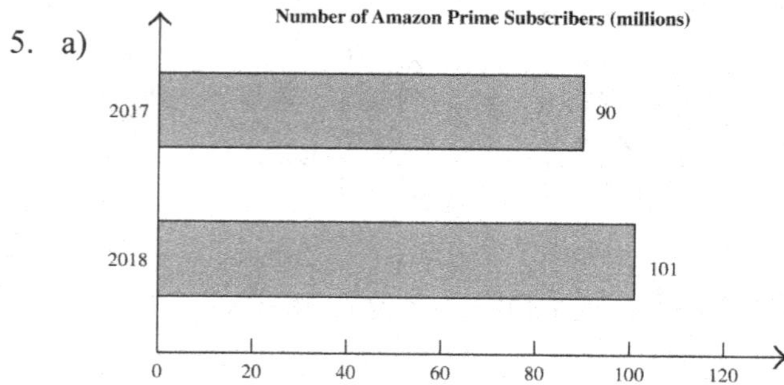

5. (continued)

b)

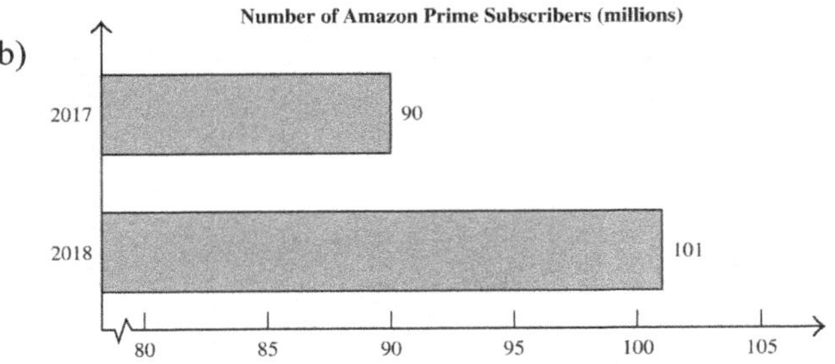

6. a)

Class	Frequency
35	1
36	3
37	6
38	2
39	3
40	0
41	4
42	1
43	3
44	1
45	1

b) and c)

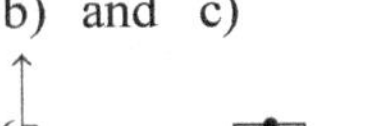

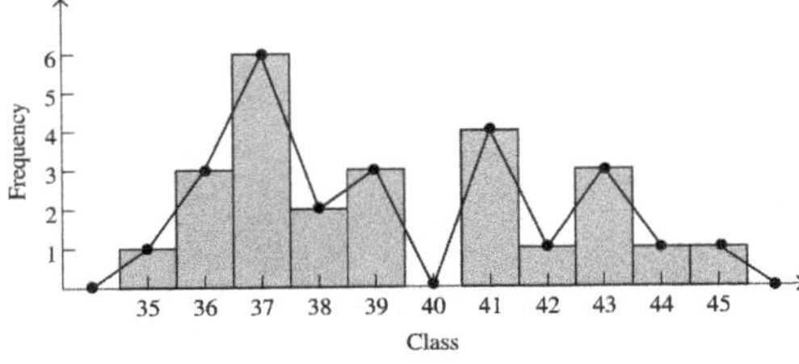

7. a)

High Temperature	Number of Cities
58–62	1
63–67	4
68–72	9
73–77	10
78–82	11
83–87	4
88–92	1

b) and c)

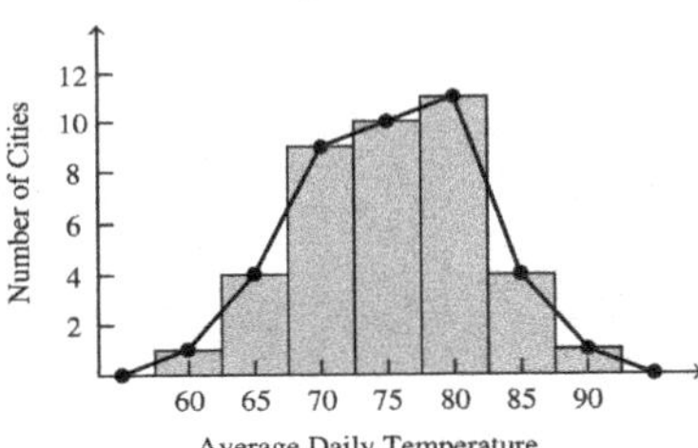

d) 5 | 8 represents 58

5	8
6	3 6 6 7 8 8 9
7	0 1 1 1 2 2 3 3 3 4 5 5 5 6 6 7 9 9 9
8	0 0 0 0 1 2 2 2 3 4 4 7
9	1

8. $\bar{x} = \frac{\Sigma x}{n} = \frac{468}{6} = 78$

9. Median: $\frac{77+81}{2} = 79$

10. Mode: none

11. Midrange: $\frac{65+91}{2} = 78$

12. Range: $91 - 65 = 26$

13. $s = \sqrt{\frac{\Sigma(x-\bar{x})^2}{n-1}} = \sqrt{\frac{400}{5}} = \sqrt{80} \approx 8.94$

Table for Exercises 8–13

x	$x-\bar{x}$	$(x-\bar{x})^2$
65	–13	169
72	–6	36
77	–1	1
81	3	9
82	4	16
91	13	169
468	0	400

14. $\overline{x} = \frac{\Sigma x}{n} = \frac{216}{12} = 18$

15. Median: $\frac{17+19}{2} = 18$

16. Mode: 12 and 17

17. Midrange: $\frac{9+28}{2} = 18.5$

18. Range: $28 - 9 = 19$

19. $s = \sqrt{\frac{\Sigma(x-\overline{x})^2}{n-1}} = \sqrt{\frac{440}{11}} = \sqrt{40} \approx 6.32$

Table for Exercises 14–19

x	$x - \overline{x}$	$(x-\overline{x})^2$
9	−9	81
10	−8	64
12	−6	36
12	−6	36
17	−1	1
17	−1	1
19	1	1
20	2	4
22	4	16
24	6	36
26	8	64
28	10	100
216	0	440

20. $z_7 = \frac{7-9}{2} = -1.00$; $z_{11} = \frac{11-9}{2} = 1.00$
(area between -1 and 1) = (area to the left of 1.00) − (area to the left of -1.00)
$= 0.8413 - 0.1587 = 0.6826 = 68.26\%$

21. $z_5 = \frac{5-9}{2} = -2.00$; $z_{13} = \frac{13-9}{2} = 2.00$
(area between -2.00 and 2.00) = (area to the left of 2.00) − (area to the left of -2.00)
$= 0.9772 - 0.0228 = 0.9544 = 95.44\%$

22. $z_{12.2} = \frac{12.2-9}{2} = 1.60$; (area to the left of 1.60) $= 0.9452 = 94.52\%$

23. $z_{12.2} = \frac{12.2-9}{2} = 1.60$
(area to the right of 1.60) = 1 − (area to the left of 1.60)
$= 1 - 0.9452 = 0.0548 = 5.48\%$

24. $z_{7.8} = \frac{7.8-9}{2} = -0.60$
(area to the right of -0.60) = 1 − (area to the left of -0.60)
$= 1 - 0.2743 = 0.7257 = 72.57\%$

25. $z_{20} = \frac{20-20}{5} = 0$; $z_{25} = \frac{25-20}{5} = 1.00$
(area between 0 and 1.00) = (area to the left of 1.00) − (area to the left of 0)
$= 0.8413 - 0.5 = 0.3413 = 34.13\%$

26. $z_{18} = \frac{18-20}{5} = -0.40$; (area to the left of -0.40) $= 0.3446 = 34.46\%$

27. $z_{22} = \frac{22-20}{5} = 0.40$; $z_{28} = \frac{28-20}{5} = 1.60$
(area between 0.40 and 1.60) = (area to the left of 1.60) − (area to the left of 0.40)
$= 0.9452 - 0.6554 = 0.2898 = 28.98\%$

28. $z_{30} = \dfrac{30-20}{5} = 2.00;$

(area to the right of 2.00) = 1 − (area to the left of 2.00)

$= 1 - 0.9772 = 0.0228 = 2.28\%$

29. a)

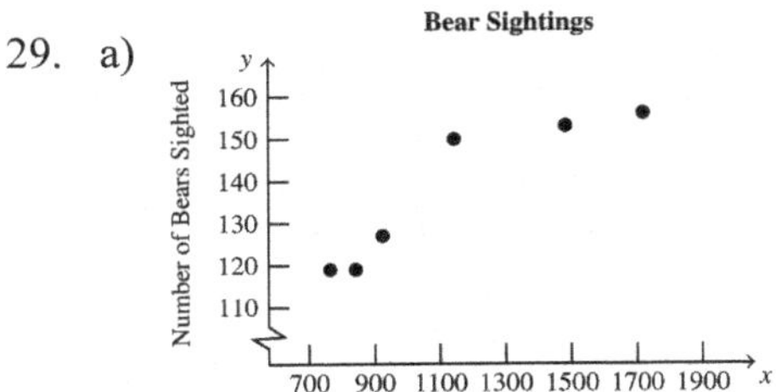

b) Yes; positive

c) $r = \dfrac{6(973,302) - (6865)(824)}{\sqrt{6(8,564,719) - (6865)^2}\sqrt{6(114,696) - (824)^2}} \approx 0.925$

x	y	x^2	y^2	xy
765	119	585,225	14,161	91,035
926	127	857,476	16,129	117,602
1145	150	1,311,025	22,500	171,750
842	119	708,964	14,161	100,198
1485	153	2,205,225	23,409	227,205
1702	156	2,896,804	24,336	265,512
6865	824	8,564,719	114,696	973,302

d) Yes; $|0.925| > 0.811$

e) $m = \dfrac{6(973,302) - 6865(824)}{6(8,564,719) - (6865)^2} \approx 0.04;\ b = \dfrac{824 - 0.043(6865)}{6} \approx 88.17;\ y = 0.04x + 88.17$

f) $y = 0.04(1500) + 88.17 = 148.17 \approx 148$ bears

30. a)

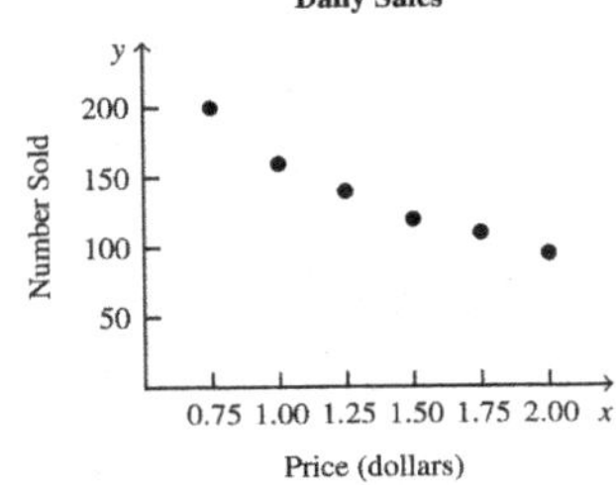

b) Yes; negative

30. (continued)

c) $r = \dfrac{6(1047.5)-(8.25)(825)}{\sqrt{6(12.4375)-(8.25)^2}\sqrt{6(120,725)-(825)^2}} \approx -0.973$

x	y	x^2	y^2	xy
0.75	200	0.5625	40,000	150.0
1.00	160	1.0000	25,600	160.0
1.25	140	1.5625	19,600	175.0
1.50	120	2.2500	14,400	180.0
1.75	110	3.0625	12,100	192.5
2.00	95	4.0000	9025	190.0
8.25	825	12.4375	120,725	1047.5

d) Yes; $|-0.973| > 0.811$

e) $m = \dfrac{6(1047.5)-8.25(825)}{6(12.4375)-(8.25)^2} \approx -79.4;\ b = \dfrac{825-(-79.429)(8.25)}{6} \approx 246.7;\ y = -79.4x + 246.7$

f) $y = -79.4(1.60) + 246.7 = 119.66 \approx 120$ sold

31. 180 lb

32. 185 lb

33. 25%

34. 25%

35. $100 - 86 = 14\%$

36. $100(192) = 19,200$ lb

37. $192 + 2(23) = 238$ lb

38. $192 - 1.8(23) = 150.6$ lb

39. $\bar{x} = \dfrac{\Sigma x}{n} = \dfrac{40}{20} = 2$

40. Mode: 0 and 2

41. Median: $\dfrac{2+2}{2} = 2$

42. Midrange: $\dfrac{0+7}{2} = 3.5$

43. Range: $7 - 0 = 7$

44. $s = \sqrt{\dfrac{\Sigma(x-\bar{x})^2}{n-1}} = \sqrt{\dfrac{72}{19}} \approx \sqrt{3.79} \approx 1.95$

Table for Exercises 39–44

x	$x-\bar{x}$	$(x-\bar{x})^2$
0	–2	4
0	–2	4
0	–2	4
0	–2	4
0	–2	4
1	–1	1
1	–1	1
1	–1	1
1	–1	1
2	0	0
2	0	0
2	0	0
2	0	0
2	0	0
3	1	1
3	1	1
3	1	1
4	2	4
6	4	16
7	5	25
40	0	72

45.

Number of Children	Number of Families
0	5
1	4
2	5
3	3
4	1
5	0
6	1
7	1

46.

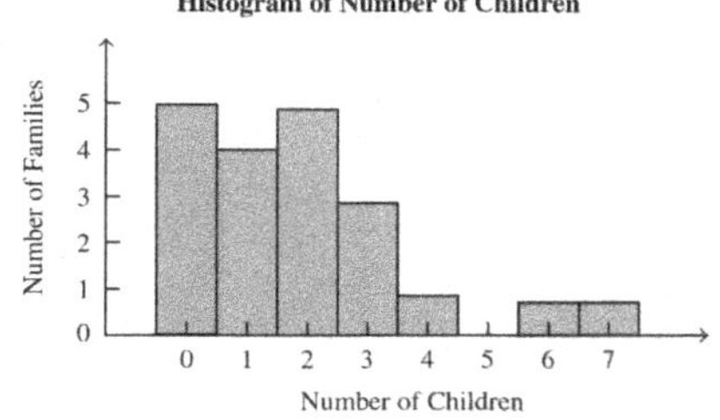

47.

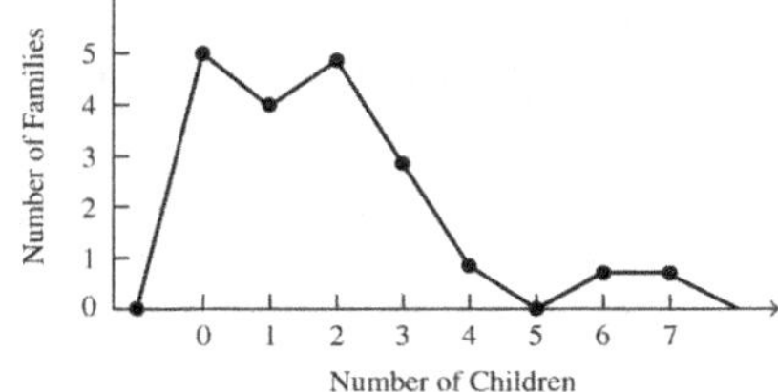

48. No, it is skewed to the right.

49. Answers will vary.

50. Answers will vary.

Chapter Test

1. $\overline{x} = \frac{\Sigma x}{n} = \frac{185}{5} = 37$

2. Median: 38

3. Mode: 38

4. Midrange: $\frac{22+47}{2} = 34.5$

5. Range: $47 - 22 = 25$

6. $s = \sqrt{\frac{\Sigma(x-\overline{x})^2}{n-1}} = \sqrt{\frac{336}{4}} = \sqrt{84} \approx 9.17$

Table for Exercises 1–6

x	$x-\overline{x}$	$(x-\overline{x})^2$
22	–15	225
38	1	1
38	1	1
40	3	9
47	10	100
185	0	336

7.

Class	Frequency
25–30	7
31–36	5
37–42	1
43–48	7
49–54	5
55–60	3
61–66	2

8.

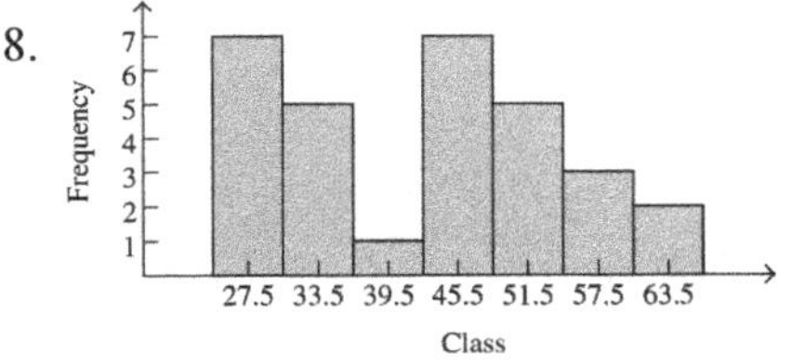

9.

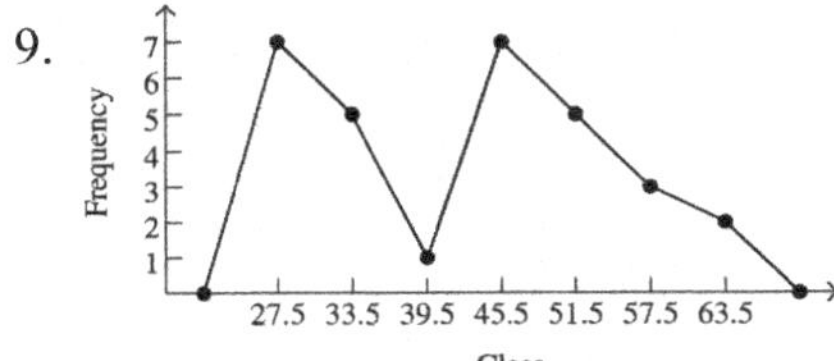

10. \$815

11. \$790

12. 75%

13. 79%

14. $100(820) = \$82,000$

15. $820 + 1(40) = \$860$

16. $z_{36} = \frac{36-42}{5} = -1.20; \quad z_{53} = \frac{53-42}{5} = 2.20$

(area between -1.20 and 2.20) = (area to the left of 2.20) – (area to the left of -1.20)

$= 0.9861 - 0.1151 = 0.871 = 87.1\%$

17. $z_{35.75} = \frac{35.75-42}{5} = -1.25;$

(area to the right of -1.25) = 1 – (area to the left of -1.25) $= 1 - 0.1056 = 0.8944 = 89.44\%$

18. $z_{48.25} = \dfrac{48.25 - 42}{5} = 1.25;$

(area to the right of 1.25) = 1 − (area to the left of 1.25) = 1 − 0.8944 = 0.1056 = 10.56%

19. $z_{50} = \dfrac{50 - 42}{5} = 1.60;$ (area to the left of 1.60) = 0.9452 = 94.52%

20. a) 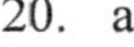

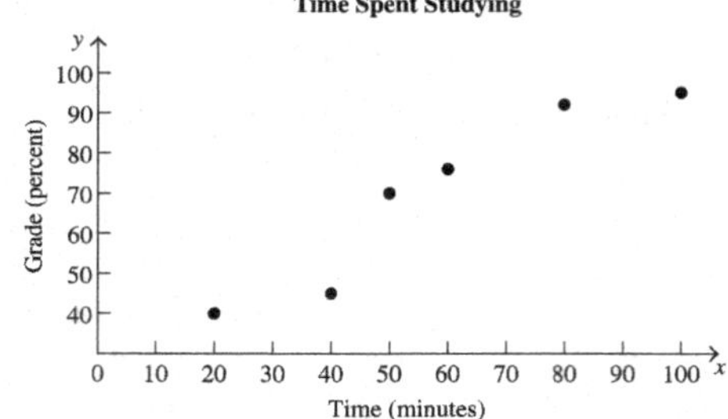

b) Yes

c) $r = \dfrac{6(27{,}520) - (350)(418)}{\sqrt{6(24{,}500) - (350)^2}\sqrt{6(31{,}790) - (418)^2}} \approx 0.950$

x	y	x^2	y^2	xy
20	40	400	1600	800
40	45	1600	2025	1800
50	70	2500	4900	3500
60	76	3600	5776	4560
80	92	6400	8464	7360
100	95	10,000	9025	9500
350	418	24,500	31,790	27,520

d) Yes, $|0.950| > 0.917$

e) $m = \dfrac{6(27{,}520) - 350(418)}{6(24{,}500) - (350)^2} \approx 0.77;\ b = \dfrac{418 - 0.768}{6} \approx 24.86$

$y = 0.77x + 24.86$

f) $y = 0.77(75) + 24.86 \approx 83$

Chapter Thirteen: Graph Theory

Section 13.1: Graphs, Paths, and Circuits

1. Graph
3. Edge
5. Path
7. Degree
9. *B* and *C* are even; *A* and *D* are odd.

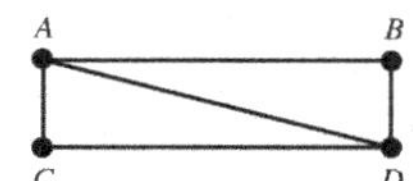

11. 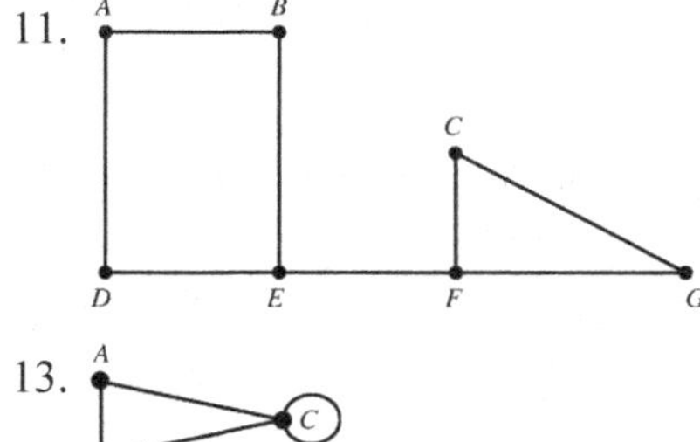

13. A B C

15. a) No. There is no edge connecting vertices *C* and *D*.
    b) Yes, there are edges connecting vertex *A* to vertex *B*, vertex *B* to vertex *C*, vertex *C* to vertex *E*, and vertex *E* to vertex *D*.
    c) No. The path does not begin and end with the same vertex.
    d) Yes, it is a path that begins and ends with the same vertex.

17 a) Yes. One example is *B*, *A*, *C*, *E*, *D*, *B*, *C*.
   b) No
   c) No

19. Yes. One example is *B*, *A*, *C*, *E*, *D*, *B*.

21. 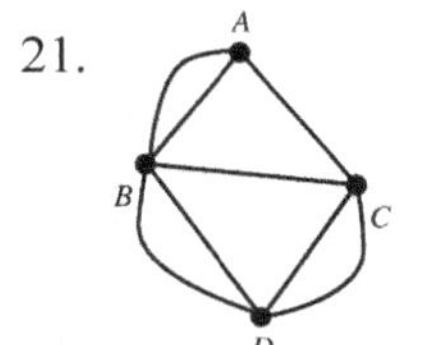
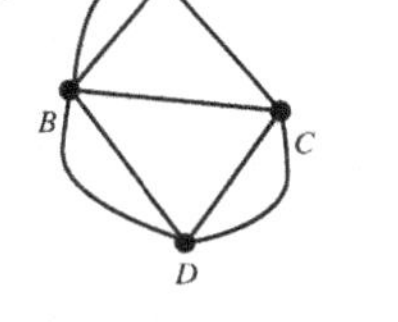

23. WA ID MT OR WY

25. 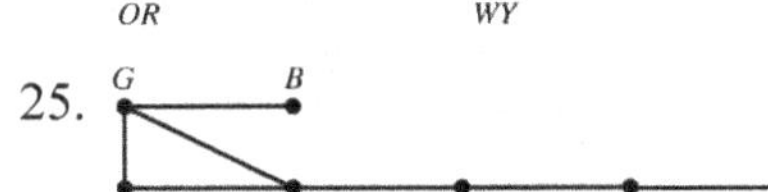

27. 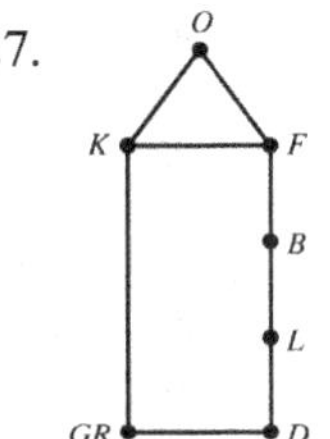

29. 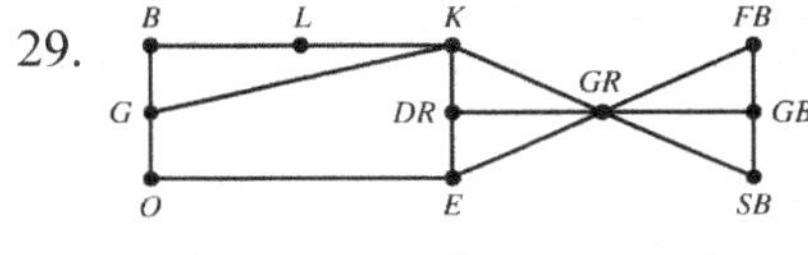

31. A B C D E F G

33. Disconnected.
    There is no path that connects *A* to *D*.

35. Connected

37. a) Edge *BC*
    b) Loop *FF*

39. a) Edge *BC*
    b) No loop

41. Answers will vary.

43. a) Answers will vary.
    b) Answers will vary.
    c) Answers will vary.
    d) The sum of the degrees is equal to twice the number of edges. This is true since each edge must connect two vertices. Each edge then contributes two to the sum of the degrees.

45. a) Answers will vary.
    b) Answers will vary.

Section 13.2: Euler Paths and Euler Circuits

1. Euler
3. No
5. Odd
7. a) Yes; one example is *A, B, D, E, C, A, D, C*.
   b) No. This graph has exactly two odd vertices, *A* and *C*. Each Euler path must begin at vertex *A* and end at vertex *C* or vice versa.
9. No. A graph with exactly two odd vertices has no Euler circuits.
11. No. A graph with more than two odd vertices has neither an Euler path nor an Euler circuit.
13. No. A graph with more than two odd vertices has neither an Euler path nor an Euler circuit.
15. *A, B, C, D, E, F, B, D, F, A*; Other answers are possible.
17. *C, D, E, F, A, B, D, F, B, C*; Other answers are possible.
19. *E, F, A, B, C, D, F, B, D, E*; Other answers are possible.
21. a) Yes. Each island would correspond to an odd vertex. According to item 2 of Euler's Theorem, a graph with exactly two odd vertices has at least one Euler path, but no Euler circuit.
    b) They could start on either island and finish at the other.
23. a) Other graphs are possible.

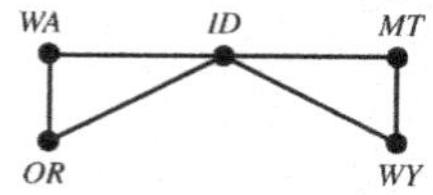

    b) Yes; WA, ID, MT, WY, ID, OR, WA
25. a) Other graphs are possible.

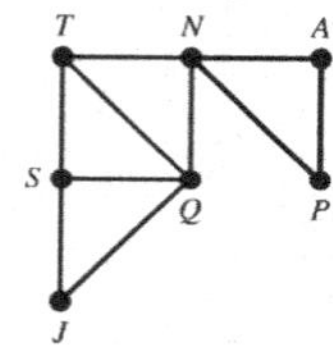

    b) Vertices *S* and *T* are both odd. According to item 2 of Euler's Theorem, since there are exactly two odd vertices, at least one Euler path, but no Euler circuits exist. Yes; *S, T, N, A, P, N, Q, J, S, Q, T*
    c) No; see part (b)
27. a) Other graphs are possible.

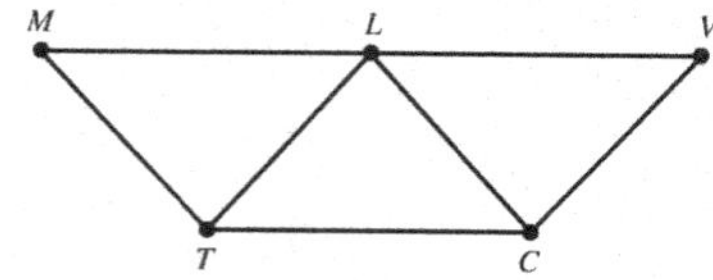

    b) Vertices *T* and *C* are both odd. According to item 2 of Euler's Theorem, since there are exactly two odd vertices, at least one Euler path, but no Euler circuits exist. Yes; *T, M, L, V, C, L, T, C*
    c) No; see part (b)

29. a) The graph for the floor plan is shown.

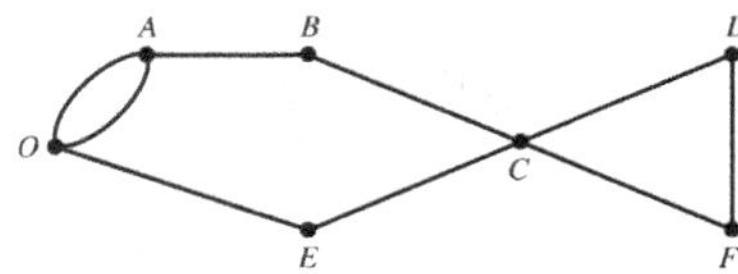

b) Yes

c) *O*, *A*, *B*, *C*, *D*, *F*, *C*, *E*, *O*, *A*

31. a) The graph for the floor plan is shown.

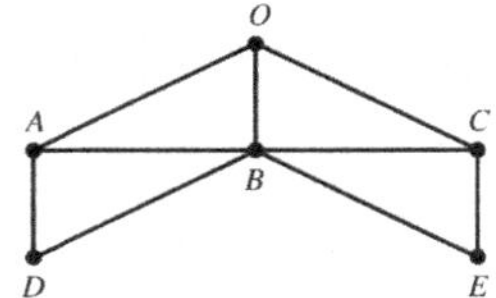

b) No; the graph has four odd vertices, and by Euler's theorem a graph with more than two odd vertices has neither an Euler path nor an Euler circuit.

33. a) Yes. The graph representing the map is shown below.

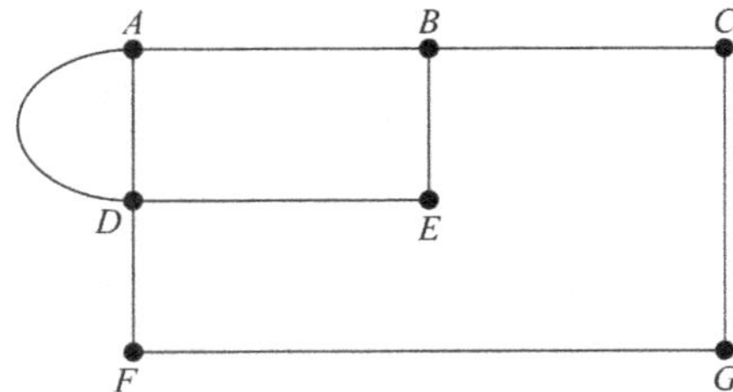

They are seeking an Euler path or an Euler circuit. Note that vertices *A* and *B* are both odd. According to item 2 of Euler's Theorem, since there are exactly two odd vertices, at least one Euler path, but no Euler circuits exist.

b) The residents would need to start at the intersection of Maple Circle, Walnut St., and Willow St. or at the intersection of Walnut St. and Oak St.

35. *A*, *B*, *E*, *D*, *C*, *A*, *D*, *B*; Other answers are possible.

37. *A*, *B*, *H*, *E*, *B*, *C*, *E*, *F*, *H*, *G*, *F*, *D*, *C*, *A*, *G*; Other answers are possible.

39. *A*, *B*, *C*, *D*, *I*, *H*, *G*, *B*, *E*, *H*, *C*, *E*, *G*, *F*, *A*; Other answers are possible.

41. *A*, *C*, *D*, *G*, *H*, *F*, *C*, *F*, *E*, *B*, *A*; Other answers are possible.

43. *A*, *B*, *C*, *E*, *B*, *D*, *E*, *F*, *I*, *E*, *H*, *D*, *G*, *H*, *I*, *J*, *F*, *C*, *A*; Other answers are possible.

45. *UT*, *CO*, *NM*, *AZ*, *CA*, *NV*, *UT*, *AZ*, *NV*; Other answers are possible.

47. *B*, *A*, *E*, *H*, *I*, *J*, *K*, *D*, *C*, *G*, *G*, *J*, *F*, *C*, *B*, *F*, *I*, *E*, *B*; Other answers are possible.

49. *J*, *G*, *G*, *C*, *F*, *J*, *K*, *D*, *C*, *B*, *F*, *I*, *E*, *B*, *A*, *E*, *H*, *I*, *J*; Other answers are possible.

51. a) Yes. There are no odd vertices.
b) Yes. There are no odd vertices.

53. a) No. There are more than two odd vertices.
b) No. There is at least one odd vertex.

55. a) No.
b) California, Nevada, and Louisiana (and others) have an odd number of states bordering them. Since a graph of the United States would have more than two odd vertices, no Euler path and no Euler circuit exists.

57. a) 

b) A ●——● B

c) 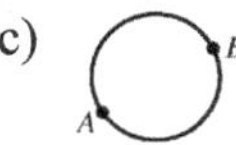

Section 13.3: Hamilton Paths and Hamilton Circuits

1. Salesman

3. Hamilton

5. Euler

7. Force

9. *B*, *A*, *E*, *F*, *G*, *H*, *D*, *C* and *H*, *D*, *C*, *G*, *F*, *E*, *A*, *B*; Other answers are possible.

11. *A*, *B*, *C*, *D*, *G*, *F*, *E*, *H* and *E*, *H*, *F*, *G*, *D*, *C*, *A*, *B*; Other answers are possible.

13. *A*, *G*, *J*, *D*, *E*, *B*, *C*, *F*, *I*, *H* and *J*, *D*, *A*, *G*, *F*, *I*, *H*, *E*, *B*, *C*; Other answers are possible.

15. *A, B, D, G, E, H, F, C, A* and *D, B, A, C, F, H, E, G, D*; Other answers are possible.

17. *A, B, C, F, I, E, H, G, D, A* and *A, E, B, C, F, I, H, G, D, A*; Other answers are possible.

19. 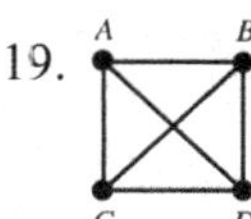

21. The number of unique Hamilton circuits within the complete graph with 6 vertices representing this situation is $(6-1)! = 5! = 5 \cdot 4 \cdot 3 \cdot 2 \cdot 1 = 120$ ways.

23. The number of unique Hamilton circuits within the complete graph with thirteen vertices representing this situation is $(13-1)! = 12! = 12 \cdot 11 \cdot 10 \cdot 9 \cdot 8 \cdot 7 \cdot 6 \cdot 5 \cdot 4 \cdot 3 \cdot 2 \cdot 1 = 479{,}001{,}600$ ways.

25. a) (Other graphs are possible.)

b)

Hamilton Circuit	First Leg Distance	Second Leg Distance	Third Leg Distance	Fourth Leg Distance	Total Distance
O, D, S, L, O	150	125	250	400	925 feet
O, D, L, S, O	150	450	250	100	950 feet
O, L, S, D, O	400	250	125	150	925 feet
O, L, D, S, O	400	450	125	100	1075 feet
O, S, D, L, O	100	125	450	400	1075 feet
O, S, L, D, O	100	250	450	150	950 feet

The shortest route is *O, D, S, L, O* or *O, L, S, D, O*

c) 925 feet (But note that if Mary goes back to her office from the library by revisiting the student center, her total trip will be 875 feet.)

27. a) (Other graphs are possible.)

b)

Hamilton Circuit	First Leg Distance	Second Leg Distance	Third Leg Distance	Fourth Leg Distance	Total Distance
T, C, S, A, T	342	521	1188	582	2633 mi
T, C, A, S, T	342	917	1188	663	3110 mi
T, S, C, A, T	663	521	917	582	2683 mi
T, S, A, C, T	663	1188	917	342	3110 mi
T, A, C, S, T	582	917	521	663	2683 mi
T, A, S, C, T	582	1188	521	342	2633 mi

The shortest route is *T, C, S, A, T* or *T, A, S, C, T*

c) 2633 miles

29. a) Other graphs are possible.

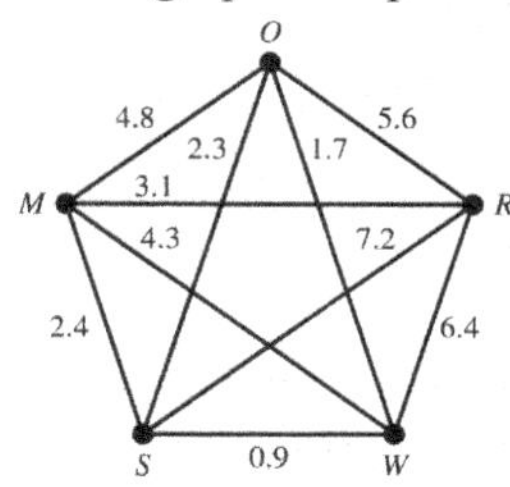

b) O, W, S, M, R, O for
$1.7 + 0.9 + 2.4 + 3.1 + 5.6 = 13.7$ miles

c) Answers will vary.

31. a) Other graphs are possible.

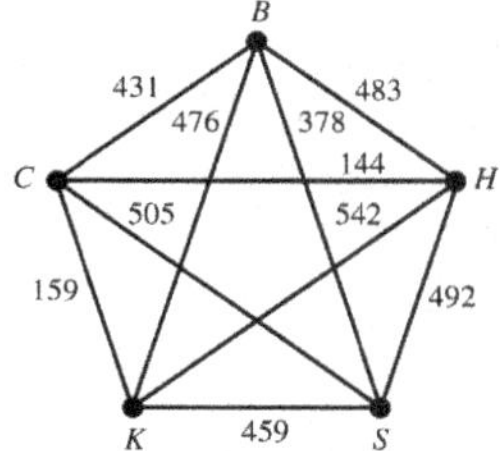

b) S, B, C, H, K, S for
$378 + 431 + 144 + 542 + 459 = \$1954$

c) Answers will vary.

33. a) Answers will vary.
 b) Answers will vary.
 c) Answers will vary.
 d) Answers will vary.

35. $A, E, D, N, O, F, G, Q, P, T, M, L, C, B, J, K, S, R, I, H, A$; Other answers are possible.

Section 13.4: Trees

1. Tree

3. Circuits

5. Minimum-cost

7.

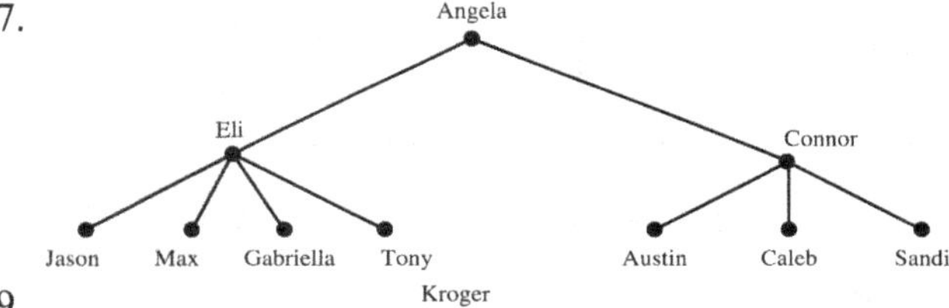

9.

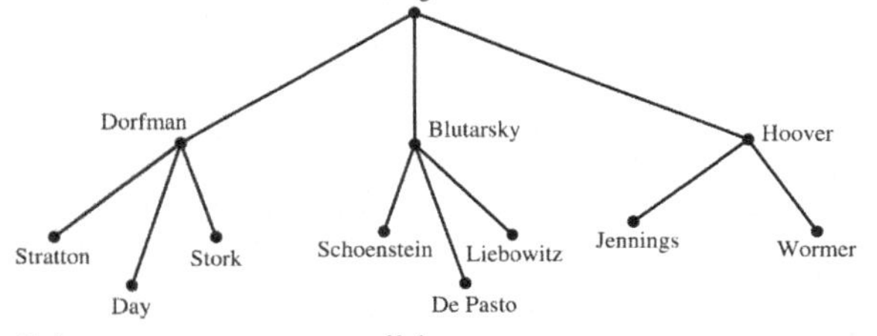

11. Other answers are possible.

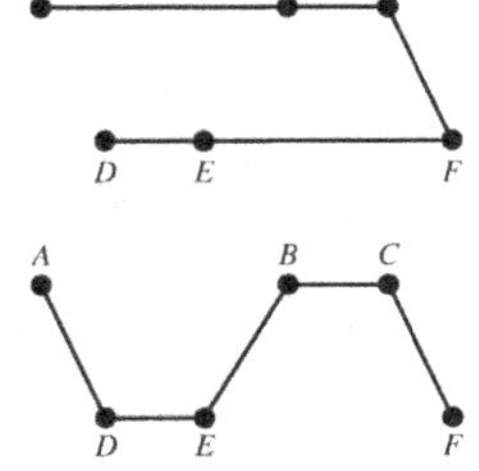

13. Other answers are possible.

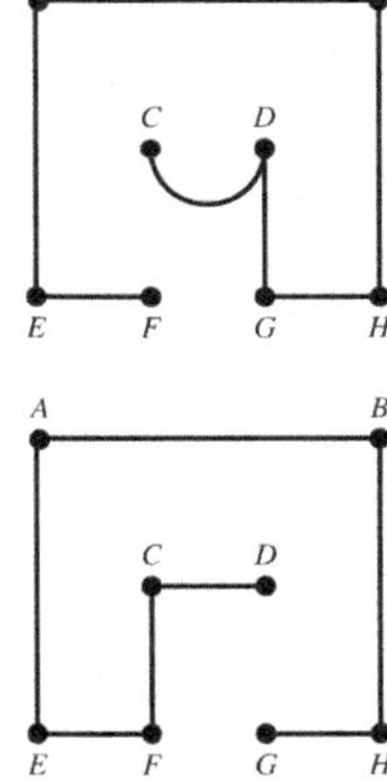

15. Other answers are possible.

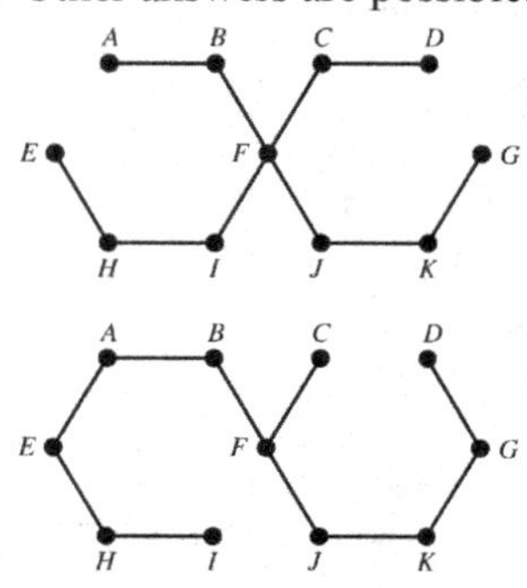

17. Choose edges in the following order.
AB, *AD*, *AC*, *BE*

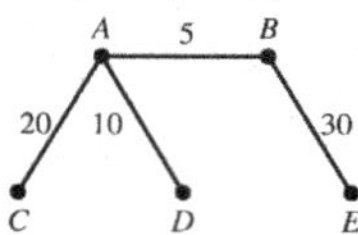

19. Choose edges in the following order.
GB, *BC*, *BA*, *AH*, *DE*, *CF*, *FE*

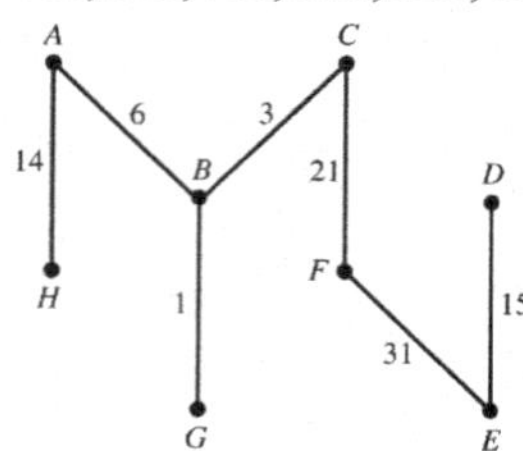

21. Choose edges in the following order.
BC, *CF*, *EB*, *AB*, *DE*

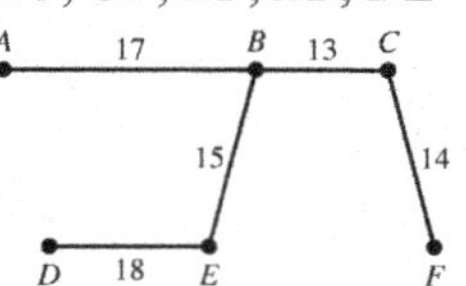

23. Choose edges in the following order.
BE, *FD*, *AH*, *EF*, *FG*, *HE*, *CF*

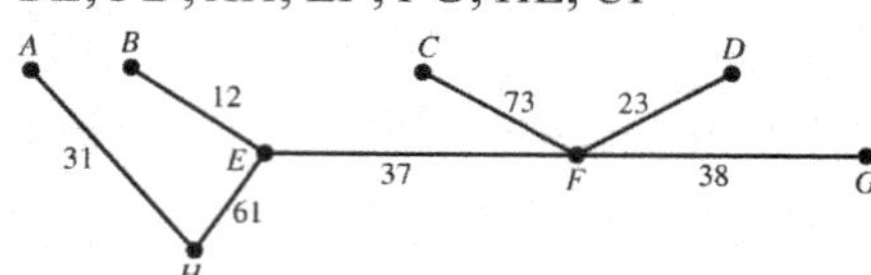

25. a) Other answers are possible.

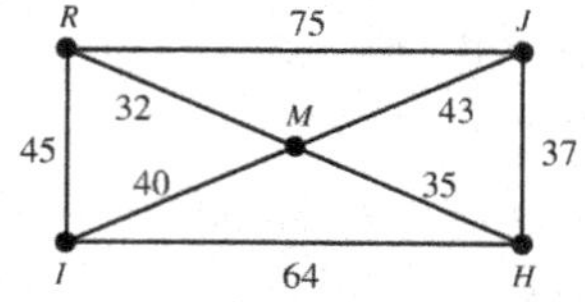

b) Choose edges in the following order.
AD, *EF*, *AB*, *DE*, *BC*

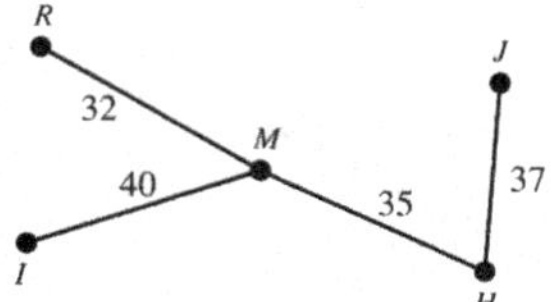

c) $59(32+35+37+40) = 59(144) = \$8496$

27. a) Choose edges in the following order: *HY*, *YL*, *LR*, *RA*, *AP*.

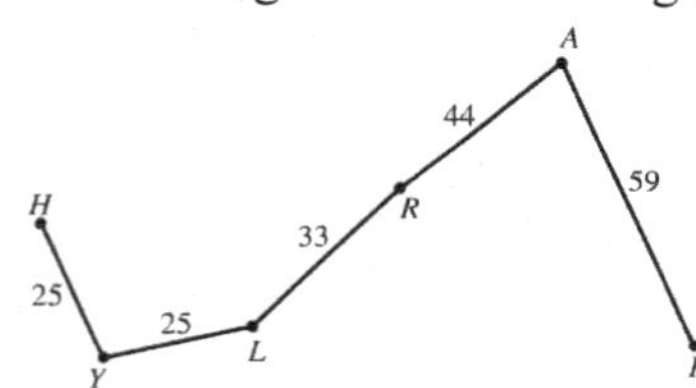

b) $1,300,000(25+25+33+44+59) = 1,300,000(186) = \$241,800,000$

29. a) Choose edges in the following order: *DH*, *HL*, *DS*.

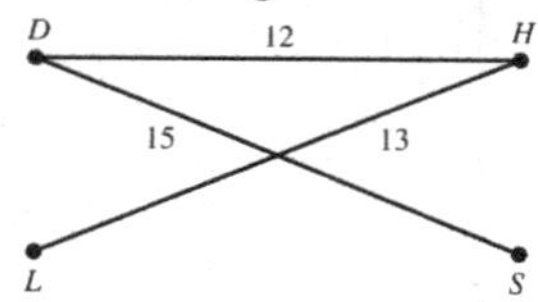

b) $11,500(12+13+15) = 11,500(40) = \$460,000$

31. a) Choose edges in the following order: *KC-SJ*, *JC-SL*, *JC-SP*, *JC-KC*.

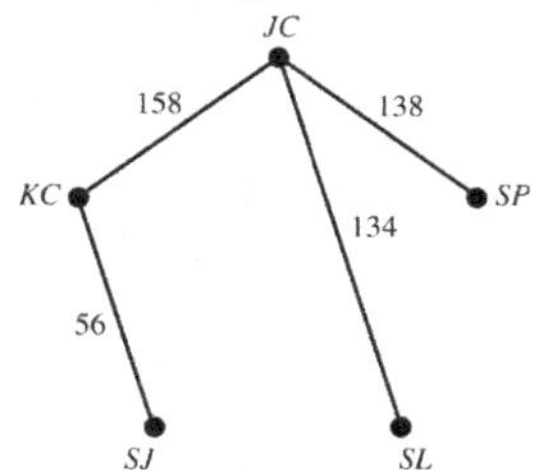

b) $9500(56+134+138+158)=9500(486)=\$4,617,000$

33. Answers will vary.

35. Answers will vary.

Review Exercises

1. Other answers are possible.

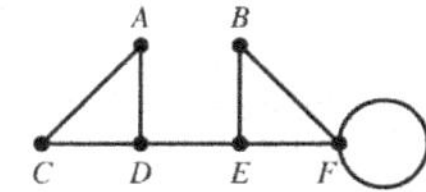

2. Other answers are possible.

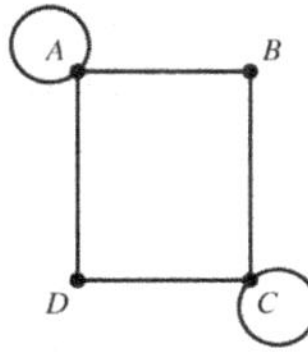

3. *E*, *C*, *D*, *F*, *E*, *G*, *A*, *B*, *H*, *G*; Other answers are possible.

4. No. A path that includes each edge exactly one time would start at vertex *E* and end at vertex *G*, or vice versa.

5.

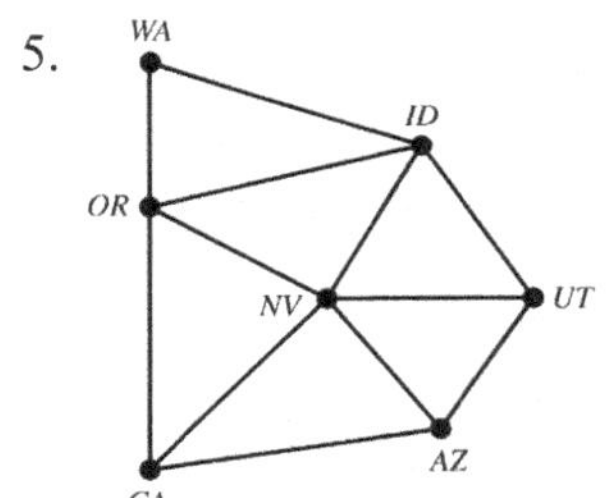

6.

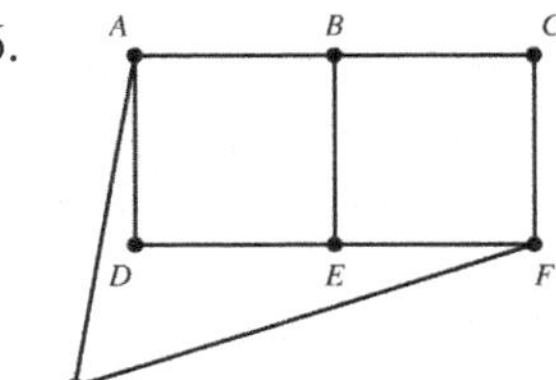

7. Connected

8. Disconnected. There is no path that connects *A* to *C*.

9. Edge *BD*

10. *C*, *A*, *B*, *F*, *H*, *G*, *C*, *D*, *E*, *D*, *E*, *F*; Other answers are possible.

11. *F*, *E*, *D*, *E*, *D*, *C*, *G*, *H*, *F*, *B*, *A*, *C*; Other answers are possible.

12. *A*, *B*, *E*, *G*, *F*, *D*, *C*, *A*, *D*, *E*, *A*; Other answers are possible.

13. *E*, *D*, *C*, *A*, *E*, *B*, *A*, *D*, *F*, *G*, *E*; Other answers are possible.

14. a)

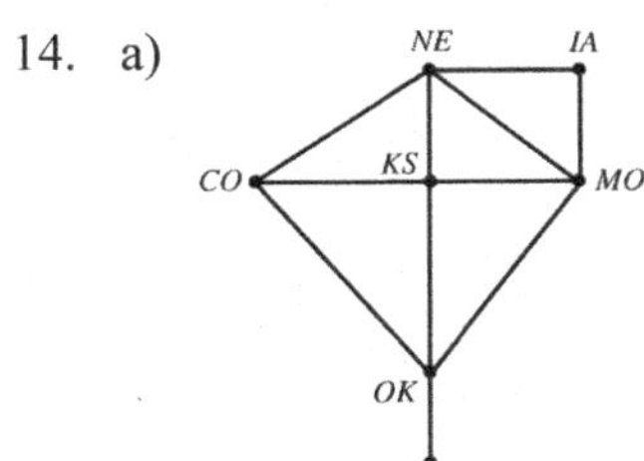

b) Vertices *CO* and *TX* are both odd. According to item 2 of Euler's Theorem, since there are exactly two odd vertices, at least one Euler path, but no Euler circuits exist. Yes; *CO*, *NE*, *IA*, *MO*, *NE*, *KS*, *MO*, *OK*, *CO*, *KS*, *OK*, *TX*; Other answers are possible.

c) No; see part (b)

15. a) 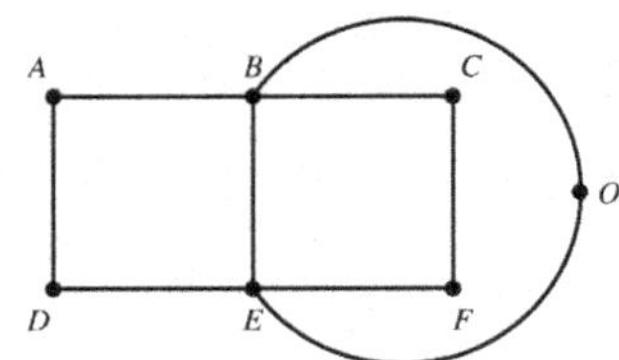

b) Yes; the graph has no odd vertices, so there is at least one Euler path, which is also an Euler circuit.

c) The person may start in any room or outside and will finish where he or she started.

16. a) Yes. The graph representing the map is shown below.

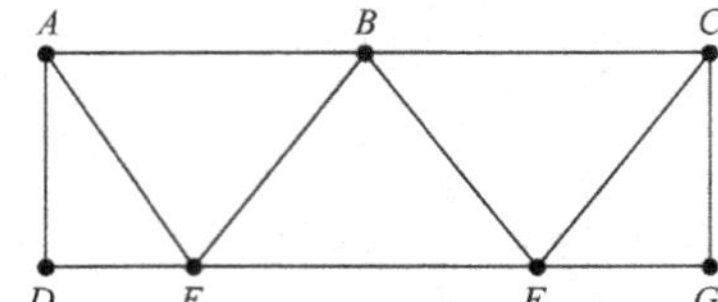

The officer is seeking an Euler path or an Euler circuit. Note that vertices A and C are both odd. According to item 2 of Euler's Theorem, since there are exactly two odd vertices, at least one Euler path but no Euler circuits exist.

b) The officer would have to start at either the intersection of Dayne St., Gibson Pl., and Alvarez Ave. or at the intersection of Chambers St., Fletcher Ct., and Alvarez Ave.

17. *C, A, B, G, F, A, D, C, F, D, B, E, D, G, E*; Other answers are possible.

18. *A, B, D, E, I, J, O, N, L, K, G, H, L, M, I, H, D, C, G, F, A*; Other answers are possible.

19. *C, A, E, K, I, F, G, J, L, H, B, D* and *D, B, H, L, J, G, F, C, A, E, K, I*; Other answers are possible.

20. *A, B, E, F, J, I, L, K, H, G, C, D, A* and *I, J, F, E, B, A, D, C, G, H, K, L, I*; Other answers are possible.

21. 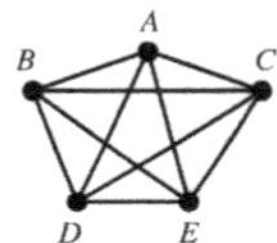

22. The number of unique Hamilton circuits within the complete graph with 5 vertices representing this situation is $(5-1)! = 4! = 4 \cdot 3 \cdot 2 \cdot 1 = 24$ ways

23. a) 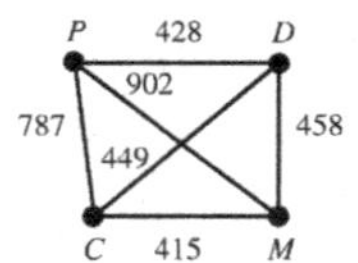

b)

Hamilton Circuit	First Leg Cost	Second Leg Cost	Third Leg Cost	Fourth Leg Cost	Total Cost
P, D, C, M, P	428	449	415	902	$2194
P, D, M, C, P	428	458	415	787	$2088
P, C, M, D, P	787	415	458	428	$2088
P, C, D, M, P	787	449	458	902	$2596
P, M, D, C, P	902	458	449	787	$2596
P, M, C, D, P	902	415	449	428	$2194

The least expensive route is *P, D, M, C, P* or *P, C, M, D, P*.

c) $2088

24. a)

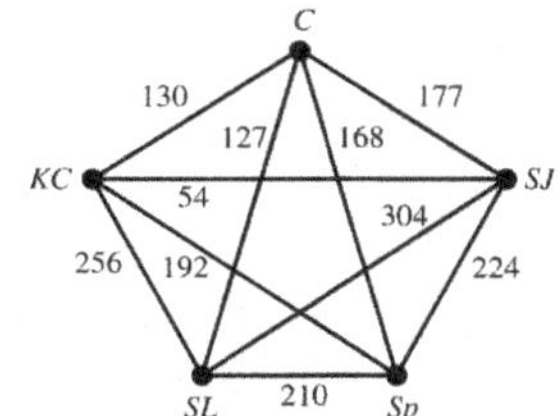

b) *SJ*, *KC*, *C*, *SL*, *Sp*, *SJ*; traveling a total of $54+130+127+210+224=745$ miles

c) *Sp*, *C*, *SL*, *KC*, *SJ*, *Sp*; traveling a total of $168+127+256+54+224=829$ miles

25.

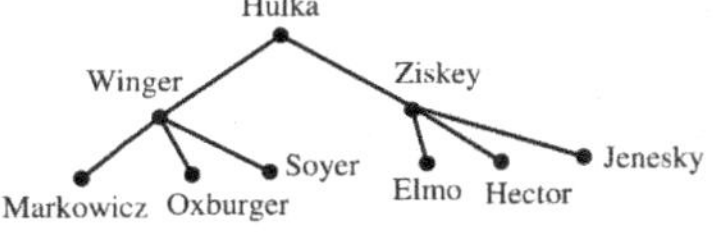

26.

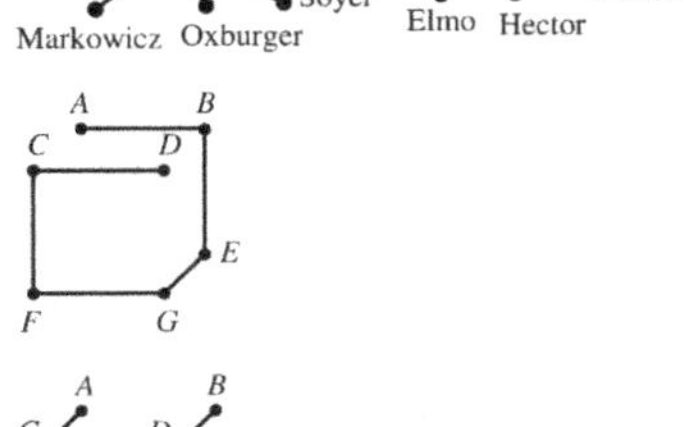

27. Choose edges in the following order. *AB*, *AD*, *DE*, *BC*, *BF*

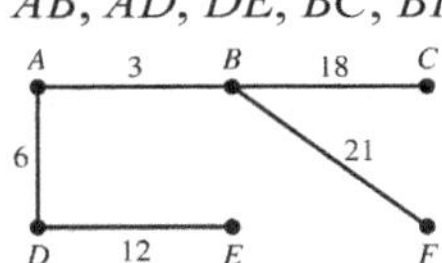

28. a) Choose edges in the following order: *O-GCJ*, *O-PF*, *J-GCJ*, *FA-O*, *GCJ-B*.

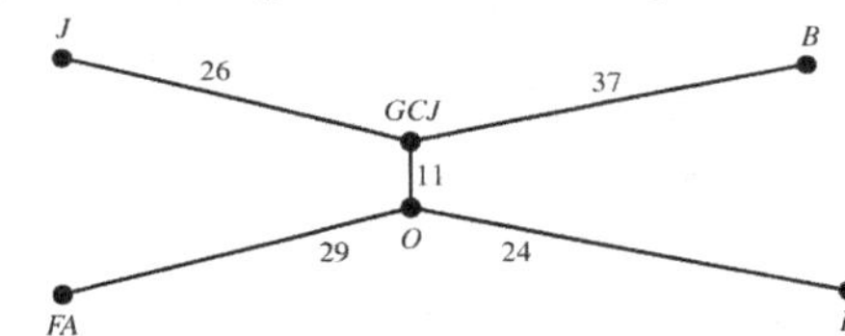

b) $56(26+29+11+37+24)=56(127)=\$7112$

Chapter Test

1. Edge *AB* is a bridge. There is a loop at vertex *G*. Other answers are possible.

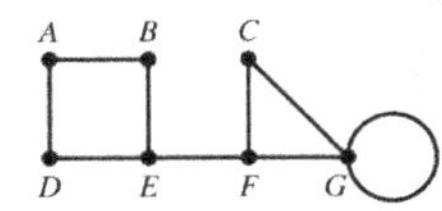

2.

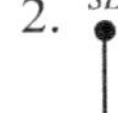
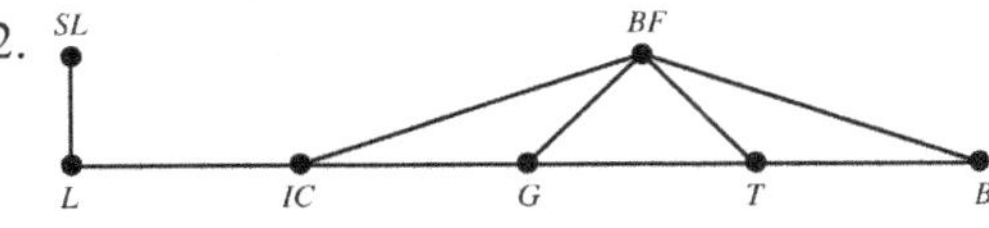

3.

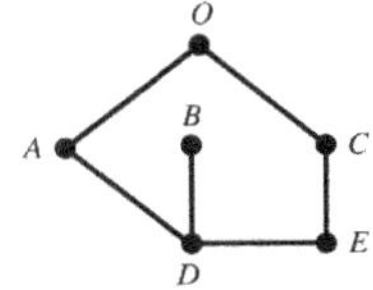

4. Other answers are possible.

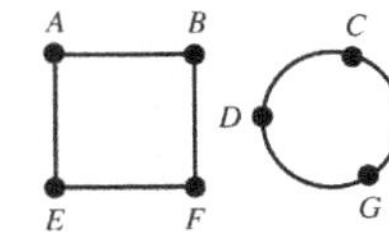

5. *A*, *B*, *E*, *D*, *C*, *A*, *D*, *B*; Other answers are possible.
6. *A*, *B*, *D*, *G*, *H*, *F*, *C*, *B*, *E*, *D*, *F*, *E*, *C*, *A*; Other answers are possible.
7. Yes. The person may start in room *A* and end in room *B* or vice versa.
8. *A*, *D*, *E*, *A*, *F*, *E*, *H*, *F*, *I*, *G*, *F*, *B*, *G*, *C*, *B*, *A*; Other answers are possible.
9. *B*, *A*, *D*, *E*, *F*, *C*, *G*; Other answers are possible.
10. *A*, *B*, *C*, *G*, *E*, *D*, *H*, *I*, *K*, *J*, *F*, *A*; Other answers are possible.

11.

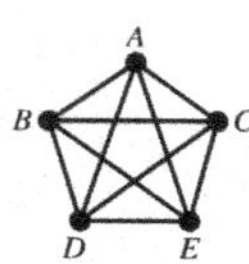

12. The number of unique Hamilton circuits within the complete graph with 6 vertices representing this situation is $(6-1)! = 5! = 5 \cdot 4 \cdot 3 \cdot 2 \cdot 1 = 120$ ways

13.

C 253 N
122
114 199 183
R 112 U

14. *C, R, N, U, C* or *C, U, N, R, C* for $114 + 199 + 183 + 122 = \$618$

15. *C, R, U, N, C* for $114 + 112 + 183 + 253 = \$662$

16.

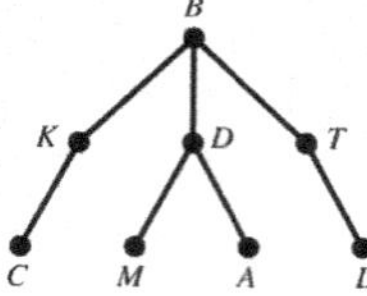

17. Other answers are possible.

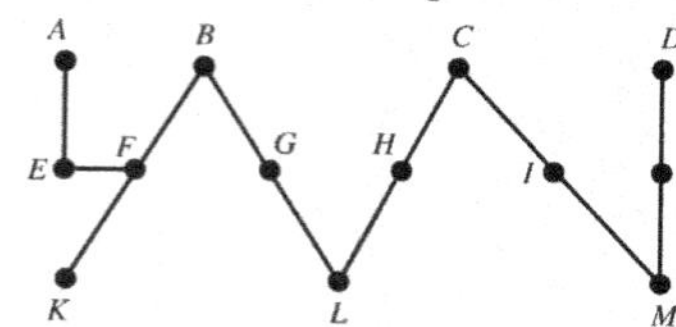

18. Choose edges in the following order.
AB, AC, CE, BF, BD.

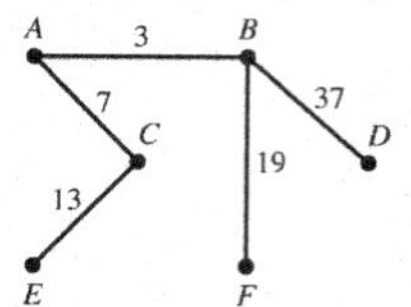

19. Choose edges in the following order.
V2-V4, V3-V4, V4-V5, V1-V2.

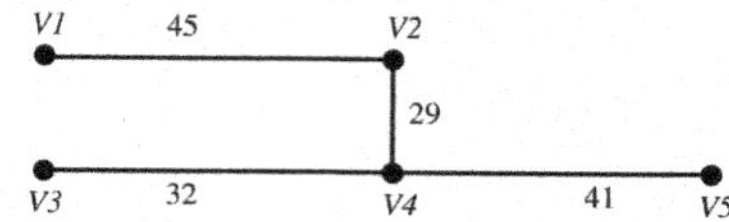

20. $1.25(29 + 32 + 41 + 45) = 1.25(147) = \$183.75$

Chapter Fourteen: Voting and Apportionment

Section 14.1: Voting Methods

1. Majority

3. $\frac{n(n-1)}{2}$

5. Plurality

7. Pairwise comparison

9. a) Li is the winner; he received the most votes using the plurality method.

   b) No. $\frac{102,503}{76,431+91,863+102,503}=\frac{102,503}{270,797}\approx 0.39$ is not a majority.

   A majority is more than 135,399 votes.

11.

Number of votes	3	1	2	2	1
First	B	D	H	H	D
Second	D	B	B	D	H
Third	H	H	D	B	B

13. $10+5+4+2=21$ members voted.

15. Guitar wins with the most votes (10).

17. A majority out of 21 votes is 11 or more votes.

    First choice votes: G: 10, K: 9, N: 2

    None receives a majority, thus N with the least votes is eliminated.

    Second round: G: 10, K: 11

    Keyboard wins with a majority of 11 votes.

19. C: $3(4)+2(2)+1(3)=19$

    N: $3(2)+2(4)+1(3)=17$

    O: $3(3)+2(3)+1(3)=18$

    Chicago wins with 19 points.

21. C (6) vs. N (3); C gets a point.
    C (4) vs. O (5); O gets a point.
    N (5) vs. O (4); N gets a point.

    C has 1 point.
    N has 1 point.
    O has 1 point.

    Three-way tie, there is no winner.

23. First choice votes: H: 9, P: 13, S: 9, T: 1

    Portland wins with the most votes.

25. A majority out of 32 votes is 17 or more votes.

    First choice votes: H: 9, P: 13, S: 9, T: 1

    None receives a majority, thus T with the least votes is eliminated.

    Second round: H: 10, P: 13, S: 9

    No majority, thus eliminate S.

    Third round: H: 16, P: 16

    Since L and H tied, there is no winner.

27. B: $3(5)+2(4)+1(3)=26$

    M: $3(6)+2(1)+1(5)=25$

    S: $3(1)+2(7)+1(4)=21$

    Brownstein wins with 26 points.

29. B (5) vs. M (7); M gets a point.
    B (9) vs. S (3); B gets a point.
    M (6) vs. S (6); M and S each get a half point.

    B has 1 point.
    M has 1.5 points.
    S has 1 point.

    Marquez wins with 1.5 points.

31. Votes: E: 2, L: 5, O: 4.

    Lehigh Road wins with the most votes.

33. A majority out of 11 votes is 6 or more votes.

    First choice votes: L: 5, E: 2, O: 4

    None receives a majority, thus E with the least votes is eliminated.

    Second round: L: 5, O: 6

    Ontario Road wins with a majority of 6 votes.

35. a) First choice votes: B: 10, C: 3, D: 2, S: 0
Bishara wins with the most votes.

b) B: $4(10)+3(5)+2(0)+1(0)=55$
C: $4(3)+3(6)+2(6)+1(0)=42$
D: $4(2)+3(4)+2(0)+1(9)=29$
S: $4(0)+3(0)+2(9)+1(6)=24$
Bishara wins with 55 points.

c) A majority out of 15 votes is 8 or more votes.
First choice votes: B: 10, C: 3, D: 2, S: 0
Because B already has a majority, Bishara wins.

d) B (12) vs. C (3); B gets a point.
B (13) vs. D (2); B gets a point.
B (15) vs. S (0); B gets a point.
C (9) vs. D (6); C gets a point.
C (15) vs. S (0); C gets a point.
D (6) vs. S (9); S gets a point.
B has 3 points.
C has 2 points.
D has 0 points.
S has 1 point.
Bishara wins with 3 points.

37. a) G: $4(8)+3(3)+2(4)+1(14)=63$
I: $4(3)+3(7)+2(19)+1(0)=71$
P: $4(14)+3(8)+2(3)+1(4)=90$
Z: $4(4)+3(11)+2(3)+1(11)=66$
Petunia wins with 91 points.

b) Votes: G: 8, I: 3, P: 14, Z: 4
Petunias win with the most votes.

c) A majority out of 29 votes is 15 or more votes.
First choice votes: G: 8, I: 3, P: 14, Z: 4
None receives a majority, thus I with the least votes is eliminated.
Second round: G: 11, P: 14, Z: 4
No majority, thus eliminate Z.
Third round: G: 15, P: 14
Geraniums win with 15 votes.

d) G (8) vs. I (21); I gets a point.
G (15) vs. P (14); G gets a point.
G (11) vs. Z (18); Z gets a point.
I (7) vs. P (22); P gets a point.
I (14) vs. Z (15); Z gets a point.
P (25) vs. Z (4); P gets a point.
G has 1 point.
I has 1 point.
P has 2 points.
Z has 2 points.
Petunias and Zinnias tie with 2 points.

39. By ranking their choices, voters are able to provide more information with the Borda count method.

41. A majority out of 12 votes is 7 or more votes.
Most last place votes: B: 3, S: 4, M: 5
Thus, M with the most last place votes is eliminated.
Second round using the most last place votes: B: 3, S: 9
Brownstein wins with the least last place votes.

43. a) If there were only two columns then only two of the candidates were the first choice of the voters. If each of the 15 voters cast a ballot, then one of the voters must have received a majority of votes because 15 cannot be split evenly.

b) An odd number cannot be divided evenly so one of the two first choice candidates must receive more than half of the votes.

45. a) C: $4+1+1=6$
R: $4+4+3=11$
W: $3+3+2+2+1+1=12$
T: $4+3+2+2=11$
The Wildcats finished 1st, the Rams and the Tigers tied for 2nd, and the Comets were 4th.

b) C: $5+0=5$
R: $5+5+3=13$
W: $3+3+1+1+0+0=8$
T: $5+3+1+1=10$
Rams finished 1st, Tigers finished 2nd, Wildcats finished 3rd, and Comets finished 4th.

47. a) Each voter assigns $4+3+2+1=10$ points: $(15)(10)=150$ points.

    b) $150-(35+40+25)=50$ points

    c) Yes. Candidate D has more points than each of the other 3 candidates.

49. One possible answer is shown below.

Number of votes	5	5	2
First	A	B	C
Second	B	A	B
Third	C	C	A

Section 14.2: Flaws of the Voting Methods

1. Majority
3. Head-to-head
5. Borda count
7. Plurality with elimination
9. The plurality method yields Orlando as the winner with a majority of 10 first place votes.

   N: $3(9)+2(10)+1(0)=47$

   O: $3(10)+2(4)+1(5)=43$

   P: $3(0)+2(5)+1(14)=24$

   The winner is New Orleans using the Borda count method, thus violating the majority criterion.

11. a) A (6) vs. B (5); A gets a point.
    A (6) vs. C (5); A gets a point.
    B (6) vs. C (5); B gets a point.

    A has 2 points.
    B has 1 point.
    C has 0 points.

    Plan A wins all head-to-head comparisons.

    b) Plan C wins by a plurality of 5 votes. No, the head-to-head criterion is not satisfied.

13. L: $3(3)+2(5)+1(1)=20$

    P: $3(4)+2(2)+1(3)=19$

    S: $3(2)+2(2)+1(5)=15$

    L wins with 20 points.

    L (4) vs. P (5); P gets a point.
    L (7) vs. S (2); L gets a point.
    P (5) vs. S (4); P gets a point.

    L has 1 point.
    P has 2 points.
    S has 0 points.

    Plan P wins all head-to-head comparisons.

    Because parking wins its head-to-head comparisons and lounge areas win by Borda count method, the head-to-head criterion is not satisfied.

15. A majority out of 27 votes is 14 or more votes.

    Plurality with elimination:

    First choice votes: A: 11, B: 3, C: 8, D: 5

    None receives a majority, thus B with the least votes is eliminated.

    Second round: A: 11, C: 11, D: 5

    Still no majority, thus eliminate D.

    Third round: A: 11, C: 16

    C wins with a majority of 16 votes.

    Pairwise comparison:

    A (11) vs. B (16); B gets a point.
    A (11) vs. C (16); C gets a point.
    A (22) vs. D (5); A gets a point.
    B (19) vs. C (8); B gets a point.
    B (22) vs. D (5); B gets a point.
    C (11) vs. D (16); D gets a point.

    A has 1 point.
    B has 3 points.
    C has 1 point.
    D has 1 point.

    B wins with 3 points.

    C wins by plurality with elimination, but B is favored over each of the other candidates using head-to-head comparisons. Therefore, the head-to-head criterion is not satisfied.

17. First place votes: A: 10, B: 4, C: 6; A wins.
If B drops out, we get the following:
First place votes: A: 10, C: 11; C wins.
The irrelevant alternatives criterion is not satisfied.

19. A: $3(8)+2(10)+1(9)=53$
B: $3(10)+2(9)+1(8)=56$
C: $3(9)+2(8)+1(10)=53$
B wins using the Borda count method.
After C drops out:
A: $2(8)+1(19)=35$
B: $2(19)+1(8)=46$
Thus, B still wins. The irrelevant alternatives criterion is satisfied.

21. A majority out of 29 voters is 15 or more votes.
First place votes: A: 11, B: 8, C: 10; none has a majority, thus eliminate B.
Second round votes: A: 11, C: 18, thus C wins.
If the three voters who voted for A, C, B change to C, A, B, the new set of votes becomes:
First place votes: A: 7, B: 8, C: 14; none has a majority, thus eliminate A.
Second round votes: B: 15, C:14, thus B wins.
Thus, the monotonicity criterion is not satisfied.

23. Original preference table:
A (13) vs. B (13); A and B each get a half point.
A (13) vs. C (13); A and C each get a half point.
A (13) vs. D (13); A and D each get a half point.
B (13) vs. C (13); B and C each get a half point.
B (13) vs. D (13); B and D each get a half point.
C (12) vs. D (14); D gets a point.
A has 1.5 points.
B has 1.5 points.
C has 1 point.
D has 2 points.
D wins with 2 pts.

After the change in votes:
A (8) vs. B (18); B gets a point.
A (13) vs. C (13); A and C each get a half point.
A (13) vs. D (13); A and D each get a half point.
B (18) vs. C (8); B gets a point.
B (13) vs. D (13); B and D each get a half point.
C (7) vs. D (19); D gets a point.
A has 1 point.
B has 2.5 points.
C has 1 point.
D has 2 points.
B wins with 2.5 pts.

The monotonicity criterion is not satisfied.

25. A (1) vs. B (4); B gets a point.
A (3) vs. C (2); A gets a point.
A (3) vs. D (2); A gets a point.
A (2) vs. E (3); E gets a point.
B (4) vs. C (1); B gets a point.
B (2) vs. D (3); D gets a point.
B (4) vs. E (1); B gets a point.
C (3) vs. D (2); C gets a point.
C (4) vs. E (1); C gets a point.
D (2) vs. E (3); E gets a point.

A has 2 points.
B has 3 points.
C has 2 points.
D has 1 point.
E has 2 points.
B wins by pairwise comparison.

After A, C and E drop out, the new set of first place votes is: B: 2 and D: 3, thus D wins.
The irrelevant alternatives criterion is not satisfied.

27. Total votes: $4+2+1=7$
A wins with a majority of 4 votes.
A: $3(4)+2(0)+1(3)=15$
B: $3(2)+2(5)+1(0)=16$
C: $3(1)+2(2)+1(4)=11$
B wins with 21 points. The majority criterion is not satisfied.

29. a) Savannah holds a majority with 12 out of 23 votes.
    b) Savannah will win the plurality method since it holds a majority.
    c) B: $4(3)+3(0)+2(8)+1(12)=40$
       C: $4(0)+3(20)+2(0)+1(3)=63$
       P: $4(8)+3(0)+2(15)+1(0)=62$
       S: $4(12)+3(3)+2(0)+1(8)=65$
       Savannah wins.
    d) Savannah will win the plurality with elimination since it holds a majority.
    e) B (3) vs. C (20); C gets a point.
       B (3) vs. P (20); P gets a point.
       B (11) vs. S (12); S gets a point.
       C (12) vs. P (11); C gets a point.
       C (8) vs. S (15); S gets a point.
       P (8) vs. S (15); S gets a point.
       B has 0 points.
       C has 2 points.
       P has 1 point.
       S has 3 points.
       Savannah wins.
    f) None of them violate the majority criterion.

31. a) A majority out of 82 votes is 42 or more votes.
       First choice votes: C: 24, H: 28, L: 30; None receives a majority, thus C is eliminated.
       Second round: H: 52, L: 30; Steve Harvey is selected.
    b) First choice votes: C: 24, H: 36, L: 22; None receives a majority, thus L is eliminated.
       Second round: C: 46, H: 36; Bradley Cooper is selected.
    c) Yes, the monotonicity criterion is violated.

33. A candidate who holds a plurality will only gain strength and holds an even larger lead if more favorable votes are added.

35. Answers will vary.

37. Answers will vary.

Section 14.3: Apportionment Methods

1. Divisor

3. Upper

5. Quota

7. Hamilton's

9. a) Webster's
   b) Adams'
   c) Jefferson's

11. a) Standard divisor: $\dfrac{7,500,000}{150}=50,000$
    b)

State	A	B	C	D	Total
Population	1,220,000	2,730,000	857,000	2,693,000	7,500,000
Standard Quota	24.40	54.60	17.14	53.86	150.00

13. a) and b) Modified divisor: 49,300

State	A	B	C	D	Total
Population	1,220,000	2,730,000	857,000	2,693,000	7,500,000
Modified quota	24.75	55.38	17.38	54.62	152.13
Modified lower quota	24	55	17	54	150

15. a) and b) Modified divisor: 50,700

State	A	B	C	D	Total
Population	1,220,000	2,730,000	857,000	2,693,000	7,500,000
Modified quota	24.06	53.85	16.90	53.12	147.93
Modified upper quota	25	54	17	54	150

17. Standard divisor: 50,000

State	A	B	C	D	Total
Population	1,220,000	2,730,000	857,000	2,693,000	7,500,000
Standard Quota	24.40	54.60	17.14	53.86	150.00
Modified quota	24.40	54.60	17.14	53.86	148.00
Modified rounded quota	24	55	17	54	150

19. a) Standard divisor: $\dfrac{675}{25} = 27$

b)

Hotel	A	B	C	Total
Number of rooms	306	214	155	675
Standard Quota	11.33	7.93	5.74	25.00

21. a) and b)

Hotel	A	B	C	Total
Number of rooms	306	214	155	675
Modified quota	11.86	8.29	6.01	26.16
Modified lower quota	11	8	6	25

23. a) and b)

Hotel	A	B	C	Total
Number of rooms	306	214	155	675
Modified quota	10.55	7.38	5.34	23.27
Modified upper quota	11	8	6	25

25.

Hotel	A	B	C	Total
Number of rooms	306	214	155	675
Standard Quota	11.33	7.93	5.74	25.00
Modified quota	11.33	7.93	5.74	23.00
Modified rounded quota	11	8	6	25

27. a) Standard divisor: $\dfrac{550}{50} = 11$

b) and c)

Resort	A	B	C	D	Total
Number of rooms	86	102	130	232	550
Standard Quota	7.82	9.27	11.82	21.09	50.00
Lower quota	7	9	11	21	48
Hamilton's apportionment	8	9	12	21	50

29. Modified divisor: 11.5

Resort	A	B	C	D	Total
Number of rooms	86	102	130	232	550
Modified quota	7.48	8.87	11.30	20.17	47.82
Modified upper quota	8	9	12	21	50

31. a) Standard divisor: $\frac{13,000}{250} = 52$

b) and c)

School	A & S	Business	Engineering	Education	V & P Arts	Total
Enrollment	1746	7,95	2131	937	1091	13,000
Standard Quota	33.58	136.44	40.98	18.02	20.98	250.00
Lower quota	33	136	40	18	20	247
Hamilton's apportionment	34	136	41	18	21	250

33. Modified divisor: 51.5

School	A & S	Business	Engineering	Education	V & P Arts	Total
Enrollment	1746	7095	2131	937	1091	13,000
Modified quota	33.90	137.77	41.38	18.19	21.18	252.42
Modified lower quota	33	137	41	18	21	250

35. a) Standard divisor: $\frac{10,800}{120} = 90$

b) and c)

Dealership	A	B	C	D	Total
Sales	3840	2886	2392	1682	10,800
Standard Quota	42.67	32.07	26.58	18.69	120.01
Lower quota	42	32	26	18	118
Hamilton's apportionment	43	32	26	19	120

37. A divisor of 90.3 was used.

Dealership	A	B	C	D	Total
Sales	3840	2886	2392	1682	10,800
Standard Quota	42.67	32.07	26.58	18.69	120.01
Modified quota	42.52	31.96	26.49	18.63	118.00
Modified rounded quota	43	32	26	19	120

39. a) Standard divisor: $\frac{75,000}{100} = 750$

b) and c)

Precinct	A	B	C	D	E	F	Total
Population	9070	15,275	12,810	5720	25,250	6875	75,000
Standard Quota	12.09	20.37	17.08	7.63	33.67	9.17	90.84
Lower quota	12	20	17	7	33	9	89
Hamilton's apportionment	12	20	17	8	34	34	91

41. The divisor 765 was used.

Precinct	A	B	C	D	E	F	Total
Population	9070	15,275	12,810	5720	25,250	6875	75,000
Modified quota	11.86	19.97	16.75	7.48	33.01	8.99	89.07
Modified upper quota	12	20	17	8	34	9	91

43. a) Standard divisor: $\frac{2400}{200} = 12$

b) and c)

Shift	A	B	C	D	Total
Room Calls	751	980	503	166	2400
Standard Quota	62.58	81.67	41.92	13.83	200.00
Lower quota	62	81	41	13	197
Hamilton's apportionment	62	82	42	14	200

45. The divisor 11.9 was used.

Shift	A	B	C	D	Total
Room Calls	751	980	503	166	2400
Modified quota	63.11	82.35	42.27	13.95	201.68
Modified lower quota	63	82	42	13	200

47. Standard divisor: $\frac{3,615,920}{105} = 34,437.33$

a) Hamilton's Apportionment: 7, 2, 2, 2, 8, 14, 4, 5, 10, 10, 13, 2, 6, 2, 18

b) Jefferson's Apportionment: 7, 1, 2, 2, 8, 14, 4, 5, 10, 10, 13, 2, 6, 2, 19

c) States that benefited: Virginia; States Disadvantaged: Delaware

49. Answers will vary. One possible answer is A; 743, B: 367, C: 432, D: 491, E: 519, F: 388

Section 14.4: Flaws of the Apportionment Methods

1. Population
3. Alabama
5. Small

7. New divisor: $\frac{1080}{61} \approx 17.70$

Clinic	A	B	C	D	E	Total
Patients	246	201	196	211	226	1080
Standard Quota	13.90	11.36	11.07	11.92	12.77	61.02
Lower quota	13	11	11	11	12	58
Hamilton's apportionment	14	11	11	12	13	61

No. No clinic suffers a loss so the Alabama paradox does not occur.

9. a) Standard divisor: $\frac{900}{30} = 30$

State	A	B	C	Total
Population	161	250	489	900
Standard Quota	5.37	8.33	16.30	30.00
Lower quota	5	8	16	29
Hamilton's apportionment	6	8	16	30

b) New divisor: $\frac{900}{31} \approx 29.03$

State	A	B	C	Total
Population	161	250	489	900
Standard Quota	5.55	8.61	16.84	31.00
Lower quota	5	8	16	29
Hamilton's apportionment	5	9	17	31

Yes. When the number of seats increases, states B and C gain a seat and state A loses a seat.

11. a) Standard divisor: $\frac{30,000}{200} = 150$

City	A	B	C	Total
Employees	9130	6030	14,840	30,000
Standard Quota	60.87	40.20	98.93	200.00
Lower quota	60.00	40.00	98.00	198.00
Hamilton's apportionment	61	40	99	200

b) New divisor: $\frac{30,125}{200} \approx 150.63$

City	A	B	C	Total
Employees	9150	6030	14,945	30,125
Standard Quota	60.74	40.03	99.22	199.99
Lower quota	60.00	40.00	99.00	199.00
Hamilton's apportionment	61	40	99	200

No. None of the cities loses a promotion.

13. a) Standard divisor: $\frac{5400}{54} = 100$

Division	A	B	C	D	E	Total
Population	733	1538	933	1133	1063	5400
Standard Quota	7.33	15.38	9.33	11.33	10.63	54.00
Lower quota	7.00	15.00	9.00	11.00	10.00	52.00
Hamilton's apportionment	7	16	9	11	11	54

b) New divisor: $\frac{5454}{54} = 101$

Division	A	B	C	D	E	Total
Population	733	1539	933	1133	1116	5454
Standard Quota	7.26	15.24	9.24	11.22	11.05	54.01
Lower quota	7.00	15.00	9.00	11.00	11.00	53.00
Hamilton's apportionment	8	15	9	11	11	54

13. (continued)

Yes. Division B loses a printer to Division A even though the population of division B grew faster than the population of division A.

15. a) Standard divisor: $\frac{4800}{48} = 100$

Cynergy Telecom.	A	B	Total
Employees	844	3956	4800
Standard Quota	8.44	39.56	48.00
Lower quota	8.00	39.00	47.00
Hamilton's apportionment	8	40	48

b) New divisor: $\frac{5524}{55} \approx 100.44$

Cynergy Telecom.	A	B	C	Total
Employees	844	3956	724	5524
Standard Quota	8.40	39.39	7.21	55.00
Lower quota	8.00	39.00	7.00	54.00
Hamilton's apportionment	9	39	7	55

Yes. Group B loses a manager.

17. a) Standard divisor: $\frac{3300}{33} = 100$

State	A	B	Total
Population	744	2556	3300
Standard Quota	7.44	25.56	33.00
Lower quota	7.00	25.00	32.00
Hamilton's apportionment	7	26	33

b) New divisor: $\frac{4010}{40} = 100.25$

State	A	B	C	Total
Population	744	2556	710	4010
Standard Quota	7.42	25.50	7.08	40.00
Lower quota	7.00	25.00	7.00	39.00
Hamilton's apportionment	7	26	7	40

No. The apportionment is the same.

Review Exercises

1. a) Comstock wins with the most votes (20).
   b) A majority out of 42 voters is 22 or more votes. Comstock does not have a majority.
2. a) Michelle wins with the most votes (231).
   b) Yes. A majority out of 413 voters is 207 or more votes.

3.

Number of Votes	3	2	1	1	3
First	B	A	D	D	C
Second	A	C	C	A	B
Third	C	D	A	B	A
Fourth	D	B	B	C	D

4.

Number of Votes	2	2	2	1
First	C	A	B	C
Second	A	C	A	B
Third	B	B	C	A

5. Number of votes: $5+3+1+2=11$

6. Chipotle Mexican Grill wins with a plurality vote of 5.

7. B: $4(2)+3(4)+2(5)+1(0)=30$

   C: $4(5)+3(2)+2(1)+1(3)=31$

   D: $4(4)+3(5)+2(0)+1(2)=33$

   J: $4(0)+3(0)+2(5)+1(6)=16$

   Domino's Pizza wins with 33 points.

8. A majority out 11 voters is 6 or more votes.

   First place votes: B: 2, C: 5, D: 4

   None has a majority, thus eliminate B.

   Second round votes: C: 7, D: 4

   Chipotle Mexican Grill wins.

9. B (6) vs. C (5); B gets a point.
   B (2) vs. D (9); D gets a point.
   B (11) vs. J (0); B gets a point.
   C (7) vs. D (4); C gets a point.
   C (8) vs. J (3); C gets a point.
   D (9) vs. J (2); D gets a point.

   B has 2 points.
   C has 2 points.
   D has 2 points.
   J has 0 points.

   There is a three-way tie. There is no winner.

10. First round last place votes: Votes: J: 6, C: 3, D: 2, B: 0, so J is eliminated.
    Second round last place votes: Votes: B: 5, C: 4, D: 2 so B is eliminated.
    Third round first place votes: Votes: C: 7, D: 4

    Chipotle Mexican Grill wins.

11. $38+30+25+7+10=110$ students voted

12. Volleyball wins with a plurality of 40 votes.

13. B: $3(32)+2(48)+1(30)=222$

    S: $3(38)+2(37)+1(35)=223$

    V: $3(40)+2(25)+1(45)=215$

    Soccer wins.

14. A majority out of 110 voters is 56 or more votes.

    First round votes: B: 32, S: 38, V: 40

    None has a majority, thus eliminate B.

    Second round: S: 45, V: 65

    Volleyball wins.

15. B (42) vs. S (68); S gets a point.
    B (70) vs. V (40); B gets a point.
    S (45) vs. V (65); V gets a point.

    B has 1 point.
    S has 1 point.
    V has 1 point.

    There is a three-way tie. There is no winner.

16. A majority out of 110 voters is 56 or more votes.

    First place votes: B: 32, S: 38, V: 40; None has a majority, thus eliminate V with the most last place votes.

    Second round: S: 68, B: 42.

    Soccer wins.

17. a) Yes. A majority out of 372 voters is 187 or more votes. American Music receives a majority.

    b) First place votes: A: 295, T: 65, W: 12, S: 0; American Music wins.

    c) A: $4(295)+3(65)+2(0)+1(12)=1387$

       S: $4(0)+3(173)+2(134)+1(65)=852$

       T: $4(65)+3(0)+2(173)+1(134)=740$

       W: $4(12)+3(134)+2(65)+1(161)=741$

       American Music wins.

17. (continued)

d) 187 or more votes is needed for a majority.

First place votes: A: 295, T: 65, W: 12, S: 0

American Music wins.

e) A (360) vs. S (12); A gets a point.
A (295) vs. T (77); A gets a point.
A (360) vs. W (12); A gets a point.
S (307) vs. T (65); S gets a point.
S (161) vs. W (211); W gets a point.
T (226) vs. W (146); T gets a point.

A has 3 points.
S has 1 point.
T has 1 point.
W has 1 point.

American Music wins.

18. a) A majority out of 200 voters is 101 or more votes.

First place votes: C: 25, S: 80, D: 45, L: 50; None of the cities has a majority.

b) First place votes: C: 25, S: 80, D: 45, L: 50; Seattle wins.

c) C: $4(30)+3(45)+2(115)+1(10)=495$

D: $4(45)+3(65)+2(30)+1(60)=495$

L: $4(55)+3(90)+2(55)+1(0)=600$

S: $4(70)+3(0)+2(0)+1(130)=410$

Las Vegas wins.

d) A majority out of 200 voters is 101 or more votes.

First place votes: C: 30, D: 45, L: 55, S: 70; No city has a majority so eliminate C.

Second round: D: 45, L: 85, S: 70; no city has a majority so eliminate D.

Third round: L: 130, S: 70; Las Vegas wins.

e) C (90) vs. D (110); D gets a point.
C (75) vs. L (125); L gets a point.
C (130) vs. S (70); C gets a point.
D (55) vs. L (145); L gets a point.
D (130) vs. S (70); D gets a point.
L (130) vs. S (70); L gets a point.

C has 1 point.
D has 2 points.
L has 3 points.
S has 0 points.

Las Vegas wins with 3 points.

19. a) A majority out of 16 voters is 9 or more votes.

First place votes: A: 0, C: 7, F: 2, W: 7; None has a majority, thus eliminate A.

Second round: C: 7, F: 2, W: 7; None has a majority, thus eliminate F.

Third round: C: 8, W: 8

Where on Earth is Carmen San Diego and *Chronicles of Narnia* tie.

b) Use the Borda count method to break the tie.

A: $4(0)+3(10)+2(1)+1(5)=37$

C: $4(7)+3(1)+2(7)+1(1)=46$

F: $4(2)+3(1)+2(3)+1(10)=27$

W: $4(7)+3(4)+2(5)+1(0)=50$

Where on Earth is Carmen San Diego wins.

c) A (7) vs. C (9); C gets a point.
A (10) vs. F (6); A gets a point.
A (4) vs. W (12); W gets a point.
C (13) vs. F (3); C gets a point.
C (8) vs. W (8); C and W each get a half point.
F (2) vs. W (14); W gets a point.

A has 1 point.
C has 2.5 points.
F has 0 points.
W has 2.5 points.

W wins with 3 points.

Where on Earth is Carmen San Diego and *Chronicles of Narnia* tie.

20. A: $4(10)+3(0)+2(0)+1(9)=49$

B: $4(2)+3(17)+2(0)+1(0)=59$

C: $4(7)+3(0)+2(12)+1(0)=52$

D: $4(0)+3(2)+2(7)+1(10)=30$

Using the Borda count, candidate B wins. However, A has a majority of first place votes, thus the majority criterion is not satisfied.

21. A wins all its head-to-head comparisons, but B wins using the Borda count method. The head-to-head criterion is not satisfied.

22. a) A majority out of 50 voters is 26 or more votes.

First place votes: A: 14, B: 20, C: 16; None has the majority, thus eliminate A.

Second round: B: 20, C: 30; C wins.

b) First place votes: A: 14, B: 12, C: 24; None has a majority, thus eliminate B.

Second round: A: 26, C: 24; A wins.

When the order is changed A wins. Therefore, the monotonicity criterion is not satisfied.

c) First place votes: A: 26, C: 24; A wins.

Since C won the first election and then after B dropped out A won, the irrelevant alternatives criterion is not satisfied.

23. a) Ragu wins again all other brands.

b) Prego with a plurality of 34.

c) B: $4(33)+3(47)+2(0)+1(34)=307$

N: $4(23)+3(33)+2(58)+1(0)=307$

P: $4(34)+3(0)+2(21)+1(59)=237$

R: $4(24)+3(34)+2(35)+1(21)=289$

Newman's Own and Barilla are tied.

d) A majority out of 114 voters is 58 or more votes.

First place votes: B: 33, N: 23, P: 34, R: 24

None has a majority, thus eliminate N.

Second round: B: 56, P: 34, R: 24

None has a majority, thus eliminate R.

Third round: B: 80, P: 34; B wins.

e) B (57) vs. N (57); B and N each get a half point.
B (80) vs. P (34); B gets a point.
B (56) vs. R (58); R gets a point.
N (80) vs. P (34); N gets a point.
N (56) vs. R (58); R gets a point.
P (55) vs. R (59); R gets a point.

B has 1.5 points.
N has 1.5 points.
P has 0 points.
R has 3 points.

Ragu wins

f) The plurality method, Borda count method, and plurality with elimination method all violate the head-to-head criterion.

24. a) Yes, Fleetwood Mac is favored when compared to each of the other bands.

b) First place votes: B: 34, F: 25, J: 13, R: 15; Boston wins.

c) B: $4(34)+3(0)+2(9)+1(44)=198$

F: $4(25)+3(34)+2(19)+1(9)=249$

J: $4(13)+3(40)+2(0)+1(34)=206$

R: $4(15)+3(13)+2(59)+1(0)=217$

Fleetwood Mac wins.

d) A majority out of 87 voters is 44 or more votes.

First place votes: B: 34, F: 25, J: 13, R: 15; None has a majority, thus eliminate J.

Second round: B: 34, F: 25, R: 28; None has a majority, thus eliminate F.

Third round: B: 34, R: 53; Boston wins.

24. (continued)

e) B (43) vs. F (44); F gets a point.
B (34) vs. J (53); J gets a point.
B (34) vs. R (53); R gets a point.
F (59) vs. J (28); F gets a point.
F (59) vs. R (28); F gets a point.
J (38) vs. R (49); R gets a point.

B has 0 points.
F has 3 points.
J has 1 point.
R has 2 points.

Fleetwood Mac wins.

f) The plurality and plurality with elimination methods violate the head-to-head criterion.

25. A majority out of 70 votes is 36 or more votes, which A has.

Using the Borda count method:

A: $4(36)+3(0)+2(26)+1(8)=204$

B: $4(20)+3(50)+2(0)+1(0)=230$

C: $4(8)+3(20)+2(0)+1(42)=134$

D: $4(6)+3(0)+2(44)+1(20)=132$

B wins with 230 points. The majority criterion is not satisfied by the Borda count method.

26. A majority out of 82 votes is 42 or more votes.

First election:

Plurality with elimination:

First place votes: A: 28, B: 24, C: 30

None has a majority, thus eliminate B.

Second round: A: 52, C: 30; A wins.

Borda Count:

A: $3(28)+2(34)+1(20)=172$

B: $3(24)+2(20)+1(38)=150$

C: $3(30)+2(28)+1(24)=170$

A wins.

Pairwise comparison:

A (38) vs. B (44); B gets a point.
A (52) vs. C (30); A gets a point.
B (24) vs. C (58); C gets a point.

A has 1 point.
B has 1 point.
C has 1 point.

Three-way tie, no winner.

Second election:

Plurality with elimination:

First place votes: A: 36, B: 24, C: 22

None has a majority, thus eliminate C.

Second round: A: 38, C: 44; B wins.

Borda Count:

A: $3(36)+2(26)+1(20)=180$

B: $3(24)+2(20)+1(38)=150$

C: $3(22)+2(36)+1(24)=162$

A wins.

Pairwise comparison:

A (38) vs. B (44); B gets a point.
A (60) vs. C (22); A gets a point.
B (24) vs. C (58); C gets a point.

A has 1 point.
B has 1 point.
C has 1 point.

Three-way tie, no winner.

The plurality with elimination method does not satisfy the monotonicity criterion.

27. Election including candidate D:

Plurality Method:

B wins with 40 first place votes.

Plurality with elimination:

A majority out of 89 voters is 45 or more votes.

First place votes: A: 8, B: 40, C: 16, D: 9, E: 16

None has a majority, thus eliminate A.

Second round: B: 40, C: 16, D: 9, E: 24

None has a majority, thus eliminate D.

Third round: B: 44, C: 16, E: 29

None has a majority, thus eliminate C.

Fourth round: B: 44, E: 45; E wins.

27. (continued)

The Borda count :

A: $5(8)+4(56)+3(9)+2(0)+1(16)=307$

B: $5(40)+4(4)+3(0)+2(29)+1(16)=290$

C: $5(16)+4(0)+3(16)+2(16)+1(41)=201$

D: $5(9)+4(16)+3(32)+2(16)+1(16)=253$

E: $5(16)+4(13)+3(32)+2(28)+1(0)=284$

A wins with 307 points.

Pairwise comparisons:

A (29) vs. B (60); B gets a point.
A (73) vs. C (16); A gets a point.
A (64) vs. D (25); A gets a point.
A (52) vs. E (37); A gets a point.
B (57) vs. C (32); B gets a point.
B (40) vs. D (49); D gets a point.
B (44) vs. E (45); E gets a point.
C (48) vs. D (41); C gets a point.
C (16) vs. E (73); E gets a point.
D (49) vs. E (40); D gets a point.

A has 3 points.
B has 2 points.
C has 1 point.
D has 2 points.
E has 2 points.

A wins.

Election excluding candidate D:

Plurality Method:

B wins with 44 first place votes.

Plurality with elimination:

A majority out of 89 voters is 45 or more votes.

First place votes: A: 8, B: 44, C: 16, E: 21

None has a majority, thus eliminate A.

Second round: B: 44, C: 16, E: 29

None has a majority, thus eliminate C.

Third round: B: 44, E: 45; E wins.

A: $4(8)+3(65)+2(0)+1(16)=243$

B: $4(44)+3(0)+2(29)+1(16)=250$

C: $4(16)+3(0)+2(16)+1(57)=153$

E: $4(21)+3(24)+2(44)+1(0)=244$

B wins with 250 points.

Pairwise comparisons:

A (29) vs. B (60); B gets a point.
A (73) vs. C (16); A gets a point.
A (52) vs. E (37); A gets a point.
B (57) vs. C (32); B gets a point.
B (44) vs. E (45); E gets a point.
C (16) vs. E (73); E gets a point.

A has 2 points.
B has 2 points.
C has 0 points.
E has 2 points.

Three way tie, there is no winner.

The pairwise comparison method and the Borda count method violate the irrelevant alternative criterion.

28. Standard divisor: $\dfrac{6000}{10}=600$

Region	A	B	C	Total
Number of houses	2592	1428	1980	6000
Standard Quota	4.32	2.38	3.30	10.00
Lower quota	4	2	3	9
Hamilton's apportionment	4	3	3	10

29. Using the modified divisor 500.

Region	A	B	C	Total
Number of houses	2592	1428	1980	6000
Modified quota	5.18	2.86	3.96	12.00
Modified lower quota	5	2	3	10

30. Using the modified divisor 700.

Region	A	B	C	Total
Number of houses	2592	1428	1980	6000
Modified quota	3.70	2.04	2.83	8.57
Modified upper quota	4	3	3	10

31. Using the modified divisor 575.

Region	A	B	C	Total
Number of houses	2592	1428	1980	6000
Standard Quota	4.32	2.38	3.30	10.00
Modified quota	4.51	2.48	3.44	9.00
Modified rounded quota	5	2	3	10

32. New divisor: $\dfrac{6000}{11} \approx 545.45$

Region	A	B	C	Total
Number of houses	2592	1428	1980	6000
Standard Quota	4.75	2.62	3.63	11.00
Lower quota	4	2	3	9
Hamilton's apportionment	5	2	4	11

Yes. Hamilton's apportionment becomes 5, 2, 4. Region B loses one truck.

33. Standard divisor: $\dfrac{690}{23} = 30$

Course	A	B	C	Total
Number of students	311	219	160	690
Standard Quota	10.37	7.30	5.33	23.00
Lower quota	10	7	5	22
Hamilton's apportionment	11	7	5	23

34. Use the modified divisor 28.

Course	A	B	C	Total
Number of students	311	219	160	690
Modified quota	11.11	7.82	5.71	24.64
Modified lower quota	11	7	5	23

35. Use the modified divisor 31.4

Course	A	B	C	Total
Number of students	311	219	160	690
Modified quota	9.90	6.97	5.10	21.97
Modified upper quota	10	7	6	23

36. Use the modified divisor 29.5

Course	A	B	C	Total
Number of students	311	219	160	690
Standard Quota	10.37	7.30	5.33	23.00
Modified quota	10.54	7.42	5.42	22.00
Modified rounded quota	11	7	5	23

37. New divisor: $\frac{698}{23} \approx 30.35$

Course	A	B	C	D	E	Total
Number of students	317	219	162	0	0	698
Standard Quota	10.44	7.22	5.34	0.00	0.00	23.00
Lower quota	10	7	5	0	0	22
Hamilton's apportionment	11	7	5	0	0	23

No. The apportionment remains the same.

38. Standard divisor: $\frac{50,000}{50} = 1000$

State	A	B	Total
Population	4420	45,580	50,000
Standard Quota	4.42	45.58	50.00
Lower quota	4	45	49
Hamilton's apportionment	4	46	50

39. Using a divisor of 990:

State	A	B	Total
Population	4420	45,580	50,000
Modified quota	4.46	46.04	50.50
Modified lower quota	4	46	50

40. Using a divisor of 1020:

State	A	B	Total
Population	4420	45,580	50,000
Modified quota	4.33	44.69	49.02
Modified upper quota	5	45	50

41. Using the standard divisor of 1000:

State	A	B	Total
Population	4420	45,580	50,000
Standard Quota	4.42	45.58	50.00
Modified quota	4.42	45.58	49.00
Modified rounded quota	4	46	50

42. New divisor: $\frac{55,400}{55} \approx 1007.27$

State	A	B	C	Total
Population	4420	45,580	5400	55,400
Standard Quota	4.39	45.25	5.36	55.00
Lower quota	4	45	5	54
Hamilton's apportionment	5	45	5	55

Yes. State B loses a seat.

Chapter Test

1. $4+3+3+2=12$ members voted.
2. No lunch has a majority of 7 or more votes.
3. Pizza wins with a plurality of 5 votes.

4. B: $3(3)+2(3)+1(6)=21$

   D: $3(4)+2(5)+1(3)=25$

   P: $3(5)+2(4)+1(3)=26$

   Pizza wins.

5. B is eliminated and D then has a plurality of 7; deli sandwiches wins.

6. B (6) vs. D (6); B and D each get a half point.
   B (3) vs. P (9); P gets a point.
   D (7) vs. P (5); D gets a point.

   B has 1 point.
   D has 1.5 points.
   P has 1 point.

   Deli sandwiches wins.

7. Votes: $43+30+29+26+14=142$

8. Salamander wins with 43 votes.

9. H: $4(40)+3(59)+2(0)+1(43)=380$

   I: $4(29)+3(40)+2(73)+1(0)=382$

   L: $4(30)+3(43)+2(43)+1(26)=361$

   S: $4(43)+3(0)+2(26)+1(73)=297$

   The iguana wins.

10. A majority out of 142 voters is 72 or more votes.

    First place votes: H: 40, I: 29, L: 30, S: 43

    None has a majority, thus eliminate I.

    Second round: H: 69, L: 30, S: 43

    None has a majority, thus eliminate L.

    Third round: H: 99, S: 43

    The hamster wins.

11. H (70) vs. I (72); I gets a point.
    H (69) vs. L (73); L gets a point.
    H (99) vs. S (43); H gets a point.
    I (69) vs. L (73); L gets a point.
    I (99) vs. S (43); I gets a point.
    L (73) vs. S (69); L gets a point.

    H has 1 point.
    I has 2 points.
    L has 3 points.
    S has 0 points.

    The lemming wins with 3 points.

12. Plurality method:

    First place votes: W: 86, X: 80, Y: 60, Z: 58; W wins.

    Borda count:

    W: $4(86)+3(0)+2(52)+1(146)=594$

    X: $4(80)+3(118)+2(0)+1(86)=760$

    Y: $4(60)+3(86)+2(86)+1(52)=722$

    Z: $4(58)+3(80)+2(146)+1(0)=764$

    Z wins

    Plurality with elimination:

    A majority out of 284 voters is 143 or more votes.

    First place votes: W: 86 X: 80, Y: 60, Z: 58

    None has a majority, thus eliminate Z.

    Second round: W: 86 X: 138, Y: 60

    None has a majority, thus eliminate Y.

    Third round: W: 86 X: 198; X wins.

    Head-to-Head:

    W (86) vs. X (198); X gets a point.
    W (138) vs. Y (146); Y gets a point.
    W (86) vs. Z (198); Z gets a point.
    X (138) vs. Y (146); Y gets a point.
    X (140) vs. Z (144); Z gets a point.
    Y (146) vs. Z (138); Y gets a point.

    W has 0 points.
    X has 1 point.
    Y has 3 points.
    Z has 2 points.

    Y wins.

The plurality, Borda count, and plurality with elimination methods each violate the head-to-head criterion. The pairwise method never violates the head-to-head criterion.

13. A majority out of 35 voters is 18 or more votes.

E: $4(18)+3(0)+2(13)+1(4)=102$

M: $4(10)+3(25)+2(0)+1(0)=115$

S: $4(3)+3(0)+2(22)+1(10)=66$

W: $4(4)+3(10)+2(0)+1(21)=67$

El Capitan has a majority of first-place votes, but the mule deer wins using the Borda count method with 115 points. Thus, the majority criterion is violated.

14. Standard divisor: $\frac{33,000}{30}=1100$

State	A	B	C	Total
Population	6933	9533	16,534	33,000
Standard Quota	6.30	8.67	15.03	30.00
Lower quota	6.00	8.00	15.00	29.00
Hamilton's apportionment	6	9	15	30

15. The divisor 1040 was used.

State	A	B	C	Total
Population	6933	9533	16,534	33,000
Modified quota	6.67	9.17	15.90	31.74
Modified lower quota	6	9	15	30

16. The divisor used was 1160

State	A	B	C	Total
Population	6933	9533	16,534	33,000
Modified quota	5.98	8.22	14.25	28.45
Modified upper quota	6	9	15	30

17. The standard divisor of 1100 was used.

State	A	B	C	Total
Population	6933	9533	16,534	33,000
Standard Quota	6.30	8.67	15.03	30.00
Modified quota	6.30	8.67	15.03	29.00
Modified rounded quota	6	9	15	30

18. Standard divisor: $\frac{33,000}{31}\approx 1064.52$

State	A	B	C	Total
Population	6933	9533	16,534	33,000
Standard Quota	6.51	8.96	15.53	31.00
Lower quota	6.00	8.00	15.00	29.00
Hamilton's apportionment	6	9	16	31

The Alabama paradox does not occur, since none of the states loses a seat.

19. Standard divisor: $\frac{33,826}{30} \approx 1127.53$

State	A	B	C	Total
Population	7072	9724	17,030	33,826
Standard Quota	6.27	8.62	15.10	29.99
Lower quota	6.00	8.00	15.00	29.00
Hamilton's apportionment	6	9	15	30

The Alabama paradox does not occur, since none of the states loses a seat.

20. Standard divisor: $\frac{38,100}{35} \approx 1088.57$

State	A	B	C	D	Total
Population	6933	9533	16,534	5100	38,100
Standard Quota	6.37	8.76	15.19	4.69	35.01
Lower quota	6.00	8.00	15.00	4.00	33.00
Hamilton's apportionment	6	9	15	5	35

The new states paradox does not occur, since none of the existing states loses a seat.